The Reference Guide to Famous Engineering Landmarks of the World

Bridges, Tunnels, Dams, Roads, and Other Structures

by Lawrence H. Berlow

ORYX
1998

The rare Arabian oryx is believed to have inspired the myth of the unicorn. This desert antelope became virtually extinct in the early 1960s. At that time several groups of international conservationists arranged to have 9 animals sent to the Phoenix Zoo to be the nucleus of a captive breeding herd. Today the oryx population is over 1,000, and over 500 have been returned to the Middle East.

© 1998 by The Oryx Press
4041 North Central at Indian School Road
Phoenix, Arizona 85012-3397

Published simultaneously in Canada
Printed and bound in the United States of America

Cover Photographs: (Front, clockwise from top) Brooklyn Bridge, Hoover Dam, Neuschwanstein Castle, and Parthenon; (Back) Stonehenge.

∞ The paper used in this publication meets the minimum requirements of the American National Standard for Information Sciences—Permanence of Paper for Printed Library Materials, ANSI Z39.48-1984.

Library of Congress Cataloging-in-Publication Data

Barlow, Lawrence H., 1945–
    Reference guide to famous engineering landmarks of the world : bridges, tunnels, dams, roads, and other structures / by Lawrence H. Berlow.
        p. cm.
    Includes bibliographical references and index.
    ISBN 0-89774-966-9 (alk. paper)
    1. Engineering—History. I. Title.
TA15.B42 1997
620'.009—dc21   97-36051
                                                                    CIP

For Barbara, Amy, and Corey

CONTENTS

PREFACE

The Reference Guide to Famous Engineering Landmarks of the World provides readers with brief, factual entries for a selection of over 600 of the most famous and noteworthy human engineering achievements of the last 5,000 years. High school and community college students, university undergraduates in engineering and related programs, reference librarians, and interested public library patrons can quickly and easily retrieve such facts as location, purpose, dimensions, date, designer/builder, and construction history for any particular structure. *The Reference Guide* includes landmarks from around the world and of many different types—bridges, dams, roads, tunnels, canals and waterways, skyscrapers, palaces, castles, walls, temples, squares and plazas, cathedrals and churches, mausoleums, and other monumental structures. Information was collected from numerous sources, including books, magazine and journal articles, websites, and encyclopedias and other reference sources.

Over 100 of the alphabetically arranged entries are illustrated. Entries begin with the location and date of construction of the landmark, and also give any former or alternate names, for which cross-references are supplied in the listing. Most of the entries in this book are followed by an "Additional Information" listing. Readers wishing to pursue information about a particular structure may find the additional information entry to be a good starting place; discussions of the structure usually are more extensive than space allowed here. Also, the referenced information is sometimes more technical. Matthys and Salvadori's discussion of the failure of the Tacoma Narrows Bridge, for example, is a comprehensive one set against the overall theme of their book *Why Buildings Fall Down.* In some cases, only lim-

ited information is available about a structure, or the amount of information available is summed up in the entry. In these situations, such as in many of the biographies, the reference given at the bottom of the entry is designated as a "Source." While a researcher may wish to take a look at the particular source, he or she may not find it a particularly rich reference and will probably want to continue research elsewhere.

The Reference Guide also contains a series of brief biographies of important builders and designers, a chronology of important events in civil engineering from the third millennium B.C. to the present, a glossary of important civil engineering and construction terms used throughout the book, a bibliography of additional sources of information, and geographical and subject indexes. Appendix A contains illustrations of bridge types and bridge truss designs mentioned frequently throughout the book and described in the glossary. Appendix B is a photographic portfolio of Ohio covered bridges, which are examples of many of the nineteenth-century bridge truss designs shown in Appendix A. Appendix C lists the 20 tallest buildings, 20 longest suspension bridges, and 10 highest dams in 1997. When reading an entry, users are alerted to terms defined in the glossary by **boldface type**. Landmarks with their own entry elsewhere in the alphabetical listing are highlighted by being set in SMALL CAPS.

What Is a Landmark?

Although nations, states, cities, towns, and villages throughout the world have designated thousands of structures as landmarks, the term still needs definition. In part, defining a landmark involves size or sociohistorical importance. The biggest bridge, longest tunnel, highest tower, and tallest dam all qualify as landmarks. A site is also a landmark if it has historical significance; the "Seven Wonders of the Ancient World" fall into this category. This book began with the intention of presenting only landmarks that could be described with those terms—biggest, highest, longest, tallest, most historically significant, etc.

That definition of landmark expanded with the realization that certain structures are incredibly special or new or different, even though they required no special knowledge or new tools. These structures had builders, engineers, and architects who knew their craft well and were willing to push the limits of their capability. The Brooklyn Bridge certainly falls into this category, as do several other well-known structures. For the builders of these structures, the project took on a life of its own, and acquired a unique importance.

Perhaps this revised definition is too inclusive, embracing structures that may not seem, at first glance, to be landmarks. But I am certain that the builders and designers of the structures presented in this book were convinced that they were working on something important, small and mean though some of the projects may have seemed to the outside world. Thus, Alfred Ely Beach's short subway tunnel underneath New York City is a landmark, not only because of its innovative engineering, but also because of its social significance. Those who labored at the Winchester Mansion, building staircase after staircase, room after room, hidden passage after hidden passage, also labored in the construction of a landmark. Knowledge of such unusual landmarks as the Winchester Mansion can open for us a wide window to the politics, society, and technology of other times.

Users of this book who are looking for information on landmarks will find listed magnificent bridges, tunnels, dams, and other structures of such size or significance that their builders knew full well they were engaged in historic work. But also included here are the structures that presented little technical challenge but came to acquire significance over time or through layers of expansion and restoration—the Statue of Liberty, the Paris Opera House, the Kremlin in Moscow. This category also includes structures that became significant because they were well done. The builders of the Bronx River Parkway in New York probably did not set out to build "the first modern parkway"; they did aim, however, to build the best road possible to protect the nearby environment and still handle local traffic demands.

Finally, *The Reference Guide* lists structures that were so well done they became landmarks by their very existence or by overcoming great obstacles to their construction. Although Timothy Palmer was understandably proud of the "Permanent Bridge" (1806) he would have been surprised to learn that it lasted far longer than its expected life of 30 or 40 years. The Grubenmann brothers, about whom little is known, must have taken justifiable pride in their wooden bridges, structures built according to accepted principles but far surpassing what was expected of them.

This book is an amalgam of facts and histories about structures that could be considered landmarks. Because the modern world has so many remarkable bridges, tunnels, towers, and buildings, some landmarks have been overlooked. My apologies in advance for those oversights, but inclusion or exclusion from this book is far from the only determining factor in calling something a landmark.

Construction workers have no time for romanticism. These workers knew that their chances of being injured or killed were high on certain projects. Those who labored on projects such as the Mont Cenis Tunnel knew that they were engaged in hard, dirty, exhausting, and dangerous work. They knew as they began their day's labor that an accident might leave them incapacitated or dead before the day was over. We cannot even conceive of what slaves or medieval prisoners of war thought about as they were put to work constructing an edifice that was meant to please or honor nobility with whom they had no contact and toward whom they were probably hostile. This book salutes those workers—the grimy, tired, and helpless; the laborer, the skilled craftsperson, and the master of a craft; the knowledgeable, the genius, and the dreamer.

Internet Sources

A portion of the research for this book was done on the Internet—specifically on the World Wide Web (WWW). Although some of the WWW sites may be ephemeral, I am convinced that most will continue to exist, or, at worst, will be replaced by successors with the same or even more information. Although formatting for the columns in this book has sometimes caused WWW addresses to break over a new line, remember that WWW addresses are to be input in a continuous string without spaces.

In many cases, WWW information is probably available in print as well as electronically. "Lonely Planet" webpages, for example, are excerpts from currently popular travel books. I am satisfied that the hard information these sites present is generally accurate; if these sites disappear, others will surely take their place. This is true not only for travel book pages, but for government sites and travel bureaus as well. If a reader wishes to follow a citation and determines that a site no longer exists, he or she can be assured that similar sites on the Web can provide much the same information. Using a listed reference site along with a WWW search engine will almost assure a reader of some information on the topic. Although this may provide a minor challenge to researchers following my references, it is the modern equivalent of an older dilemma with which researchers have learned to deal—referencing the only copy of a newspaper or magazine that exists only in the archives of a single, far-away library or repository.

ACKNOWLEDGMENTS

Very many people helped with this book, and I gratefully acknowledge their cooperation, assistance, and interest.

I must thank my friends and relatives who provided encouragement and maintained their patience during what must have seemed an interminable process. Many volunteered ideas, suggestions, background material, information, and research leads that helped bring the book to its conclusion.

Maria Allen, Chip and Barbara Antonelli, Bob Baird, Jean Guyader, Bob Johnson, David Karchov, Marilyn Kelso, Jan P. Phillips, Rich Preiss, Mary Scott, Gilbert Siegel, Andrew Shelofsky, Gary Skeels, Bob and Eva Wingate, Mel and Evelyn Wittenstein, James Wen, Michelle Zmajkovic, and a host of others provided assistance, encouragement, support, patience, and large measures of needed good humor. Over a two-year period, so many people have contributed in one way or another to this book's progress that I fear I may fail to thank them here by name, although I know they will forgive me and that their good wishes and hopes for the book are still with me.

Special acknowledgments and thanks are due to the people who assisted in the creation of this book by contributing their knowledge, experience, and time. Very special thanks are due to Donald Jackson for clarifying large sections of my work at the preliminary stage and for being a valuable resource. Scott Nelson of the National Performance of Dams Program provided speedy assistance, under deadline pressure, with several important questions about dams. Aileen McGuire of the engineering firm of Lawler, Matusky & Skelly in Pearl River, New York, graciously granted me access to necessary publications in her firm's library. George Yansick of the Suffern Free Library assisted me with more than a few difficult research challenges.

Numerous people stopped what they were doing at their library/town hall/historical society/place of business to answer a sudden, obscure question from an unknown researcher calling on the telephone, or to put together a group of photographs or a batch of notes for me to use. Very few declined to assist. I'm proud to consider myself a part of that community of research, inquiry, and cooperation.

Very special thanks are in order to the wonderful people at Oryx Press, especially John Wagner, who not only edited my manuscript but has also dealt calmly and professionally with a nearly endless stream of e-mails, phone calls, letters, and overnight packages filled with fixes, changes, insertions, deletions, and rearrangements.

Most importantly, the book would never have been completed without the support and urging of Barbara, Corey, and Amy. Barbara, my wife, provided invaluable research, copyediting, and organizational skills and a near-infinite amount of encouragement and patience, as well as tolerance for what the demands on my time did to our daily lives. Corey's willingness to take on research projects and to assiduously pursue information for me by telephone, as well as his constant support and interest, made a difficult task much easier than it would have otherwise been. Amy's encouragement, and her assistance in helping to locate some essential research material, was an important contribution in bringing this book to fruition.

INTRODUCTION
A Brief Historical Background

A full history of civil engineering is far beyond the scope of this book. The practice of engineering—the conscious planning of structures for a specific purpose—extends backward in time into prehistory. The pyramids at Giza and other extant Egyptian structures go back to the third millenium B.C., and recent evidence has suggested the possible one-time existence of a rope and masonry bridge constructed by the Maya of Central America, probably for the purpose of encouraging commerce. It took many years until the civil engineering profession as we know it today came into being.

The English builder, John Smeaton, was the first to use the term "civil engineering." Smeaton, himself a distinguished practitioner, was referring to projects constructed for civilian, rather than military, purposes. Today the term has come to mean the building of structures for a particular purpose, usually utilitarian, and, often, military. Despite a structure's utilitarian inspiration, however, many of these structures have an overwhelming sense of awesomeness about them because of their beauty, magnificence, uniqueness, or individuality.

The development of the technology for constructing the type of landmarks discussed in this book reflects changes in both the engineering and social environment, and frequently in the political environment. The major technologies are briefly reviewed below.

Bridges

The earliest bridges were probably made of fallen trees that lay across opposite banks of a river or stream. The next development would have been the purposeful moving of a tree to a location where it could serve as a crossing. Later, some early engineer considered, in advance of construction, what size of crossing would work best, be safest, last longest, and be easiest to construct.

Once simple beam bridges appeared, it was a short leap to constructing bridges and their supports of stone. Such construction might involve purposely building a tower of stone on either side of a proposed crossing. If the rocks were well chosen, they would fit together tightly without any other medium to hold them. American farmers, for example, traditionally made stone fences from rocks plowed up in a new field. Without any holding medium, some of these fences have stood for 200 years or more with no sign of deterioration.

A simple beam bridge, mounted on piles of rocks on either side, would soon lead to a beam bridge with two sections and three supports, the third support placed under the center. If the center area was underwater, it was possible to fill the area under the water with rocks until the pile broke the surface.

The development of mortar and other holding media, beginning with the Roman development of pozzolino, enabled bridge building to take a major leap forward. Combined with a knowledge of the strength of the arch, bridges were constructed of such size and durability that they are still respected today as examples of solid technology.

Structures in the United States tended, in the eighteenth century, to be made of timber, a readily available material. With the coming of the age of industrialization, iron and steel made possible, in both the United States and elsewhere, the development of structures that took advantage of the new material. But the timber trusses developed early on in the United States and elsewhere were still used in the steel and cast-iron bridges of a later date.

Eugene Freyssinet, Robert Maillart, and others made wonderful use of prestressed concrete during the early years of the twentieth century. Their works, combining both up-to-date engineering with an almost artistic sense of design, are examples of the fusion of developments in technology with a keen awareness of the importance of how a project appears. In recent years, further developments in material science have made possible the widespread use of such materials as aluminum and even ceramics. Developments in the capabilities of engineers and other builders have led to even greater structures than their predecessors would have imagined.

Tunnels

Through soft dirt and gravel, tunnels can be built easily if proper attention is paid to roof supports and other supporting structures. The oldest known tunnel is the Euphrates Tunnel (c. 3000 B.C.), which is no longer in existence. Because the task of tunnelling underwater is so onerous, no other subaqueous tunnel was attempted until the early 1800s (Brunel's Thames Tunnel). Tunnelling through dirt and rock for the purpose of making war or bringing water to a particular location seems to be an ancient technology. Similarly, the creation of a tunnel by the cut and cover technique is not one that creates many challenges, at least not today.

Cutting through rock, though done in the distant past, was an onerous job until the nineteenth century. Rock had to be hammered out by hand; power tools, and even good quality hand tools, are relatively modern luxuries. Large rocks were broken by heating them until they cracked, sometimes aided by pouring water on a hot section to allow steam to enlarge a fault.

The major developments in tunnel construction occurred in only a few technical areas. The development and use of gunpowder, followed by the development of nitroglycerine, dynamite, and other explosives, made it possible to blast through hard rock for long distances, something that was nearly impossible previously. In the nineteenth century, the construction of underwater tunnels was greatly enhanced by the invention of the Greathead Shield, a device that allowed several workers to dig or cut away at the face of a tunnel while working underwater. Added to the capabilities that the shield gave workers were the understanding of how to deal with the bends (a knowledge also useful for those excavating bridge foundations), and how to properly supply air to a long, underground tunnel. Holland's achievement in air conditioning the Holland Tunnel was amazing in its time, and it provided a method and a model that has been used again in several places since the construction of that historic tunnel.

Dams and Reservoirs

Water management has always been a human concern. Small dams were probably made as far back in time as people began to live in a single location for an extended period of time. The Roman contribution to water delivery, primarily through long, well-built aqueducts, is well known. Major dams began to appear in the nineteenth century. The major obstacle to constructing a large dam was in analyzing how big a dam should be, and what type of dam would be best for a particular purpose, a mathematical and engineering challenge that is perhaps not completely understood even today. As a result, early dams tended to be massive affairs, usually rubble filled. Knowledge of the strength of the arch made it possible for dams to become thinner; although, in some cases, dams were built too thin, or too weak, to withstand the forces they held back. When that happened, the results were disastrous.

Additionally, dams are subject to several severe stresses. In the Johnstown Flood, for example, the collapse of the dam was caused not by the original structure of the dam but by modifications that had been made to the dam over the years; these changes, cumulatively, prevented the dam from standing against floodwaters. Earthen dams need the special protection of a spillway, or channel for high waters, since water coming over the top of an earthen dam would rapidly erode the structure.

Interesting technical challenges occur in the building of dams. At Hoover Dam, for instance, the concrete poured into the dam's base would not cool and set, under normal conditions, for decades. The solution, innovative at the time and still impressive, was to pipe cold brine (cold salted water) through the setting concrete to speed the cooling process.

As pressures on water supply grew, so did the number of dams being constructed. The height of dam building in the U.S. was in the 1930s, when public works projects provided the money, time, and manpower to actually construct large water storage facilities. Dams are still being proposed and built today, and some of them, like the Tehri High Dam in India, will be superb pieces of engineering when completed. However, large dams engender opposition, on both environmental and social grounds. The future will likely see the construction of many smaller dams with less threat of displacing or flooding culturally important areas, or of affecting the environment.

Roads

As with water supply systems, the Romans built magnificent roads that served the empire well. The Roman roads, apart from their careful construction, are also known for their amazing straightness; the Roman legions marched straight, and obstacles were removed or overlaid, rather than detoured around. The very existence of the New Testament epistles is an indication of the quality of Roman roads, upon which an adequate mail delivery system could function.

The growth of cities, and the increased use of horse-coaches to move about, prompted the development of roads in the eighteenth and nineteenth centuries. Although simple dirt roads were not sufficient for heavy traffic, the roads of gravel were more durable, and knowledge of grading kept gravel roads relatively dry. Early roads often linked centers of trade, or were built to encourage or support commerce. As commerce grew, the demand for better roads grew also. In eighteenth-century England, during the Industrial Revolution, the famous engineers Thomas Telford and John MacAdam developed methods of road construction in times that called for innovative engineering technologies.

With the coming of the automobile in the twentieth century, even gravel roads were insufficient. Roads had to meet the stresses of high speeds and accommodate the fragile rubber tire. The most advanced roads, by the middle of the twentieth century, were those of Germany, the "autobahns" constructed by the Third Reich before and during World War II. In later years, the autobahns were expanded, and they are among the finest roads in existence today.

In the United States, pressure to develop roads came early in the twentieth century from automobile manufacturers. Realizing that good roads meant more customers for their cars, the manufacturers organized lobbies and fostered associations to press for better thoroughfares. After World War II, the United States began to build an interstate highway system, suitable both for the modern automobile traveler and, not coincidentally, for moving American armed forces through the country to seaports and airports.

Canals

Canals were built to support commerce. The oldest dam and associated canal in England, the Alresford Dam and Canal, was constructed in 1189. In the United States, Albert Gallatin's early report called for several canals to be built in the east to spur the development of trade. In other countries, canals also supported trade—the Bridgewater Canal in England, for example, was a result of the need to stoke the fires of the Industrial Revolution, moving coal from the mines into the rest of the country.

The two largest canals in the world, the Suez and the Panama, were both built with the improvement of trade in mind. Both canals are legendary, and both changed history. The Suez Canal was constructed largely with forced or slave labor, although the political and economic benefits were many. The Panama Canal was a major accomplishment; built by American engineers and United States armed forces, the canal met many of the expected challenges of a major construction project in the middle of the tropics. Both canals are serving well today, and the picture of the world as we now know it would be critically altered if these canals did not exist.

The Reference
Guide

Terms set in **boldface type** are described in the "Glossary" at the back of the book; terms set in SMALL CAPS have their own entries elsewhere in the alphabetical listing.

Accademia Bridge. *See* Grand Canal [Venice].

Allegheny Tunnels. *See* Pennsylvania Turnpike.

Akashi Straits Bridge. *See* Akashi-Kaikyo Bridge.

Akashi-Kaikyo Bridge (near Kobe, Japan; expected completion 1998)

The Akashi-Kaikyo Bridge will connect the Japanese city of Kobe on Honshu with the island of Awaji. When finished, the structure will be the longest **cable-stayed bridge** in the world, with a total length of 3,910 meters (12,825 feet), and a center span of 1,990 meters (6,527 feet). Using two towers, this three-span bridge is being built to withstand the winds, earthquakes, and other natural phenomenon it is likely to encounter at its location. The bridge will be able to withstand winds up to 80 meters (262 feet) per second, and earthquakes as great as 8.5 on the Richter scale. The Akashi-Kaikyo Bridge will exceed the length of Japan's current longest **suspension bridge,** the Akashi Straits Bridge built in 1991. The Akashi Straits Bridge has a span of 1,780 meters (5,838 feet).

Additional Information: Lay, M.G. *Ways of the World.* New Brunswick, NJ: Rutgers University Press, 1992; http://www.kobe-u.ac.jp/hyogo/akashi.html.

Aki Tunnel. *See* Hokuriku Railway Tunnel.

Alaska Highway (Dawson Creek, British Columbia, to Fairbanks, Alaska; 1942-1943)

This 1,422-mile (2,294-kilometer) road is jointly maintained by Canada and the United States. Originally constructed in 1942 for military purposes, the Alaska Highway is now a major civilian thoroughfare. Beginning in Dawson Creek, British Columbia, the road runs for 1,221 miles (1,969 kilometers) through Canada, crossing the province of British Columbia and the Yukon Territory; in Alaska, the highway's 201-mile run (325 kilometers) proceeds through Delta Junction, where it joins the Richardson Highway for the final 98 miles (158 kilometers) into Fairbanks. Most of the road is either paved or a tar and gravel mix; during the winter months, hard, frozen ice serves as an excellent roadway cover for drivers daring the cold, arduous trip. Although the possibility of such a road had been discussed since the beginning of the twentieth century, the Alaska Highway, which was originally called the Alcan Highway, only became reality during World War II when the United States feared that the Pacific coast, and especially Alaska, was vulnerable to Japanese attack. In about 20 months, an army of 14,000 American engineers, contractors, and soldiers constructed the highway across rough and mountainous terrain to keep American forces in Alaska supplied. In 1946, the United States transferred control of the Canadian part of the highway to Canada.

Additional Information: The excellent Alaska Highway Web page [http://alaskan.com/bells/alaska_highway.html] offers a description of the road, driving advice, lists of campgrounds and motels along the highway, and telephone numbers for emergency aid services along the route; *see also* Rose, Albert C. *Historic American Roads*. New York: Crown Publishers, 1976.

Albert Canal (between Antwerp and Liege, Belgium; 1939)

Named for Belgium's King Albert I (reigned 1909-1934), who took an active part in his country's industrial reconstruction after World War I, the Albert Canal runs about 80 miles (130 kilometers) across northeastern Belgium between Antwerp on the west and Liege on the east. From Antwerp, the River Schelde offers access to the sea. The canal has a minimum bottom width of 80 feet (24 meters) and can be navigated by vessels of 2,000 tons with a maximum draft of 9 feet (2.7 meters). The waterway passes through a highly industrialized region of northwestern Europe. The canal has six sets of triple locks and one single lock on the Liege end.

Additional Information: Hadfield, Charles. *World Canals*. Newton Abbot, England: David & Charles, 1986.

Albert-Louppe Bridge. *See* Plougastel Bridge.

Alcan Highway. *See* Alaska Highway.

Alcantara Bridge (Alcantara, Spain; A.D. 98)

Caius Julius Lacer built this Roman masonry **arch bridge** across the Tagus River near Alcantara in the Estremadura region of western Spain. Built about A.D. 98 to honor Roman Emperor Trajan, who was born in Spain, the bridge rises to a height of 175 feet (about 53 meters). Each of its six arches spans a distance of 98 feet (30 meters), allowing the bridge to carry a roadway of almost 600 feet (183 meters). The central span is 89 feet (27 meters) wide, perhaps the widest bridge span constructed by the Romans. The granite **voussoirs** were assembled without mortar. Lacer must have had no doubt about the majesty of his accomplishment. A temple facing the southern approach to the bridge has a tablet over the doorway with the justifiably proud epitaph: "I leave a bridge forever to the generations of the world." Although 1,900 years old, the bridge is still in use today. The bridge has held up well under flooding conditions, but has been damaged several times by war, and portions of the bridge were restored in about 1543 and 1778. One span was completely destroyed in 1809 during the Napoleonic wars, and was rebuilt between 1858 and 1860. The bridge is located away from major roads; it may have been built not to facilitate everyday traffic, but to allow Roman forces to move quickly westward toward Portugal.

Additional Information: *Academic American Encyclopedia* (electronic version). Danbury, CT: Grolier Electronic Publishing, 1991; Brown, David J. *Bridges*. New York: Macmillan, 1993; Kirby, Richard Shelton et al. *Engineering in History*. New York: Dover Publications, 1990; O'Connor, Colin. *Roman Bridges*. Cambridge, England: Cambridge University Press, 1993.

Alex Fraser Bridge, formerly the Annacis Bridge (Vancouver, British Columbia; 1986)

This fascinating composite bridge of **reinforced concrete** and steel crosses the Fraser River in British Columbia. The main span of this **cable-stayed bridge** is an impressive 1,526 feet (465 meters) in length, currently the maximum span achieved for a cable-stayed bridge. The bridge's two concrete towers soar 506 feet (154 meters) into the sky, and its cross beams run below the deck and about halfway up above the deck. The deck itself is made of precast panels and is only 8.7 inches (22 centimeters) thick, although it is supported by a steel frame that is 6.6 feet (2 meters) thick and cantilevered out from the towers. Cables from the bridge deck that run close to the tower are attached to the towers about half way up, while cables from the ends of the bridge attach to the tops of the towers.

Additional Information: Brown, David J. *Bridges*. New York: Macmillan, 1993; Lay, M.G. *Ways of the World*. New Brunswick, NJ: Rutgers University Press, 1992.

Alexander Column (Palace Square, St. Petersburg, Russia; 1834)

This 47.5-meter (156-foot) column was erected on orders of Czar Alexander I in honor of the Russian victory over Napoleon in 1812-13. The column is surrounded by major historical structures, such as the Russian **Baroque** WINTER PALACE. The column is not secured to the ground, and depends solely upon gravity to keep its 704 tons in place.

Additional Information: http://www.spb.su/fresh/sights/palacesq.html; Glickman, Yevgenia, "Where Only Angels Dare," 1995, *St. Petersburg Press*, reprinted in http://www.spb.su/lifestyl/136/where.html

Allegheny Portage Tunnel (near Johnstown, Pennsylvania; 1833)

This 701-foot-long (214-meter) tunnel was the first railroad tunnel built in the United States. The tunnel was built and used as a railway portage to transport barges across the Allegheny Mountains to the Ohio-Mississippi waterway system in the Midwest. The tunnel cost $37,798 to construct. Although the tunnel still exists, it is no longer in use.

Additional Information: Sandstrom, Gosta E. *Tunnels*. New York: Holt, Reinhart and Winston, 1963.

Allt-na-Lairige Dam (Allt-na-Lairige, Scotland; 1958)

Allt-na-Lairige is the first **prestressed concrete**-buttress dam in the world; it uses only 60 percent of the concrete that an ordinary concrete dam would need. Allt-na-Lairige Dam is located about 50 miles (80 kilometers)

north of the industrial city of Glasgow in western Scotland. The dam is 73 feet (22 meters) high and 1,360 feet (415 meters) long. It is made of prestressed concrete on its water face with groups of iron rods sunk into its rock foundations.

Additional Information: Smith, Norman. *A History of Dams.* Secaucus, NJ: Citadel Press, 1972.

Alsea Bay Bridge (Waldport, Oregon; 1936, replaced 1992)

The Alsea Bay Bridge carries the Oregon Coast Highway (U.S. Route 101) over Alsea Bay at Waldport in western Oregon. The original bridge, built by engineer Conde B. McCullough in 1936, was over 3,000 feet (915 meters) long. The bridge was a combination of **tied arches** and a plain deck. Deterioration of the concrete doomed the original bridge, which was replaced in 1992 with a bridge of the same design.

Additional Information: DeLony, Eric. *Landmark American Bridges.* Boston: Bullfinch Press, 1993.

Alvord Lake Bridge (Golden Gate Park, San Francisco, California; 1889)

Engineer Ernest L. Ransome, a successful popularizer of **reinforced concrete**, used steel reinforcing bars to build this first concrete **arch bridge** in the United States. An ornamental bridge with an imitation stone finish, the Alvord Lake Bridge has concrete stalactites and stalagmites inside its 20-foot (6.1-meter) span, and appears to be a vault rather than an arch.

Additional Information: DeLony, Eric. *Landmark American Bridges.* Boston: Bullfinch Press, 1993.

Ambassador Bridge (Detroit, Michigan, to Windsor, Ontario, Canada; 1927-1929)

At completion, the Ambassador Bridge, which spans the Detroit River, became the longest **suspension bridge** in the world. Jonathan Jones was chief engineer on the Ambassador, and the architectural firm of Smith, Hinchman, and Gryllis designed the bridge. The Ambassador runs 1,850 feet (564 meters), 100 feet (30.5 meters) longer than the PHILADELPHIA-CAMDEN BRIDGE over the Delaware River, the previous record holder. The Ambassador was itself superseded in 1931 by the 3,500-foot-long (1,067.5-meter) GEORGE WASHINGTON BRIDGE between New York and New Jersey. The construction of the George Washington ushered in the age of giant suspension bridges.

Source: Steinman, David B. and Watson, Sara Ruth. *Bridges.* New York: Dover Publications, 1941, 1957.

An Ji Bridge, also known as the Zhaozhou Bridge (Zhao Xian, Hebai Province, China; c. A.D. 590-605)

Li Chun, the founder of a famous Chinese bridge design school, designed and built this 121-foot-long (37-meter) span over the Jiaohe River in Hebai Province in central China. Built in the late sixth century during the rule of the Sui dynasty, the An Ji Bridge, which is known locally as the Dashiqiao or "Great Stone" Bridge, is one of the most remarkable stone bridges in the world. Legend says the bridge was built in one night by Lu Ban, the first Chinese stonemason. Such a feat performed by a human alarmed some of the immortals, who tried to destroy the bridge by crossing it with an overloaded donkey and wheelbarrow. Lu Ban held up the bridge with his own hands, and gouges still visible in the bridge's stone deck are seen as evidence of the donkey's passing. The main arch of the bridge is supported on the abutment ends by two smaller arches on each side. Because the main arch is segmental, forming only a portion of a circle rather than a full semicircle, it rises only 23 feet (7 meters) and uses only 28 **courses** of masonry. The bridge is almost 32 feet (9.6 meters) wide and uses four pairs of bow-shaped iron connectors to keep adjacent stones together and make the relatively flat arch of the bridge possible. Five transverse iron ties hold the arches together. Li Chun's use of iron ties and connectors was highly innovative and much in advance of bridge design in sixth-century Europe.

Additional Information: Brown, David J. *Bridges.* New York: Macmillan, 1993; Lay, M.G. *Ways of the World.* New Brunswick, NJ: Rutgers University Press, 1992.

Annacis Bridge. *See* Alex Fraser Bridge.

Antioch Bridge (Antioch, California; 1978)

The Antioch Bridge in northern California carries Highway 160 across the San Joaquin River, connecting Contra Costa and Sacramento counties. The strikingly beautiful Antioch Bridge replaces a 1926 two-lane lift span that had been the first toll bridge built across a tributary of San Francisco Bay. With a length of 2,877 meters (9,437 feet), this new steel-plate girder bridge, which is still a toll bridge, carries almost 11,000 vehicles per day. The older bridge was a serious impediment to traffic because its lift was raised and lowered an average of 95 times a month. The older bridge was also too low, and ship collisions with the structure in 1958, 1963, and 1970 nearly destroyed the bridge. The new bridge is 135 feet (41 meters) high, more than enough for the nautical traffic in San Francisco Bay.

Additional Information: http://www.dot.ca.gov/dist4/calbridgs.htm#ab.

The Antioch Bridge crosses the San Joaquin River in northern California. Photo by Bob Colin, courtesy of the State of California Department of Transportation.

Antonine Wall. *See* Hadrian's Wall.

Apennine Railroad Tunnel. *See* Great Apennine Railway Tunnel.

Appian Way. *See* Via Appia.

Aqua Appia (Rome, Italy; c. 312 B.C.)

Appius Claudius Caecus, the Roman censor (a city magistrate) who was responsible for the construction of the VIA APPIA, also built this **aqueduct**, the first to supply Rome with water. Appius skillfully played off the plebeians (the non-noble classes in Rome) against the aristocrats to win Senate approval for the projects. A majority of Roman citizens had opposed both projects as too expensive. Neither the aqueduct nor the road were complete on the expiration of Appius's term of office. Fearing that his successors would refuse to complete the projects, Appius convinced the Senate to extend his term until both projects had been finished. Appius then secured his place in history by naming the two projects after himself. The Aqua Appia, none of which remains in existence today, ran underground for 10.6 miles (17 kilometers) from the springs from which it drew water to the east wall of Rome. Within the city, the aqueduct was carried on arches for about 100 feet (30.5 meters) and then dropped underground again to run southwest across the city until turning sharply northwest. At its end, the aqueduct was again carried on arches for a short stretch to a terminal reservoir on the south bank of the River Tiber. From the terminal reservoir, water was piped to 20 distribution reservoirs, which each received over 75,000 cubic meters (over 97,000 cubic yards) of water per day. The water duct of the Aqua Appia measured 5 feet (1.5 meters) by 1.75 feet (0.5 meters); the tunnel through the rock that carried the water duct measured 5 feet by 5 feet (1.5 meters).

Additional Information: O'Connor, Colin. *Roman Bridges.* Cambridge, England: Cambridge University Press, 1993; Sandstrom, Gosta E. *Man the Builder.* New York: McGraw-Hill, 1970.

Arc de Triomphe (Paris, France; 1806-1836)

Napoleon I ordered construction of this "arch of triumph" to commemorate his 1805 victory over the Austrians at Austerlitz. J.F. Chalgrin began construction of the arch in 1806 from his own designs, and the work was carried on by several other builders until the arch's completion in 1836. The arch stands in the center of the Place de l'Étoile, which is formed by the intersection of 12 avenues, including the CHAMPS ÉLYSÉES, the main tree-lined avenue of Paris. The Arc de Triomphe is 160 feet (49 meters) high, 150 feet (46 meters) wide, and 72 feet (22 meters) deep. In 1920, the body of an unknown French soldier from World War I was interred beneath the arch and an eternal flame was lighted. Engraved at the top of the arch are the names of the major battles won by French arms during the Revolutionary and Napoleonic periods. On the inside walls are engraved the names of 558 generals; the names of those who died in battle are underlined.

Additional Information: http://www.focusmm.com.au/~focus/fr_re_02.html; http://www.paris.org/Monuments/Arc.html.

The Arc de Triomphe stands at the center of the Place de l'Étoile in Paris, France.

Arch of Constantine (Rome, Italy; A.D. 315)

The Arch of Constantine is the largest and best-preserved Roman triumphal arch in the world. Emperor Constantine built this monument to commemorate his victory at Milvian Bridge in A.D. 312 over Maxentius, a rival claimant for the throne. Portions of the arch were taken from imperial monuments built by earlier emperors, including Trajan (reigned A.D. 98-117) and Hadrian (reigned A.D. 117-138), the builder of HADRIAN'S WALL in northern Britain. Constantine, who converted to Christianity around A.D. 312, supposedly had a vision before the battle at Milvian Bridge of a cross and the words *"in hoc signo vinces"* ("by this sign you shall conquer") appearing in the heavens. Consequently, the inscription on the triumphal arch reads "Constantine overcame his enemies by divine inspiration."

Additional Information: http://harpy.uccs.edu/roman/html/archconslides.html.southern England. Photo courtesy of Ray Farrar.

Arlesford Dam and Reservoir (south central England; 1189)

Bishop Godfrey de Lucy of Winchester built Arlesford Dam, England's first recorded dam, in the late twelfth century. The dam was designed to make the River Itchen

The canal-like locks of Arlesford waterway in southern England. Photo courtesy of Ray Farrar.

navigable between Arlesford and Southampton. As a part of a series of canal-like locks, the reservoir was required for water. The dam and reservoir still exist, although the waterway is abandoned. The earth dam is about 250 feet (76 meters) long, 60 feet (18.3 meters) thick at its base, 30 feet (9.2 meters) thick at its crest, and 20 feet (6.1 meters) high. Originally, the reservoir covered about 200 acres (81 **hectares**). The dam still provides power to mills in Arlesford, and a still usable road runs along its crest.

Additional Information: Smith, Norman. *A History of Dams.* Secaucus, NJ: Citadel Press, 1972.

Arlington Memorial Bridge (Washington, DC; 1926-1932)

The Arlington Memorial Bridge carries U.S. Route 50 across the Potomac River between the Lincoln Memorial in Washington, DC, and ARLINGTON NATIONAL CEM-

The Arlington Memorial Bridge, looking across the Potomac from Virginia toward the Lincoln Memorial in Washington, DC. Photo courtesy of the Library of Congress, Farm Security Administration-Office of War Information Collection.

ETERY in northern Virginia. This **neoclassical** bridge with a double-leaf **bascule** in the center was first suggested by President Andrew Jackson in the 1830s as a symbol of the link between the North and the South. The design of this granite and concrete bridge was controversial because some critics considered the neoclassical style to be pompous and not in keeping with the principles of republicanism. The bridge is decorated with golden equestrian figures created by Leo Friedlander; the figures were cast in Italy as a gift from that country to the United States. The **reinforced concrete** span is 2,138 feet long (652 meters), and the draw span is 216 feet (66 meters) long, making it one of the longest draw spans in the world.

Additional Information: Jackson, Donald C. *Great American Bridges and Dams.* Washington, DC: The Preservation Press, 1988; Steinman, David B. *Famous Bridges of the World.* London: Dover Publications, 1953, 1961.

Arlington National Cemetery (Arlington, Virginia; 1864, enlarged 1889 and 1897)

In 1864, during the Civil War, the federal government confiscated land in northern Virginia across the Potomac River from Washington to use as a burial place for Union war dead. The 1,100 acres (445.5 **hectares**) of land comprised the Arlington plantation, the estate of the Custis and Lee families, which was then in the possession of General Robert E. Lee, commander of the Confederate Army of Northern Virginia. In 1883, the government paid General Lee's son and heir $150,000 for the estate. Ironically, the first man buried at Arlington National Cemetery was a Confederate prisoner of war who

A view (c.1900) of the monument to the unknown dead at Arlington National Cemetery in Virginia. Photo courtesy of the Library of Congress, World's Transportation Commission Photograph Collection.

had died at a local hospital. Today, more than 60,000 American war dead are buried at Arlington. The most famous feature of this national cemetery is the Tomb of the Unknown Soldier, an 80-ton block of white marble in which rest the unidentified remains of soldiers from the two World Wars, the Korean War, and the Vietnam War. Soldiers from nearby Fort Myer (from the U.S. 3rd Infantry, "The Old Guard") guard the tomb; a formal and solemn changing of the guard occurs every hour or every half hour during the day, depending on the season. The guard is changed every two hours at night. Each of the guards marches exactly 21 steps, and then faces the tomb for exactly 21 seconds. The "21" symbolizes the 21-gun salute, the highest gun-salute used. The inscription on the tomb reads as follows: "Here rests in honored glory an American Soldier . . . known but to God." The 408-acre (165-hectare) grounds of the national cemetery also enclose a large marble amphitheater and Arlington House, Lee's mansion, which has been preserved as the Lee Mansion National Memorial since 1925. Arlington House stands on a hill overlooking the cemetery; below the hill is the eternal flame marking the grave site of President John F. Kennedy, who was buried in Arlington after his assassination in November 1963. The ARLINGTON MEMORIAL BRIDGE crosses the Potomac at Arlington, connecting the cemetery with the city of Washington across the river.

Source: http://www.dgsys.com/-mwardell/cem.html; Duffield, Judy; Kramer, William; and Shephard, Cynthia. *Washington D.C.: The Complete Guide*. New York: Vintage Books, 1988.

Armor, Swift, Burlington Bridge. *See* A.S.B. Bridge.

Arrowrock Dam (Boise, Idaho; 1912-1916)
This concrete arch-gravity dam was one of the earliest projects of the United States Bureau of Reclamation. At 354 feet (108 meters), Arrowrock Dam, which stretches across the Boise River east of the city of Boise, was the highest dam in the world until surpassed by the 400-foot (122-meter) OWHYEE DAM in Oregon in 1932. Arrowrock Dam's original height was 349 feet (106 meters); an additional 5 feet were added in late 1936 and early 1937 when some minor reconstruction was done on the dam's downstream side. The dam is 1,100 feet (335.5 meters) long and forms a 17-mile-long (27-kilometer) reservoir. In 1904, the Bureau of Reclamation Engineering Department did the preliminary survey for the dam, which was intended to irrigate southwestern Idaho. Construction began eight years later and was completed by 1916. Only about 190 feet (58 meters) of the dam's diversion tunnel still exist today along the south abutment canyon wall; the diversion tunnel was originally 30 feet (9.2 meters) wide and 25 feet (7.6 meters) high.

Additional Information: Condit, Carl W. *American Building Art: The Twentieth Century*. New York: Oxford University Press, 1961; Jackson, Donald C. *Great American Bridges and Dams*. Washington, DC: The Preservation Press, 1988; .

Artemision. *See* Temple of Artemis (Diana).

Arthur Kill Railroad Bridge (New York-New Jersey; 1959)
This **vertical-lift bridge** across the Arthur Kill connects Staten Island, New York, with Elizabeth, New Jersey. ("Kill" is a Dutch term used primarily in the states of New York, Pennsylvania, and Delaware for a creek or river.) Like other vertical-lift bridges, the Arthur Kill Bridge consists of a beam end supported from towers by vertical cables, like a giant horizontal elevator. The 558-foot (170-meter) Arthur Kill Bridge was the world's longest vertical-lift span when it was built in 1959. The prior record for length of a vertical-lift span was held by the 544-foot-long (166-meter) railroad lift bridge built over the CAPE COD CANAL in Massachusetts in 1935.

Additional Information: Overman, Michael. *Roads, Bridges, and Tunnels*. Garden City, NY: Doubleday & Company, 1968; Steinman, David B. *Famous Bridges of the World*. London: Dover Publications, 1953, 1961.

Arts and Industries Building, formerly the U.S. National Museum (Washington, DC; 1881)
Montgomery Meigs and the firm of Cluss and Schultze designed the Arts and Industries Building to showcase materials from the 1876 Centennial Exposition in Philadelphia. The building, which later became the basis for the Smithsonian Institution, is especially notable for Meigs' use of natural light and flexible space. Windows on all sides of the building, as well as on the roof, provide excellent lighting, and no uninterrupted solid walls separate any of the 17 exhibition halls. A statue of Columbia protecting the seated figures of Science and Industry stands over the entrance. The rotunda contains a working fountain surrounded by seasonal plants. The building has gone through several interior restorations, rehabilitations, and reconstructions since its opening. As the concept of various national museums grew into the Smithsonian concept, the building came to be used primarily as administrative space for the museum system. Toward the end of the nineteenth century, smaller galleries and lecture spaces were introduced. Vacated exhibition space was turned into office space in the 1960s and 1970s. In 1976, the building was partially restored for the national Bicentennial celebration. Several of the original and older displays can still be seen, including an 1876 locomotive, some Civil War artillery, and a Centennial costume display of Butterick sewing patterns.

Additional Information: http://www.si.edu/organiza/museums/artsind/homepage/artsind.html.

Arvida Bridge (Arvida, Quebec, Canada; 1950)
This 87-meter (285-foot) **arch bridge** over the Saguenay River near the southern Quebec city of Arvida is one of the world's first all-aluminum bridges. Arvida was the site of a large plant run by the Aluminum Company of Canada. Because of the local availability of aluminum, which is lighter than steel, aluminum was used to construct the Arvida Bridge. The use of lighter aluminum can cut maintenance and construction costs, thus justifying the metal's higher capital cost. The Arvida Bridge is unpainted and noncorrosive. It weighs only 193,000 kilograms (425,565 pounds); if constructed of steel, the bridge would have weighed 437,000 kilograms (963,585 pounds).

Additional Information: Overman, Michael. *Roads, Bridges and Tunnels*. Garden City, NY: Doubleday & Company, 1968.

A.S.B. Bridge, also known as the Armor, Swift, Burlington Bridge (Kansas City, Missouri; 1917, 1982)
This unique double-decker, **vertical-lift bridge** across the Missouri River was designed and built by J.A.L. Waddell. As originally designed, the 428-foot (131-meter) lower span rose up to allow river traffic to pass underneath. The lower level is designed to carry railroad traffic; when the lift needs to be raised, the lower level telescopes upward. An upper level, designed for vehicular traffic, was removed in 1982. The intent of the original design was to prevent the raising of the bridge from disrupting vehicular traffic on the upper level. Waddell first received a commission to build the bridge in the mid-1880s, but national financial crises, including the panic of 1893, delayed completion of the bridge until 1917.

Additional Information: Jackson, Donald C. *Great American Bridges and Dams*. Washington, DC: The Preservation Press, 1988.

Ascutney Mill Dam (Windsor, Vermont; 1834)
Local businessmen formed the Ascutney Mill Dam Company in 1833. Their goal was to revive the local economy by providing a continuous and dependable supply of water for the area's various mills. Earlier droughts had forced the closing of some of these mills. Simeon Cobb, the engineer, began construction of the dam in early 1834. Most of the materials for this gravity dam, including the rock and the timber, came from the local area. Most of the construction was completed by the end of November 1834, and water impoundment occurred over the following two winters. Whether or not Cobb knew of the work of Zachariah Allen, author of *Science of Mechanics*, the Ascutney Mill Dam follows Allen's suggestion that a dam be placed with its concave side facing downstream. Ice and water caused vibrations and rumbling in the wintertime, and the reservoir was drained in the spring of 1835 for inspection. Although the inspection revealed no problems, the downstream side of the dam was strengthened with a 150-foot (46-meter) wall. The wall was constructed with a battered (not smooth) profile to reduce the vibration and noise from falling water and ice. Several years later, a concrete lip was also added to the top of the dam. Despite the hopes of its backers, the dam did not regenerate the local economy until the 1860s, when the Civil War revived the machine tooling and cotton milling businesses in the area. The city of Windsor now owns the dam, which still provides the city with a reliable and cheap source of hydroelectric power.

Additional Information: Schodek, Daniel L. *Landmarks in American Civil Engineering*. Cambridge, MA: MIT Press, 1987.

Ashland Bridge (Ashland, Delaware)
Ashland Bridge is Delaware's only remaining covered bridge. Perhaps no more than 12 feet (3.6 meters) long, this **Town truss** bridge crosses Red Clay Creek and adjoins the Ashland Nature Center. The date of its construction and its builder are unknown.

Additional Information: http://nl-44-225.macip.drexel.edu/top/bridge/CBAshl.HTML; http://willian-king.www.drexel.edu/top/bridge/CBAshl/html.

Aswan Dam (1902) (Aswan, Egypt; 1902)

The first Nile dam at Aswan was completed in 1902 by the British, who then controlled Egypt. The city of Aswan stands at the First Cataract (rapids) of the Nile, at what was usually described as the southern extent of ancient Egypt. Made of locally quarried granite, the first Aswan Dam was 176.5 feet (54 meters) high and an amazing 6,400 feet (2,240 meters) long (more than a mile). It was fitted with 180 low-level **sluices** to control the river and to feed water to the barrage at Asyut, an irrigation dam 350 miles (563 kilometers) downstream. Sir William Willcocks designed and built the dam. In 1905, the engineer Sir Benjamin Baker supervised the heightening of the dam, which was completed in 1912. The dam was heightened again between 1929 and 1933. Now underwater because of the construction of a new ASWAN DAM in the 1960s, the first Aswan Dam was crossed by a roadway 26 feet (8 meters) wide, and affected the water level of the river as far upstream as Abu Simbel, 174 miles (280 kilometers) away. A system of locks allowed navigation around the dam. With the barrage at Asyut, the dam's chief purpose was to regulate the level of the Nile for irrigation, and so enlarge the food supply for the growing Egyptian population. The dam permitted the introduction of perennial irrigation, allowing five crops to be harvested in every two-year cycle.

Additional Information: Sandstrom, Gosta E. *Man the Builder*. New York: McGraw-Hill, 1970; Smith, Norman. *A History of Dams*. Secaucus, NJ: Citadel Press, 1972.

Aswan Dam (1970), also known as the High Dam at Aswan (Aswan, Egypt; 1970)

In 1953, a group of left-wing military officers overthrew the Egyptian monarchy and adopted as their own a plan the king had been considering for the construction of a new dam on the Nile at Aswan. By 1954, Gamal Abdel Nasser, who had emerged from the revolutionary leadership to become president of Egypt, began talking to a consortium of British, German, and French firms about beginning construction on the project. The United States and Great Britain offered loans of $270 million to help finance the dam. Nasser's friendly relations with the Soviet Union and other communist states caused the U.S. and the other Western nations to back out of the project, and in 1958 the Soviet Union agreed to finance construction of the dam, which by now had become a major propaganda point for the Nasser regime. The agreement stipulated that only Soviet equipment and engineering methods could be used. Construction began in 1960, after a

power station had been built at the first ASWAN DAM to supply the energy needed for operations at the upstream site for the new dam. Completed in 1970, the second Aswan Dam is a rock and earthfill dam that acts essentially as a ridge built across the river. Composed entirely of local materials, the dam used 28.6 million cubic yards (21.9 million cubic meters) of rock, 20 million cubic yards (15.3 million cubic meters) of sand, and 4 million cubic yards (over 3 million cubic meters) of clay; its total volume is 55 million cubic yards (over 42 million cubic meters) of material. The dam rises 1,542 feet (470 meters) from the river bottom to a 131-foot-wide (40-meter) crest, which is 11,810 feet (3,602 meters) long. The clay core of the dam contains three concrete inspection galleries. The dam impounds 205,000 million cubic yards (over 156,800 million cubic meters) of water and creates a reservoir that averages 6 miles (9.7 kilometers) wide and is 310 miles (499 kilometers) long, extending to the Second Cataract (rapids). The lake covers an area of 1,860 square miles (4,817 square kilometers). The benefits of the dam include several million acres of new arable land and huge amounts of hydroelectric power. The reservoir created by the dam threatened to inundate several important archeological sites, including the temples of Ramses II at Abu Simbel. These structures were dismantled piece by piece and moved to higher ground where they were painstakingly reassembled.

Additional Information: Sandstrom, Gosta E. *Man the Builder*. New York: McGraw-Hill, 1970; Smith, Norman. *A History of Dams*. Secaucus, NJ: Citadel Press, 1972.

Ataturk Dam (near Diyarbakir, Turkey; 1990s)

The 184-meter-high (604-foot) Ataturk Dam is one of the world's largest dams; it is used for both irrigation and hydroelectric power. The dam, which crosses the

Built in the 1960s, the High Dam at Aswan provides Egypt with huge amounts of hydroelectric power.

Firat River in eastern Turkey, is a focus of political interest because it could also be used, in a time of hostility, to cut off water to both Iraq and Syria.

Additional Information: http://gurukul.ucc.american.edu/ted/ataturk.htm; American University's "Trade and Environment Database" at http://www.ids.ac.uk/e/did/data/d0s1/e02135.html. Additional details were provided by Scott K. Nelson, The National Performance of Dams Program (NPDP), Stanford, California.

Austin Dam (Austin, Texas; 1893)

Joseph P. Frizell and J.T. Fanning built this dam across the Colorado River for the generation of hydroelectric power. The Austin Dam was the first large masonry dam (albeit with a concrete core) ever built specifically for hydroelectricity. It collapsed on 7 April 1900, when a heavy rain increased the flow of water over the dam to more than 11 feet (3.4 meters) above the crest. The cause of the collapse was the erosion of the dam's downstream toe and the failure of the dam's poorly constructed foundations. Siltation was also a problem. By 1897, the 53,500-**acre-foot** reservoir had 41.5 feet (12.7 meters) of sediment. The dam was eventually rebuilt as the Tom Miller Dam.

Additional Information: Jackson, Donald C. *Great American Bridges and Dams*. Washington, DC: The Preservation Press, 1988; Smith, Norman. *A History of Dams*. Secaucus, NJ: Citadel Press, 1972.

Autobahn Road System (Germany; begun September 1933)

Large, carefully engineered roads that are still the pride of Germany, the autobahns were begun by the German government early in the chancellorship of Adolf Hitler. Fritz Todt was the head of the autobahn construction program. The autobahns had several purposes besides enabling the swift movement of automobile traffic; they were used to demonstrate the efficiency of the Hitler government, to provide work for economically stricken Germans during the Great Depression of the 1930s, and to create a way to move the highly mechanized German military throughout the country. Access to the autobahn is limited, and the roads are landscaped, making them appear both attractive and orderly. The autobahn is reserved for high speed travel; it has no speed limits, and exits and entrances are usually spaced about 20 miles (32 kilometers) apart. In 1942, when World War II halted construction, 2,326 miles (3,743 kilometers) of the autobahn system had been built. Construction resumed, especially in West Germany, several years after the end of the war in 1945.

Additional Information: Patton, Phil. *Open Road: A Celebration of the American Highway*. New York: Simon and Schuster, 1986.

Avenue Road, also known as Prospective Road (Moscow to St. Petersburg, Russia; 1722-1746)

This road was begun in the eighteenth century by Czar Peter the Great of Russia to connect Moscow to his new capital at St. Petersburg, 750 kilometers (465 miles) to the west. Most of the road was of **corduroy** construction.

Source: Lay, M.G. *Ways of the World*. New Brunswick, NJ: Rutgers University Press, 1992.

B

B&O Railroad. *See* Baltimore and Ohio Railroad.

Bailey Island Bridge (Harpswell, Maine; 1928)

This fascinating bridge, which was built like an **open crib dam**, carries State Route 24 over Casco Bay in southern Maine near Portland. Engineer Llewellyn Nathaniel Edwards built the bridge on an underwater rock-ledge that extends the entire distance from Orrs Island to Bailey Island; the ledge made the sinking of piers impractical. The open arrangement of the granite slabs keeps water flowing smoothly during tides. The entire structure is 1,120 feet (342 meters) long, with a width of 18 feet (5.5 meters) and a channel span of 52 feet (16 meters).

Additional Information: DeLony, Eric. *Landmark American Bridges.* Boston: Bullfinch Press, 1993.

Baiyoke Sky Hotel (Bangkok, Thailand; 1996)

This 320-meter-high (1,050-foot) building will be the tallest in Thailand, and one of the tallest in Asia and in the world. Operated as the Best Western Baiyoke Suite Hotel, the structure has about 200 export garment showrooms on its lower floors.

Additional Information: "Edifice Complex," August 1995, *Asia, Inc.*, in Asia, Inc. Online at http://www.asia-inc.com/archive/0895edifice.html.

Baltimore and Ohio (B&O) Railroad (Baltimore, Maryland, to Ellicott's Mills, Maryland; 1830)

The B&O was the first railroad in the United States to carry public traffic, both passengers and freight. After the War of 1812, the city of Baltimore experienced tremendous growth because of its proximity to the eastern terminus of the CUMBERLAND ROAD, the first major highway to the west. The canal boom of the 1820s, particularly the opening of the ERIE CANAL in New York State in 1825, threw the city into a period of economic decline. In 1827, before construction had begun on the proposed CHESAPEAKE AND OHIO CANAL to connect Washington, D.C., with western Maryland, a group of Baltimore business leaders petitioned the state legislature for a charter authorizing the construction and operation of a railroad line to run westward out of Baltimore. Construction on the B&O Railroad began in 1828, and the first section of track, which ran 14 miles (23 kilometers) from Baltimore to Ellicott's Mills (now Ellicott City), Maryland, opened for business in 1830. Squire Whipple, a well-known bridge engineer, did the first survey for the railroad, and Jonathan Knight was the first chief engineer on the project, to be succeeded later by Benjamin Henry Latrobe II. At first, horses pulled the trains, but the successful trial in 1830 of Peter Cooper's *Tom Thumb* steam locomotive soon led the B&O to replace horses with steam-powered engines. As the line stretched west, it created a demand for large masonry bridges to carry track across rivers and streams. The B&O built the first such bridge, the CARROLLTON VIADUCT, in Baltimore in 1829. The B&O reached Cumberland, Maryland (the eastern end of the Cumberland Road) in 1842; Wheeling, Virginia (now West Virginia) in 1852; and, over connecting roads, St. Louis, Missouri, in 1857. During the Civil War, the line was vital for the transportation of troops and supplies to the west, and was frequently attacked by Confederate raiders. The B&O made connections with Chicago in 1874, Philadelphia in 1886, and New York City in 1887. Because of its central location, the B&O remained one of the nation's more important rail lines well into the twentieth century.

Additional Information: Sandstrom, Gosta E. *Man the Builder.* New York: McGraw-Hill, 1970.

Ban Dam (Saint-Chamond, France; 1867-1870)

This masonry-gravity dam near Saint-Chamond in southeastern France was constructed to be a replica of the FURENS DAM, which had been built about 25 miles (40 kilometers) to the west in Saint-Etienne in 1866. Compared to the Furens project, the Ban Dam had a higher compressive **stress limit** of 8 kilograms per centimeter squared (114 pounds per inches squared). The dam supplied water to Saint-Chamond and the surrounding area.

Additional Information: Smith, Norman. *A History of Dams.* Secaucus, NJ: Citadel Press, 1972.

Bank of China Tower (Hong Kong, China; 1989)

Designed by architect I.M. Pei, this magnificent 70-story building stands in one of the most congested parts of Hong Kong. Because of its location, the building had to be typhoon-proof; the wind-load requirements for the tower meet standards four times the earthquake standards required of buildings in Los Angeles. The base of the building is a cube, above which are four triangular shafts, each terminating at a different height with a diagonal of glass roofing. The southern shaft reaches to the building's full height. To achieve the maximum space inside (170,000 square meters of office space), the building has no internal columns. At the top of the building is a glass-enclosed penthouse that looks out over the city.

Additional Information: Hong Kong Architecture at http://bcwww.cityu.edu.hk/b-s/boc.html.

Barrackville Covered Bridge
(Barrackville, West Virginia; 1853)

In 1853, Lemuel and Eli Chenoweth built this rare two-lane covered bridge across Buffalo Creek near Barrackville in what was then northwestern Virgina (now north central West Virginia). The 146-foot (45-meter) single-span bridge is the second oldest covered bridge in West Virginia. The bridge is a classic example of the Burr arch-truss, a wooden trussing system that superimposes an arch on a rectangular truss to increase strength. The bridge is owned by the state and is currently closed to traffic for renovation.

Additional Information: DeLony, Eric. *Landmark American Bridges.* Boston: Bullfinch Press, 1993; Jackson, Donald C. *Great American Bridges and Dams.* Washington, DC: The Preservation Press, 1988; http://www.wvonline.com/post/Barrackville.htm.

BART. *See* Bay Area Rapid Transit.

Bartlett Dam (northeast of Phoenix, Arizona; 1936-1939)

Bartlett Dam is part of the SALT RIVER PROJECT, a large project undertaken by the U.S. Bureau of Reclamation in 1905 to harness the Salt and Verde rivers of Arizona for power and irrigation. Bartlett Dam was named after a government surveyor. This multiple-arch dam is 283 feet (86 meters) high and 800 feet (244 meters) long; if the spillway is included, the dam's total length is 950 feet (290 meters). It is composed of a series of cylindrical vaults and is set between massive **abutments** that are actually gravity dams. The first dam built on the Verde River, the chief tributary of the Salt River, Bartlett drew 80 percent of its construction funds from the Salt River Project and 20 percent from the Bureau of Indian Affairs. Bartlett Dam is located about 40 miles (64 kilometers) northeast of Phoenix. The Verde rises in central Arizona north of Prescott and runs south to join the Salt River east of Phoenix. Bartlett Dam might be given a hydroelectric generating capacity in the future.

Additional Information: Condit, Carl W. *American Building Art: The Twentieth Century.* New York: Oxford University Press, 1961; http://donews.do.usbr.gov/Denver/tsc/Concdams/Buttress/Bartlett.html

Barton Aqueduct (Lancashire, England; 1761)

James Brindley designed this **aqueduct** to carry water across the Irwell Valley in northwestern England as part of the BRIDGEWATER CANAL, Great Britain's first fully artificial canal. The duke of Bridgewater needed the canal

The Barton Aqueduct in northwestern England. Photo courtesy of Ray Farrar.

to transport coal from his mines to Manchester, 10 miles (16 kilometers) away. The aqueduct used three low arches to support a trough of puddled clay (a mixture of clay, sand, and water). Puddled clay, once it dried, was impervious to water and thus made the channel of the aqueduct watertight. Barton became a model for other aqueducts as Great Britain's canal system grew in the late eighteenth century. Today the aqueduct has been replaced by a swing span situated at the point where the Bridgewater Canal crosses the Irwell River and the MANCHESTER SHIP CANAL.

Additional Information: Brown, David J. *Bridges*. New York: Macmillan, 1993; Kirby, Richard Shelton et al. *Engineering in History*. New York: Dover Publications, 1990.

Basra Bridge (across the Shatt-al-Arab, Basra, Iraq; 1940s)

An Indian army engineer built the Basra Bridge across the Shatt-al-Arab, the 120-mile-long (193-kilometer) river formed by the confluence of the Tigris and Euphrates rivers. Basra, the chief port of Iraq, is the furthest point to which the river can be navigated. Because the engineer could not get the materials required to build a standard lift bridge, he built the Basra Bridge to be lowered into the water to a depth of 7 meters (23 feet) to allow water traffic to pass above it.

Additional Information: Overman, Michael. *Roads, Bridges and Tunnels*. Garden City, NY: Doubleday & Company, 1968.

Bassano Bridge (Bassano, Italy; 1570)

Andrea Palladio, an Italian architect and engineer, was one of the first builders to make use of trusses in bridge construction. Palladio built this bridge across the Cismon-Brenta River in northeastern Italy, near the medieval town of Bassano. The bridge has four spans, each about 14 meters (46 feet) long. Made of timber, and undoubtedly rebuilt several times, the Bassano Bridge lasted over 300 years, until the late nineteenth century.

Additional Information: Davison, C.S. "Bridges of Historical Importance." *Engineer* 211 (1961), pp. 196-98; Lay, M.G. *Ways of the World*. New Brunswick, NJ: Rutgers University Press, 1992.

Bath-Bath #28 Bridge (Bath, New Hampshire; 1832)

This four-span Burr truss structure with added arches is the fifth bridge built in this location to carry the Pettyboro Road over the Ammonoosuc River in northwestern New Hampshire. A single-lane bridge with an enclosed sidewalk, its four spans measure 117 feet (36 meters), 66 feet (20 meters), 62 feet (19 meters), and 80 feet (24 meters). The overall length of the bridge is 374 feet (114 meters).

Additional Information: http://vintagedb.com/guides/covered3.html.

Battle Monument, West Point Military Academy (West Point, New York; 1897)

A tall granite tower overlooking the Hudson River, the West Point Battle Monument is the only monument in the United States dedicated to the officers and enlisted men of the Regular Army who were killed in the Civil War. Designed by an architectural firm that employed the famed designer Stanford White, the monument is reported to be the largest polished granite shaft in the

The Battle Monument at West Point Military Academy in New York is the largest polished granite shaft in the western hemisphere. Photo courtesy of the Department of the Army, Public Affairs Office, U.S. Military Academy.

Western Hemisphere. The names of 2,230 Regular Army soldiers who died in the Civil War are inscribed on the base of the monument. A statue of "Victory," sometimes called "Lady Fame," crowns the monument. The statue was sculpted by Frederick MacMonnies, a well-known sculptor of the late nineteenth century who first gained fame for his fountain for the Court of Honor at the World's Columbian Exposition in Chicago in 1893. Officers and men of the Regular Army paid for the monument by contributing 6 percent of each month's pay over a period of several years.

Source: http://tamos.gmu.edu/-marcus/battle.html; "Battle Monument" press release, Public Affairs Office, United States Military Academy.

Bay Area Rapid Transit (BART) (San Francisco, California; 1970s)

The BART subway tunnels run under San Francisco Bay, connecting San Francisco and the cities to its south on the peninsula with Oakland and the cities of the East Bay. BART's trans-bay tubes are the longest underwater vehicular tunnels in North America. Financed by a sales tax, BART cost four times more than the original estimates for the project, and ridership in the system's early years was significantly less than had been expected.

Additional Information: http://www.transitinfo.org/BART/

Bay Bridge. *See* Oakland-San Francisco Bay Bridge.

Bayonne Bridge (New York-New Jersey; 1931)

The Bayonne Bridge carries State Route 40 over the Kill van Kull from Bayonne, New Jersey, to Port Richmond on Staten Island, New York. A 1,675-foot-long (511-meter) parabolic arch span, the Bayonne Bridge was the longest **arch bridge** in the world until the NEW RIVER GORGE BRIDGE opened in West Virginia in 1978. The geography of the Bayonne region made a high single-span

The Bayonne Bridge crosses the Kill van Kull to connect Bayonne, New Jersey, with Port Richmond on Staten Island, New York. Photo courtesy of The Port Authority of New York and New Jersey.

bridge necessary. Engineer Othmar Ammann decided to build an arch rather than a **suspension bridge** because the anchorages needed for a suspension bridge would have been too expensive to excavate in the thick, strong bedrock on either side of the water. Unlike other arch bridges, the Bayonne Bridge was not built by using the **cantilever** method; it is the only major arch bridge that was constructed with **falsework**. High-strength alloy steel mixed with manganese was used for the first time in bridge construction in its arch ribs and rivets. Like the SYDNEY HARBOUR BRIDGE in Australia, the Bayonne Bridge is a spiritual descendant of the EADS BRIDGE, which was built over the Mississippi River at St. Louis in 1874. The Bayonne Bridge carries two railroad tracks and six automobile lanes. It was acclaimed the "most beautiful

steel bridge" in a 1931 judging by the American Institute of Steel Construction.

Additional Information: Bishop, Gordon. *Gems of New Jersey.* Englewood Cliffs, NJ: Prentice-Hall, 1985; DeLony, Eric. *Landmark American Bridges.* Boston: Bullfinch Press, 1993; Lay, M.G. *Ways of the World.* New Brunswick, NJ: Rutgers University Press, 1992; Steinman, David B. *Famous Bridges of the World.* London, Dover Publications, Inc., 1953, 1961.

Bazacle Dams (Toulouse, France; twelfth and eighteenth centuries)

Several dams were built at this site on the Garonne River in southwestern France, including a **diversion dam**; the first dam was built in the twelfth century. The original dam was 900 feet (275 meters) long, and probably made of earth, wood, and stones. The dam continued to be rebuilt with these materials until the early 1700s, when a tremendous flood totally destroyed the dam. A new dam was then constructed on the site by an engineer known as Abeille, who built a structure about 16 feet (4.9 meters) high and almost 70 feet (21 meters) thick. Abeille's dam was built on a gigantic foundation of timber piles driven deep into the river bed. Although modifications have since been made to it, including the addition of a masonry facing, Abeille's dam still spans the Garonne at Bazacle.

Additional Information: Smith, Norman. *A History of Dams.* Secaucus, NJ: Citadel Press, 1972.

Beach's Pneumatic Underground Railroad Tunnel (New York City; 1870)

Not at all significant by tunneling standards, the story of this secret passageway deserves to be told, if only because it is an example of how progress can be made by dedicated people committed to their projects at all costs. Designer and engineer Alfred Beach, the owner of the newly established *Scientific American* magazine, proposed in 1849 to build an underground railroad below Broadway in Manhattan. The proposal was stopped dead by "Boss" Tweed, the legendarily corrupt city commissioner of public works. Because he had accepted bribes from street level transportation companies, Tweed was not about to let their business be taken away by an underground system. In 1869, Beach applied for permission to build an underground pneumatic mail delivery system, which was granted because it did not seem to threaten Tweed or his interests. On 20 February 1870, Beach opened his pneumatic railroad. His work crew had dug only at night, and the dug-out soil and rock was carried away secretly by carts. The completed 312-foot (95-meter) tunnel took 312 nights to construct. A waiting room under Warren Street had chairs, couches, chandeliers, a grandfather clock, a goldfish pond, and a grand piano because Beach wanted the waiting room to be "cozy" for women passengers. Riders were moved from one end of the tunnel to the other by the action of a fan which sucked the car backwards when reversed. Observers were entranced, and Beach had proven that it was possible to tunnel under city streets without undermining them. The money Beach collected from fares (25 cents per ride) went to charity because Beach was not licensed to run the railroad for profit. A financial panic in the 1870s prevented Beach from finding investors to extend the tunnel and the pneumatic line. Financial difficulties eventually forced Beach to close the tunnel. By 1878, elevated railroads had become the way to travel in New York City. In 1912, workers digging a tunnel for New York City's new underground railroad system accidentally broke into Beach's long forgotten tunnel. Although Beach's pneumatic plan was probably impractical for an underground system serving all of New York City, it had been the first true underground system in the city. A part of Beach's tunnel is now part of the City Hall Station in New York City, and a plaque on the wall acknowledges Beach as the father of the New York City subway system.

Additional Information: Epstein, Sam and Epstein, Beryl. *Tunnels.* Boston: Little Brown & Company, 1985; Sandstrom, Gosta E. *Tunnels.* New York: Holt, Reinhart and Winston, 1963.

Bear Mountain Bridge (Peekskill, New York; 1922-1924)

This magnificent **suspension bridge**, the first vehicular bridge to cross the Hudson River, was designed by Howard C. Baird. The bridge spans the Hudson near Peekskill, New York, about 6 miles (9.7 kilometers) south of West Point. The bridge's central span is 1,632 feet (498 meters) long; its total length is 2,257 feet (688 meters). Its two main cables have a diameter of 17 inches (43 centimeters). Built as a toll bridge in the 1920s, the two-lane structure was acquired by the state of New York in 1940. The bridge held the record for longest suspension span in the world for two years until surpassed by the 1,750-foot-long (534-meter) PHILADELPHIA-CAMDEN BRIDGE.

Additional Information: Brown, David J. *Bridges.* New York: Macmillan, 1993; Condit, Carl W. *American Building Art: The Twentieth Century.* New York: Oxford University Press, 1961.

Bear Valley Dam (San Bernadino Mountains northeast of San Bernardino, California; 1884)

F.E. Brown designed this first arch dam built in the United States; the dam was built to impound irrigation water for southern California. Because of the remoteness of the site in the 1880s, Brown used a thin, arch construction that needed a minimum of material. Made of masonry blocks set in Portland cement mortar, the dam was designed for a maximum stress of 620 pounds per inch square. It was 65 feet (20 meters) high, with a radius to its vertical air face of 335 feet (102 meters).

Only 3 feet (0.9 meters) thick at the crest, the dam's thickness increased to 8.5 feet (2.6 meters) near the base. Although the dam's strength gave concern from the start, it held up well until 1910, when the need for additional water led to the construction of its replacement (completed in 1913), the BIG BEAR VALLEY DAM, a multiple arch structure. The original dam is now covered under Big Bear Lake, the reservoir of the newer dam.

Additional Information: Condit, Carl W. *American Building Art: The Twentieth Century.* New York: Oxford University Press, 1961; Smith, Norman. *A History of Dams.* Secaucus, NJ: Citadel Press, 1972.

Belont Dam (Bolton, England; 1827)

This early water supply dam in northwestern England was originally 65 feet (20 meters) high, but was heightened to its present 81 feet (25 meters) in 1843. The earth embankment is 1,038 feet (317 meters) long. Still part of the Bolton water system, the dam was rebuilt extensively between 1923 and 1926.

Additional Information: Smith, Norman. *A History of Dams.* Secaucus, NJ: Citadel Press, 1972.

Bendorf Bridge (Coblenz, Germany; 1965)

Ulrich Finsterwalder built this bridge across the Rhine River in western Germany at a site 5 miles (8 kilometers) north of Coblenz. One of the longest **prestressed concrete cantilever bridges** in the world, the Bendorf Bridge is a three-span cement **girder** bridge, 3,378 feet (1,030 meters) in length and 101 feet (31 meters) wide, with a main span of 682 feet (208 meters). For this bridge, Finsterwalder developed a new method of "free cantilevering" to build with prestressed beams.

Additional Information: Brown, David J. *Bridges.* New York: Macmillan, 1993; Hopkins, H.J. *A Span of Bridges: An Illustrated History.* New York: Praeger, 1970.

Benjamin Franklin Bridge, formerly the Delaware River Bridge; also known as the Camden Bridge (New Jersey-Pennsylvania; 1926)

Engineers Ralph Modjeski and Leon Moisseiff built the Benjamin Franklin Bridge to connet Camden, New Jersey, with Philadelphia. When it opened, the structure was the world's largest single-span bridge, but was surpassed in 1929 by the AMBASSADOR BRIDGE in Detroit. The main span of the Benjamin Franklin Bridge is 1,750 feet (534 meters) long. At its highest point, the bridge is 380 feet (116 meters) over the Delaware River.

Additional Information: Bishop, Gordon. *Gems of New Jersey.* Englewood Cliffs, NJ: Prentice-Hall, 1985; Brown, David J. *Bridges.* New York: Macmillan, 1993.

Betsy Ross Bridge (New Jersey-Pennsylvania; 1976)

The Betsy Ross Bridge crosses the Delaware River, connecting Pennsauken, New Jersey, with Philadelphia.

Named for Betsy Ross, the revolutionary war heroine who lived in Philadelphia, the bridge is the first in the United States to be named after a woman. A continuous truss design, the bridge has an eight-lane span measuring 729 feet (222 meters) in length and 90 feet (27 meters) in width. At its highest point, the deck of the bridge is 215 feet (66 meters) above the Delaware River.

Additional Information: Bishop, Gordon. *Gems of New Jersey.* Englewood Cliffs, NJ: Prentice-Hall, 1985.

Bhakra Dam (on the Sutlej River, India; 1962)

This well-known gravity dam in the Himalayan foothills of northern India is 740 feet (226 meters) high and 1,700 feet (519 meters) long. The volume of concrete in the dam is 5.6 million cubic yards (over 4.2 million cubic meters) and its associated reservoir has a capacity of 7.1 million **acre-feet** of water. It has a power capacity of 1.8 million kilowatts. The dam's irrigation system waters over 4 million acres (over 1.6 million **hectares**) of land. Built by the International Engineering Company in the 1960s, the dam is owned by the government of the Indian state of Punjab.

Additional Information: Sandstrom, Gosta E. *Man the Builder.* New York: McGraw-Hill, 1970; Smith, Norman. *A History of Dams.* Secaucus, NJ: Citadel Press, 1972.

Bidwell Bar Suspension Bridge (Oroville, California; 1855)

The Bidwell Bar Bridge is the only surviving example of this type of **suspension bridge**, which was typical of the bridges built in northern California during the gold rush days of the mid-nineteenth century. The Bidwell originally spanned the Feather River, about 10 miles (16 kilometers) northeast of Oroville, but was reconstructed near its original site in 1964. The original bridge had four wire cables, two on each side; each cable was several inches in diameter and made up of many parallel wire strands spirally wrapped with a **wrought-iron** wire and heavily painted. Completed cables were 407 feet (124 meters) long and anchored in solid rock and concrete. The anchorages were bent bars sealed in oil and buried deep in the foundations. Each of the bridge towers consisted of four **cast-iron** posts tied together with a cross cast-iron plate at the base. The posts had specially fabricated cast-iron caps topped with cast-iron saddles that moved with the cables. The roadway was 18 feet (5.5 meters) wide, and the **abutments** were 12 feet (3.7 meters) high. The bridge

was dismantled in 1964 and reconstructed 600 feet (183 meters) above and one-half mile (0.8 kilometers) west of its original location; the reconstruction used the original cables, hangers, towers, and saddles, as well as some original stone for the abutment facings. The site is now part of the Kelly Ridge recreational area near OROVILLE DAM. The new Bidwell Bar Bridge, built in the 1960s, has a main span of 1,108 feet (338 meters) and an overall length of 1,790 feet (546 meters) from anchorage to anchorage. Initially, the new bridge carried the highway 627 feet (191 meters) above the Feather River, but the completion of Oroville Dam raised the level of the water, and the bridge is now only 47 feet (14.4 meters) above the river.

Additional Information: Schodek, Daniel L. *Landmarks in American Civil Engineering.* Cambridge, MA: MIT Press, 1987.

Big Bear Valley Dam (San Bernadino Mountains, northeast of San Bernadino, California; 1913)

This dam replaced the BEAR VALLEY DAM built nearby for irrigation in 1884. Designed by J.S. Eastwood, the Big Bear Valley Dam is 363 feet (111 meters) long at the crest and 92 feet (28 meters) high. The dam is divided into 32 vaults, each with a 32-foot (9.7-meter) span. Big Bear Lake, the reservoir created by the dam, is an important recreation area. The lake now covers the original Bear Valley Dam.

Additional Information: Condit, Carl W. *American Building Art: The Twentieth Century.* New York: Oxford University Press, 1961.

The new Bidwell Bar Bridge, pictured here, was built in the 1960s to replace the original bridge on the site, built 111 years earlier. Photo courtesy of the Bethlehem Steel Corporation.

Big Ben. *See* Westminster Palace.

"The Big Dig." *See* Boston Central Artery/Tunnel

Birchenough Bridge (Rhodesia [now Zimbabwe]; 1935)

The Birchenough Bridge, which was designed by Sir Ralph Freeman, has the reputation of being the most economical long-span steel **arch bridge** ever built. The 324-meter (1,063-foot) parabolic arch bridge was built in 1935 for only $400,000. The bridge's light weight allowed it to be the first long arch bridge built for less than the cost of a comparable multispan bridge. The bridge carries two lanes of automobile traffic and weighs a light 3,150,000 kilograms (over 6.9 million pounds); in comparison, the SYDNEY HARBOUR BRIDGE in Australia weighs 59 million kilograms (over 130 million pounds). When constructing the bridge, engineers used an aerial cableway strung from shore to shore. Parts of the **girders** were carried from one side on this ropeway and transferred in mid-air to the creeper cranes used for assembly. Because the bridge was so far from other sites, almost everything taken to the site was used in building the bridge so that little material needed to be taken away afterwards. Erection cranes were made from the parts of the bridge that would become the roadway in the final stage of construction, and the cables used as stays during the cantilever construction phases were subsequently cut to length to form the roadway hangers. These hangers were secondhand, having earlier been used as the erection stays for the Sydney Harbour Bridge. The Birchenough Bridge is one of the longest steel arch bridges in the world.

Additional Information: Overman, Michael. *Roads, Bridges and Tunnels.* Garden City, NY: Doubleday & Company. 1968.

Bixby Creek Bridge (near Carmel, California; 1933)

Designed and built by F.W. Panhorst and C.H. Purcell, this beautiful masonry arch highway bridge across Bixby Creek in west central California is built into the granite walls of a narrow V-shaped canyon. It crosses the creek near the spot where it flows into the Pacific Ocean. The bridge has open **spandrels**, a main span of 320 feet (98 meters), and a clearance of up to 280 feet (85 meters) above the water. This graceful bridge was designed to complement the beauty of the Pacific coastline.

Additional Information: Condit, Carl W. *American Building Art: The Twentieth Century.* New York: Oxford University Press, 1961; Jackson, Donald C. *Great American Bridges and Dams.* Washington, DC: The Preservation Press, 1988.

Blenheim Bridge (North Blenheim, New York; 1855)

Nicholas Montgomery Powers built this single-span covered bridge, one of the longest in the world. Blenheim Bridge, which crosses Schoharie Creek in east central New York, has a span of 232 feet (71 meters), and is a rare example of a "double-barreled" bridge using three lines of trussing. The Long truss (named after Colonel Stephen H. Long, who patented the design in 1830) was needed to span 210 feet (64 meters) of the crossing. On 8 July 1931, the Schoharie County Board of Supervisors voted to retain the bridge as a public historical site. The bridge is now a national landmark.

Additional Information: DeLony, Eric. *Landmark American Bridges.* Boston: Bullfinch Press, 1993; Schodek, Daniel L. *Landmarks in American Civil Engineering.* Cambridge, MA: MIT Press, 1987.

Blenheim Palace (Woodstock, England; 1724)

Blenheim Palace was built by Parliament between 1705 and 1724 as a national gift for John Churchill, duke of Marlborough, in gratitude for his victory over the French and Bavarians at the battle of Blenheim in Germany in August 1704. The huge palace was built on the site of an old royal hunting lodge near Woodstock in Oxfordshire in south central England. Designed by Sir John Vanbrugh, Blenheim is considered one of the finest examples of **Baroque** architecture in England. The palace was originally surrounded by formal gardens designed by Henry Wise, gardener to Queen Anne. Wise followed the style employed by Andre Le Notre in designing the gardens at VERSAILLES for Louis XIV. In the mid-eighteenth century, Lancelot Andrews was asked to redesign the gardens in the informal, natural style of woods, lawns, and waterways then in vogue. Sir Winston Churchill, a descendent of the first duke of Marlborough, was born at Blenheim in 1874.

Additional Information: Montgomery-Massingberd, Hugh. *Blenheim Revisited: The Spencer-Churchills and Their Palace.* London: Bodley Head, 1985.

Bloukrans Bridge (Van Stadens Gorge, South Africa; 1983)

This 892-foot-long (272-meter) **prestressed-concrete** arch crosses Van Stadens Gorge in the southern part of South Africa's Cape Province. The bridge's thinness, as well as its simple form and nearly invisible supports on the sides of the chasm, are reminiscent of the work of Robert Maillart, the builder of the SALGINATOBEL BRIDGE in Switzerland, although the Bloukrans Bridge is significantly longer than anything Maillart was ever able to construct.

Additional Information: Brown, David J. *Bridges.* New York: Macmillan, 1993.

Blue Mountain Tunnels. *See* Pennsylvania Turnpike.

Blue Ridge Parkway (Virginia to North Carolina; 1930s-1950s)

Stanley Abbott, who was known as a landscaper before he became a highway engineer, helped create this 469-mile (755-kilometer) stretch of scenic roadway running through the Blue Ridge Mountains (part of the Great Smoky Mountains) of western Virginia and western North Carolina. The parkway is one of the most beautifully landscaped roads in the United States; it was built specifically to show the grand vistas of the Blue Ridge to greatest advantage. The parkway was constructed according to specifications that sought to minimize damage to the environment. Landscape architects were employed to assure that trees and bushes were planted to cover the signs of road building. The parkway begins in Virginia at Shenandoah National Park and runs southwest into North Carolina to the Great Smoky Mountains National Park. Besides magnificent views, many natural and historical sights and attractions are to be found along the route.

Additional Information: "Exploring America's Scenic Highways," prepared by the Special Publications Division, National Geographic Society, Washington, DC, 1985.

Blue Ridge Tunnel. *See* Crozet Tunnel.

Boardman's Lenticular Bridge (New Milford, Connecticut; 1888)

The Berlin Iron Works Company built this **lenticular truss** bridge across the Housatonic River in 1888. The longest surviving bridge of its type in Connecticut, this 188-foot (57-meter) span was closed to traffic in the mid-1980s. Although the ornate polygonal top and bottom trusses (which form a lens shape) required more **shop work** than standard **Pratt trusses**, they required less metal. The American Bridge Company, controlled by the wealthy industrialist Andrew Carnegie, bought out the Berlin Iron Works Company in 1900. The Berlin Iron Works Company built hundreds of lenticular trusses in New York and New England at the turn of the century and made the possible but probably exaggerated claim of having constructed 90 percent of the iron bridges in the region.

Additional Information: Jackson, Donald C. *Great American Bridges and Dams*. Washington, DC: The Preservation Press, 1988.

Bollman Truss Bridge (Savage, Maryland; 1852)

Named for its builder, Wendel Bollman, the Bollman Truss Bridge carries the Baltimore and Ohio Railway across the Little Patuxent River at Savage in western Maryland, a site to which the bridge was moved in 1869. The bridge is a surviving example of the **Bollman truss**, a unique bridge type that helped fuel the expansion of the railroads in the United States in the nineteenth century. Posts on the bridge support **wrought-iron eyebars**

that are pinned into place. An iron connection is made between the tops of the post supports to each corner of each parallelogram in the bridge. The Bollman Truss Bridge was one of the first bridges to use iron in all its principal structural members.

Additional Information: DeLony, Eric. *Landmark American Bridges*. Boston: Bullfinch Press, 1993; Schodek, Daniel L. *Landmarks in American Civil Engineering*. Cambridge, MA: MIT Press, 1987.

Bonar Bridge (Dornoch, Scotland; 1811)

Thomas Telford built this 150-foot (46-meter) **arch bridge** crossing Dornoch Forth in northeastern Scotland. The four principal **girders** were precast in five sections each, rather than being assembled from individual components.

Additional Information: Brown, David J. *Bridges*. New York: Macmillan, 1993.

Bonneville Dam (Bonneville, Oregon; 1937)

Constructed by the U.S. Army Corps of Engineers, the Bonneville Dam was the first of several hydroelectric projects built in the 1930s to produce hydroelectric power from the 1,210-mile-long (1,947-kilometer) Columbia River, which forms much of the boundary between the states of Washington and Oregon. Built on a site that includes Bradford Island, the 1,230 foot-long

Ship canal gates for the Bonneville Dam in Oregon. Photo courtesy of the Bethlehem Steel Corporation.

(375-meter) concrete **overflow dam** is on one side of the island, and a large power plant, now capable of producing 500,000 kilowatts of power, is on the other side. The dam has a total crest length of 2,690 feet (820 meters) and a height of 197 feet (60 meters). Its volume is over 1.1 million cubic yards (over 840,000 cubic meters). A navigation lock can lift river traffic a height of 60 feet (18.3 meters), and an extensive series of fish ladders assists migrating fish, such as salmon, to pass the dam safely. A large fish hatchery was built next to the dam in 1936.

Additional Information: Condit, Carl W. *American Building Art: The Twentieth Century*. New York: Oxford University Press, 1961; Jackson, Donald C. *Great American Bridges and Dams*. Washington, DC: The Preservation Press, 1988.

Bosmelac Dam. *See* Nantes-Brest Canal.

Bosporus Bridge (Istanbul, Turkey; 1973)

This 3,520-foot (1,074-meter) span was the first modern road link between Europe and Asia. The bridge is one of the longest suspension spans in the world. It crosses the Bosporus near the site where the Persian Emperor Xerxes built a bridge of boats to take his army across the Bosporus for his invasion of Greece in 480 B.C. *See also* the SECOND BOSPORUS BRIDGE.

Additional Information: *Academic American Encyclopedia*, (electronic version). Danbury, CT: Grolier Electronic Publishing, 1991.

Boston Central Artery/Tunnel, also known as "The Big Dig" (Boston, Massachusetts; under construction in the 1990s)

This Massachusetts Highway Department project is the largest highway project ever undertaken in a major American city. Covering a distance of 7.5 miles (12 kilometers), half of which will be tunnel, the project will either build or reconstruct roads through and around Boston. The project has two portions: Interstate 93 in the south to Charlestown's Sullivan Square in the north, and the Interstate 93 interchange near South Station to Route 1A in East Boston. Included in the project is an extension of the Massachusetts Turnpike (I-90) to Logan International Airport through a new tunnel running under Boston Harbor. The first portion of the project, the Ted Williams Tunnel, is named after Boston's most famous baseball player. It was designed by the firms of Bechtel/Parsons Brinckerhoff of Boston and Sverdrup Civil of New York, and constructed by the firms of Morrison-Knudsen of Boise, Idaho, and Interbeton Company of The Netherlands. The tunnel was completed in mid-1996. The tunnel's construction made use of the largest circular **cofferdam** in North America; the dam is 85 feet (26 meters) deep and 250 feet (76 meters) in diameter. The world's largest clamshell dredge dug a trench for the tunnel on the bottom of Boston Harbor; the trench is 3,850 feet (1,174 meters) long, 100 feet (30.5 meters) wide, and 50 feet (15 meters) deep. Unlike other immersed tunnels, two-thirds of the trench for the Ted Williams had to be excavated through rock rather than through dirt and sediment, a circumstance that required extensive underwater blasting. During construction, special care was taken to avoid disruption in the nearby neighborhoods and to protect fish and lobster in the Charles River. The tunnel won the 1996 Outstanding Civil Engineering Achievement Award.

Additional Information: Pollak, Axel J. and Lalas, V. Peter. "Boston's Third Harbor Tunnel." *Civil Engineering* 66:3 (March 1996), pp. 3A-6A; http://www.tiac.net/users/kat/CAT.html; Robison, Rita. "Boston's Home Run." *Civil Engineering* 66:7 (July 1996), pp. 36-39.

Boston Post Road (U.S. Highway 1 between Boston and New York; 1704)

First constructed in the early eighteenth century, the road was maintained specifically for the delivery of mail between New York and Boston, although it became a major automobile route in the twentieth century. U.S. Highway 95, which runs along the east coast of the United States, is a more recent modern interstate that runs roughly parallel to U.S. 1 between New York and Boston. In many towns and villages along the way, the remains of the Boston Post Road are still designated as the Post Road or the Old Post Road.

Additional Information: Lay, M.G. *Ways of the World*. New Brunswick, NJ: Rutgers University Press, 1992; Merdinger, C.J. "Roads Through the Ages." *Journal of the Society of American Military Engineers* 44: 300 and 301 (1952), pp. 268-73, 340-44.

Boulder Canyon Project. *See* Hoover Dam.

Boulder Dam. *See* Hoover Dam.

Bouzey Dam (Épinal, France; 1881)

Built on the Moselle River in northeastern France to supply water to the Canal del Est, the Bouzey Dam was 1,700 feet (519 meters) long and 49 feet (15 meters) high. Poorly constructed, the dam had severe leakage and cracks. In 1884, a large part of the dam slipped downstream. In 1895, 155 lives were lost when a section of almost 600 feet (185 meters) was destroyed to a depth of 33 feet (10 meters) below the crest; the disaster attracted much attention to the problems of dam building. Because of its length, the dam was especially susceptible to thermal expansions and contractions. The destruction of the dam was due to shear failure, that is, too much water rushed horizontally across the water face of the dam, cutting into it like a knife.

Additional Information: Smith, Norman. *A History of Dams*. Secaucus, NJ: Citadel Press, 1972.

Bow Bridge (Central Park, New York City; 1862)
Designed by Calvert Vaux and Jacob Wrey Mould, this wrought-iron **girder** bridge is the oldest surviving example of its type in the United States; it makes use of **wrought-iron** truss rods to strengthen the deck beams. The bridge is covered with delicate masonry detailing.

Additional Information: DeLony, Eric. *Landmark American Bridges*. Boston: Bullfinch Press, 1993.

Bratsk Dam (Irkutsk, Siberia, Russia; 1964)
The Bratsk Dam stands on the Angara River, 372 miles (599 kilometers) northwest of Irkutsk. When completed in 1964, the Bratsk Dam was the largest hydropower dam in the world—417 feet (127 meters) in height and almost 3 miles (4.8 kilometers) long. It produces between 22 and 24 billion kilowatt hours of electricity annually. A gravity and earth dam, its volume is 21,180,000 cubic yards and its reservoir is capable of holding 137,230,000 **acre-feet** of water.

Additional Information: Smith, Norman. *A History of Dams*. Secaucus, NJ: Citadel Press, 1972.

Bridge of Sighs. *See* Grand Canal [Venice].

Bridge Over the River Kwai (Thailand; c.1944)
Made famous by the 1957 movie, *Bridge Over the River Kwai*, this structure is notable not for its engineering achievement but because it was built by the labor of Allied prisoners of war held by the Japanese during World War II. Although subsequently rebuilt, the original bridge was destroyed on 13 February 1945, not by the prisoners who built it, as the movie indicates, but by Allied bombing. Although the movie portrays the prisoners building a bridge of wood, the actual bridge was, and is, made of steel on nine concrete pilings. The nearby Allied cemeteries hold the remains of many of the more than 10,000 British, Dutch, Australian, and New Zealander prisoners who died in the labor camp. The 356 American soldiers who died building the bridge have been reinterred in the United States.

Additional Information: Davies, Peter N. *The Man Behind the Bridge: Colonel Toosey and the River Kwai*. London: Athlone, 1991; Harriman, Stephen. "A Day Trip to Bridge on River Kwai." *The Virginian-Pilot*, November 12, 1995, in http://scholar3.lib.vt.edu/VA-news/VA-Pilot/issues/1995/vp951112/9511100073.html.

Bridge 28 (Central Park, New York City; 1864)
The fifth **cast-iron** bridge in Central Park to be designed by Calvert Vaux and Jacob Wrey Mould, Bridge 28 was fabricated by J.B. & W.W. Cornell Ironworks of New York. Although used in Europe, the gothic arch design and art nouveau lines of this bridge hardly ever appeared in the United States at the time. The arch is just over 37 feet (11.3 meters) long, and just over 15 feet (4.6 meters) high. Vaux and Mould worked with Frederick Law Olmstead, famed designer of Central Park, to create the design for this bridge.

Additional Information: DeLony, Eric. *Landmark American Bridges*. Boston: Bullfinch Press, 1993.

Bridgeport Covered Bridge (Grass Valley, California; 1862)
Built by David Ingerfield Wood to cross the South Fork of the Yuba River in northeastern California, the 233-foot (71-meter) Bridgeport Covered Bridge is the longest single-span covered bridge in the world. The bridge was part of a 14-mile (22.5-kilometer) toll road built for the Virginia City Turnpike Company to link the port of San Francisco with the Comstock Lode silver mining area in western Nevada. The bridge is a double-intersectional truss set between two laminate arches. The vertical members are made of **wrought-iron** and the bearing blocks are **cast-iron**. The rest of the structure is Douglas fir. Although the sides are now shingled, the trace of the wooden arch can still be seen like a ghostly shadow when the bridge is viewed from the outside. The arch is superimposed on a **Howe truss**. Acquired by the California Department of Parks and Recreation in 1986, the bridge is the centerpiece of the South Yuba River Project. The bridge is a state historic landmark as well as a National Historic Landmark.

Additional Information: DeLony, Eric. *Landmark American Bridges*. Boston: Bullfinch Press, 1993; Schodek, Daniel L. *Landmarks in American Civil Engineering*. Cambridge, MA: MIT Press, 1987; http://www.websterweb.com/bridgeport.html.

Bridgewater Canal (Worsely to Liverpool, England; 1761, 1776)
On his grand tour of the Continent, the third duke of Bridgewater saw how beneficial a canal could be for improving transportation. The duke needed a better way to move coal from the mines on his estate at Worsely to Manchester, 10 miles (16 kilometers) away. The duke commissioned James Brindley to conduct a survey and report on the feasibility of building a canal to connect the mines to the city. Brindley proposed digging a level canal without locks, a scheme considered impossible by contemporary engineering opinion. The duke persuaded Parliament to pass an enactment authorizing the enterprise in 1760. Work began immediately, and the canal opened to traffic on 17 July 1761. To complete the canal, Brindley had to built the large stone BARTON AQUEDUCT over the Irwell River. To waterproof the canal bottom, the embankments, and the **aqueduct**, Brindley had his workers "heel" (tramp with their boots) the clay mixture known as puddled clay. The hooves of oxen, rather than the booted feet of men, had been the traditional instruments for heeling puddled clay. An anxious Brindley reportedly took to his bed when the time came to test the canal; he stayed there until notified that water was flowing through the canal successfully.

The canal broke the monopoly of the owners of the Irwell River, who had charged exorbitant tolls for its use, and greatly increased the supply, while reducing the transport cost, of coal for millowners in Manchester. The canal extension to Liverpool was completed in 1776. As an inexpensive way of moving coal from the mines to urban factories, the canal was a vital component of the British industrial revolution in the eighteenth century.

Additional Information: *Academic American Encyclopedia*, (electronic version). Danbury, CT: Grolier Electronic Publishing, 1991; McNeil, Ian, ed. *An Encyclopedia of the History of Technology*. London: Routledge, 1990; Sandstrom, Gosta E. *Tunnels*. New York: Holt, Rinehart and Winston, 1963.

Britannia Railway Bridge (near Bangor, Wales; 1850)

Built by Robert Stephenson, the builder of the LONDON-BIRMINGHAM RAILWAY, the Britannia Railway Bridge crosses the Menai Straits to link the Isle of Anglesey with northwestern Wales. The bridge opened to traffic on 4 March 1850. The bridge comprises four continuous spans, two of 230 feet (70 meters) and two of 460 feet (140 meters). The bridge was made of **wrought-iron** tubes put together in a rectangular shape, the prototype for the modern steel **box girder**. Each rectangular box weighed 4,680 tons (4,212 metric tons) and extended 1,511 feet (461 meters). The bridge's two **abutments** and three masonry towers hold up its beams. Orignally, traffic passed through the bridge's tubes, but today the bridge carries traffic above the tubes.

Additional Information: *Academic American Encyclopedia*, (electronic version). Danbury, CT: Grolier Electronic Publishing, 1991.

Bronx River Parkway (Bronx, New York; 1923)

The Bronx River Parkway, the first of many scenic parkways to be built in the United States in the twentieth century, was proposed by a New York State commission in 1907 as a way to clean up the banks of the Bronx River and protect the animals in the Bronx Zoo from water pollution. William White Niles, a member of the New York Zoological Society, urged the Bronx to buy the land and build the roads to make the parkway a reality and reduce pollution. A parkway was essentially a series of linear parks containing a road designed for slower speeds and noncommercial traffic. Built between 1916 and 1923, the 27-kilometer (17-mile) parkway ran from the Bronx Zoo to White Plains. The project was untertaken jointly by Westchester County and New York City. The parkway had no intersections; crossroads

passed over it. The design speed was 60 kilometers per hour (37.5 miles per hour) and no trucks were permitted. The parkway was successful, and Westchester real estate developers actively supported the extension of the concept so that by 1932 the county had over 270 kilometers (167 miles) of parkway.

Additional Information: *Academic American Encyclopedia*, (electronic version). Danbury, CT: Grolier Electronic Publishing, 1991; Lay, M.G. *Ways of the World*. New Brunswick, NJ: Rutgers University Press, 1992; Patton, Phil. *Open Road: A Celebration of the American Highway*. New York: Simon and Schuster, 1986.

Brooklyn Bridge (New York City-Brooklyn, New York; 1883)

The need for a bridge to span the East River and connect New York with Brooklyn had been recognized since the beginning of the nineteenth century, but actual con-

An early twentieth-century view of the Brooklyn Bridge looking west across the East River toward the Manhattan skyline. Photo courtesy of the Library of Congress, Detroit Publishing Company Photographic Collection.

struction of such a bridge had always been dismissed as impractical. John Roebling first proposed building such a bridge in 1857, but his original suggestion attracted little attention or support. In 1865, Roebling again proposed the bridge, but strong public support for the project only came when the severe winter of 1866-67 halted ferry traffic across the East River. In April 1867, State Senator Henry C. Murphy secured passage through the New York Legislature of an act incorporating the New York Bridge Company for the purpose of constructing and maintaining a bridge over the East River between New York and Brooklyn. Completion of the bridge was set for 1 June 1870.

The public grew skeptical of the project when it attracted the support of William "Boss" Tweed, New York's notoriously corrupt director of public works, who probably hoped to siphon off construction funds for his own purposes. Even after Tweed was removed from office, the project retained some of the stigma associated with his name. In 1869, army engineers decided the bridge needed a clearance of 130 feet (40 meters) if shipping was to pass unimpeded; this figure became standard for bridges over water with traffic. The federal government was involved in the project because the bridge would cut across waters important to the national military interest. Congress then passed and President Ulysses S. Grant signed a bill authorizing construction of the bridge.

John Roebling died on 22 July 1869, as a result of an accident. Standing on a pier surveying the site of the future bridge, Roebling failed to notice the approach of a boat coming in to dock. The boat crushed Roebling's foot, and he was taken home, where he refused further medical care. The cause of death is unknown, but Roebling probably died of either tetanus or gangrene. His son, Washington Roebling, a respected engineer in his own right, was appointed to complete the bridge.

The main piers, four 271.5-foot (83-meter) masonry towers, were built by using pneumatic **caissons** to excavate under the East River. Although this method of construction was new to the United States in the 1860s, it was being used at the same time in the construction of the EADS BRIDGE in St. Louis. Work began first on the Brooklyn side caisson, which was made of 12-inch by 12-inch yellow pine and provided six working compartments, about 50 feet (15 meters) square, separated by 2-foot-thick (0.6-meter) partitions. The overall dimensions of the caisson were 102 feet (31 meters) by 168 feet (51 meters). The working chambers were 9.5 feet (2.9 meters) high, covered by a timber roof 15 feet (4.6 meters) thick to support the load above. Timber walls were 9 feet (2.7 meters) thick at the top and tapered to 6 inches at bottom to make a cutting edge that was fitted with a rounded iron casting. Inside, the caisson was caulked with **oakum** and coated with pitch to make it watertight. Outside, the caisson had a protective coating of tin and 3-inch-thick creosote-treated planks. The caisson was towed to the site, and three 10-ton (9-metric ton) derricks were mounted on top. The caisson was loaded with limestone blocks and additional blocks were added to make the caisson sink to the bottom. **Wrought-iron** shafts were installed into the caisson to remove material and to allow workers and supplies to be sent down. Piping was added for gas and water and for pressurized air for pumping out sand. In the course of construction, workers removed more than 20,000 cubic yards (15,300 cubic meters) of earth and stone. Six compressors on the bank supplied air, pressurized to a maximum of 23 pounds per square inch above normal pressure.

The first men entered the caisson's air lock on 10 May 1870. They worked in three eight-hour shifts, using steel bars to pry material loose. Explosives were used after a while to loosen the material to be excavated. The work site was illuminated by candles, oil lamps, gas lamps, and calcium lights—electricity was not available at the time. Fire was a major danger for men in the caisson, which caught fire several times. The worst fire occurred on 2 December 1870, when a candle ignited the oakum. The only solution was to flood the caisson. After the fire was out, it took three days to pump out the water. Calcium lights and gas burners were then used to light the caisson.

When the excavation was completed, concrete was poured into the working chambers of the caisson in layers of six to eight inches. After the Brooklyn pier was completely filled, work began on the New York pier. The New York pier used similar methods as the Brooklyn pier, but was of even greater depth. The New York caisson was 4 feet (1.2 meters) longer than the Brooklyn one, and the 22-foot (6.7-meter) roof was 7 feet (2.1 meters) thicker to make up for the greater depth and weight. The maximum air pressure was 34 pounds above normal.

Washington Roebling worked regularly in the caissons. In 1872, he collapsed from decompression sickness (the **bends**). The illness resulted in the loss of his voice and partial paralysis. Roebling continued to direct the project from his home in Brooklyn Heights, sending daily messages to the workers through his wife, Emily Warren Roebling, who also helped with the technical details of the project.

The Brooklyn Bridge was the first to use galvanized steel wire in cable construction. Wire began to be drawn across the bridge on 29 May 1877. A specially built carriage moved back and forth between the piers, laying down wires that were then spliced together. In July 1878, the Roeblings discovered that rejected wire had been used with false inspection stickers. Because removing the existing cables was too expensive, the Roeblings decided to have the contractor place extra wires in the cable, making the cables much wider in diameter than had been planned. To keep the circular shape of the cables, 19 strands were used. Because of these corrections, the widths of the cables, and the number of cables, differ from section to section of the bridge. The last wire was run across the span on 5 October 1878. Each of the four main cables contains about 5,439 parallel, galvanized-steel, oil-coated wires wrapped into a cylinder about 15.75 inches (40 centimeters) in diameter.

The flooring system for the bridge was completed on 10 December 1881. The single deck is divided to provide two elevated railroad tracks, two trolley car tracks, single lane roadways on either side of the trolley tracks, and a 15-foot-wide (4.6-meter) central walkway. The total width is 80 feet (24 meters). Renovations have expanded the roadway sections. New loads, such as heavier railroad stock, eventually meant enlarging planned support members, such as floor beams, and stiffening trusses, although the well-designed cables did not need additional work.

The bridge was completed in the spring of 1883. The central span was 1,595.5 feet (487 meters) long, with two side spans of 930 feet (284 meters); with the approaches, the total length of the bridge was 5,989 feet (1,825 meters). The total cost was almost $9 million. The bridge was officially opened on 24 May 1883; President Chester A. Arthur and his cabinet, as well as many other dignitaries, attended the opening ceremonies and walked across the bridge to Brooklyn. Many of the dignitaries later assembled outside the apartment windows of the crippled Washington Roebling and his wife. The magnificent bridge inspired American poets and artists, such as Walt Whitman and Hart Crane, as well as artists from abroad.

The creation of the Brooklyn Bridge was an amazing adventure in engineering. The technical description of the bridge, one of the marvels of the nineteenth century, is almost as impressive today as it was more than 100 years ago. The bridge comprises two stone towers, four main cables, anchorages, diagonal stay cables, and four deck-stiffening trusses. The towers stand 276.5 feet (84 meters) above the mean water line. Below the waterline, the Brooklyn tower is 44.5 feet (13.6 meters) deep on the Brooklyn side, and 78.5 feet (24 meters) deep on the Manhattan side. The towers were built on pneumatic caissons made of timber and driven below the river bed by placing successive stone **courses** on their roofs. As the stone pressed down on the caissons, workers excavated rock, clay, and dirt from pressurized chambers inside the caissons. 5,434 wires make up each of the four 15.75-inch (40-centimeter) diameter steel cables. Eight 12-foot (3.7-meter) **eyebars** arched 90 degrees manage the transition between cables and anchorages. Each anchorage is made of granite and limestone, with a weight of 60,000 tons (54,000 metric tons). Diagonal stay cables provide a distinctive appearance and also assist the main cables in carrying and stiffening the deck.

Additional Information: DeLony, Eric. *Landmark American Bridges.* Boston: Bullfinch Press, 1993; McCullough, David. *The Great Bridge.* New York: Simon and Schuster, 1972; Schodek, Daniel L. *Landmarks in American Civil Engineering.* Cambridge, MA: MIT Press, 1987.

Buckingham Palace (London, England; 1702-1705)

Buckingham Palace, the official London residence of the ruling monarch of Great Britain, stands on the site of the former Buckingham House, which was built in the early eighteenth century for the duke of Buckingham, who sold the house to George III in 1762. In 1825, architect John Nash received a royal commission from

Buckingham Palace in London is the official residence of the British monarch.

George IV to convert the house into a royal residence. Nash's commission was withdrawn in 1830 because he overspent his budget. Various other architects expanded the palace in the nineteenth century, including Edward Blore, James Pennethorne, and Aston Webb. The expansions generally deepened the structure and added wings to the front. Queen Victoria (reigned 1837-1901), the first sovereign to actually live in Buckingham Palace, further remodelled the building later in the century. In 1847, Edward Blore built the east front. The facade of the palace was refaced in 1912. In 1940, a German bomb destroyed the palace chapel. Elizabeth II, the current monarch, resides in the north wing. The palace has 188 staff bedrooms, 52 royal and guest bedrooms, 92 offices, and 78 bathrooms. About 40,000 people are entertained at the palace each year. One of the most interesting tourist attractions at the palace is the changing of the guard, which occurs most days at 11 a.m.

Additional Information: http://www.londonmail.co.uk/palace.html; http://www.buckinghamgate.com/events/past_features/palace/palace.html; http://cafeonternet.co.uk/buckp.htm.

Buffalo Bill Dam, formerly the Shoeshone Dam (Cody, Wyoming; 1910)

When completed, this dam, built by F.E. Brown across the Shoeshone River about 6 miles (9.7 kilometers) from Cody, Wyoming, was the highest **concrete arch dam** in the world. The surrounding area had been settled by the legendary Colonel William "Buffalo Bill" Cody at the end of the nineteenth century. Cody constructed at least one canal to irrigate the surrounding land; in 1904, he transferred his rights to the federal Bureau of Reclamation in the Department of the Interior. Although Cody had envisioned only water canals, the Bureau decided a dam and reservoir could also provide irrigation for farming and hydroelectric power for the region. Locating the dam at a site with high canyon walls meant allowing the arch to be supported by the natural topology. Construction began in 1905, and water impoundment began in 1910. Some additional modifications were performed in the years following, and the Heart Mountain power plant was constructed between 1945 and 1948. The dam is 325 feet (99 meters) high, although the water height is kept at a maximum of 233 feet (71 meters). The dam is 10 feet (3 meters) wide at its top, 108 feet (33 meters) at its base, and set on a granite foundation. Its crest length is 200 feet (61 meters). With an operating capacity of 25,000 cubic feet per second, the dam draws on a reservoir with a capacity of 439,800 **acre-feet** and a surface of 6,710 acres (2,718 **hectares**). The power plant can produce 5,000 kilowatts. Nearly 93,000 acres (over 37,000 hectares) of farmland are irrigated from the dam.

Additional Information: Condit, Carl W. *American Building Art: The Twentieth Century*. New York: Oxford University Press, 1961; Schodek, Daniel L. *Landmarks in American Civil Engineering*. Cambridge, MA: MIT Press, 1987. For information on the theoretical advances which made both this dam and the PATHFINDER DAM possible, see Smith, Norman. *A History of Dams*. Secaucus, NJ: Citadel Press, 1972.

Burgiano Bridge (Arezzo, Italy; date unknown)

This medieval stone bridge across the Arno River at Arezzo in central Italy is still in use and may well be the bridge that appears in Leonardo da Vinci's famous painting, the *Mona Lisa*. While some art historians consider that da Vinci made up the background of his famous painting, others believe that the setting is real. Carlo Starnazzi, a paleontologist, and Cesare Mafuzzi, a lawyer, argue that the bridge is the Burgiano, which stands about 40 miles (64 kilometers) southeast of Florence.

Additional Information: Williams, Daniel. "Art Sleuths Find 'Mona Lisa's' Landscape." *Washington Post* wire service, in *Rockland Journal-News*, 15 December 1995, p. A:17.

C

C & O Canal. *See* Chesapeake and Ohio Canal.

Caballo Dam. *See* Elephant Butte Dam.

Cabin John Aqueduct (Cabin John, Maryland; 1864)

For 39 years, the Cabin John Aqueduct, built by Montgomery C. Meigs over Cabin John Creek northwest of

A view (c. 1900) of Cabin John Aqueduct in Maryland, a part of the Washington, DC, water system. Photo courtesy of the Library of Congress, World's Transportation Commission Photography Collection.

Washington, DC, as part of the Washington **aqueduct** system, was the longest masonry arch in the world. It brought water from the Great Falls on the Potomac 12 miles (19 kilometers) to the distribution reservoir at Georgetown. The 220-foot-long (67-meter) arch is a segment of 110 degrees, with a rise of 57 feet (17 meters) and granite **voussoirs** 4 feet (1.2 meters) deep at the crown and 6 feet (1.8 meters) at the spring lines. The **spandrels** are made of locally quarried Seneca sandstone and the backing behind the arch ring is laid with radial joints that add to the strength of the structure.

Additional Information: DeLony, Eric. *Landmark American Bridges.* Boston: Bullfinch Press, 1993.

Caernarvon Castle (Caernarvon, Wales; begun 1284)

Designed by Master James of St. George, Caernarvon Castle, along with HARLECH CASTLE, represents a major construction achievement of the thirteenth century. Both castles were part of Edward I's attempt to use local strongpoints to control the newly conquered mountainous region of North Wales. Caernarvon lies in northwest Wales in the old Welsh principality of Gynwedd. The most distinctive feature of the castle is its banded masonry, in which different horizontal patterns of stone are set into the outer walls of the buildings. Unusual for its time, the castle features polygonal towers; the major building, the Eagle Tower, has three polygonal turrets. The castle complex was supposedly modeled on the fifth-century walls of the city of Constantinople in a deliberate attempt to echo that city's connection with the history of Christianity.

Additional Information: Platt, Colin. *The Castle in Medieval England and Wales.* New York: Barnes & Noble, 1996.

Edward I of England began construction of Caernarvon Castle in North Wales in the 1280s.

California State Water Project (begun 1960)

Two-thirds of California's population is served by this network of dams, reservoirs, and **aqueducts**. Initial construction began after approval of a $1.75-billion bond issue by California voters. Initial water delivery began in 1973, and the water project is still growing in the mid-1990s, although it is essentially complete. The project now encompasses 23 reservoirs and lakes, 17 hydroelectric pumping plants, four pumping-generating plants, six hydroelectric power plants, and approximately 600 miles (965 kilometers) of aqueducts and pipelines. The project begins in northern California on the Feather River, which is dammed by the well-known OROVILLE DAM in Butte County. The dam is capable of holding 3.5 million **acre-feet** of water; 2.7 million acre-feet are actually used for storage and the remaining 800,000 acre-feet are used to absorb water for flood control purposes. From the Oroville Dam, water flows to the Sacramento-San Joaquin Delta, where some of the water is sent through the North Bay Aqueduct to Napa and Solano Counties. Further south, water is sent into the 444-mile-long (714-kilometer) California Aqueduct; water then moves through the South Bay Aqueduct to Alameda and Santa Clara Counties. Before reaching the Perris Reservoir, the final stop in the process, water might flow through the San Luis Reservoir, the Coastal Branch Aqueduct (under construction), or the A.D. Edmonston Pumping Plant, which raises the water 1,926 feet (587 meters) in a single rise into 10 miles (16 kilometers) of tunnels and siphons across the Tehachapi mountain range to the West Branch or East Branch Aqueduct.

Additional Information: http://tiger-2.water.ca.gov/dir-state_water_project/State_Water_Project.html.

Camden Bridge. *See* Benjamin Franklin Bridge.

Campton-Blair Bridge #41 (Campton, New Hampshire; 1869)

The 292-foot-long (89-meter) Campton-Blair Bridge #41 carries Blair Road across the Pemigewasset River. This Long truss with added arches replaces an earlier bridge that was destroyed by arson in 1829. The arsonist claimed that he burned the bridge on instructions from God, but was found not guilty in court because there were no witnesses to the crime.

Additional Information: http://vintagedb.com/guides/covered3.html.

Canadian National Railroad Tower. *See* CN Tower.

Canal du Midi, also known as the Languedoc Canal (southern France; 1666-1692)

In the seventeenth century, engineer Pierre-Paul Riquet and Jean Colbert, Louis XIV's chief minister, built this canal along one of two routes suggested for a major canal by Leonardo da Vinci, who spent his last years (1516-1519) in France. A feeder canal, the Rignole de Montagne (now the Rigole de canal du Midi), was built between May and October 1665. This feeder canal connects the two stretches of the Languedoc Canal, which were begun in October 1666. The canal opened in May 1681, seven months after Riquet's death. The Languedoc Canal was not finally completed until 1692. Still operational, the Languedoc Canal (now the Canal du Midi) was the greatest engineering achievement of the seventeenth century. Connecting the Mediterranean Sea with the Atlantic Ocean, the canal is 150 miles (241 kilometers) long, rising 206 feet (63 meters) from the Garonne River at Toulouse before falling 620 feet (189 meters) to the Étang de Thau. The canal has 100 locks, three large **aqueduct** bridges, a tunnel, and many **weirs** (relatively small **diversion dams**), road bridges, and control works. Water for the canal is provided mainly by the rivers of the Montagne Noire region. Water is available mostly in winter; in summer, the supply dries up. Riquet therefore built a dam and reservoir for summer runoff across the River Audot about 2 miles (3.2 kilometers) southeast of Revel. The dam is 2,560 feet (781 meters) long at its crest, has a maximum height of 105 feet (32 meters) above river bed, and a base thickness at its center of more than 450 feet (137 meters). It creates a reservoir of 5,400 **acre-feet**. A low-level outlet allows the removal of excess water; as with most dams, water flowing over the top would ruin the earth embankment. Work began in November 1667, and took four years. Construction involved 208,000 cubic yards (159,000 cubic meters) of material, of which 155,000 cubic yards (118,500 cubic meters) was earth fill. This was the first dam ever built to supply water to a summit-level canal. For 100 years, this canal was the only one to supply water to the Languedoc region of southern France.

Additional Information: Sandstrom, Gosta E. *Man the Builder*. New York: McGraw-Hill, 1970; Smith, Norman. *A History of Dams*. Secaucus, NJ: Citadel Press, 1972.

Canterbury Cathedral (Canterbury, England; eleventh century)

Canterbury Cathedral, known officially as the Cathedral of Christchurch, is the seat of the archbishop of Canterbury, the primate of England and head of the Church of England under the ruling monarch. In A.D. 597, Augustine arrived from Rome to convert the English to Christianity; he was consecrated "bishop of the English" in France, but founded an abbey in England at Canterbury, the capital of the southeastern kingdom of Kent. Although a church has stood on the site since the seventh century, most of the present cathedral was built in the romanesque style between 1070 and 1089. The nave (the long, narrow central hall of a cruciform church) and the towers were built in the Gothic style in the fourteenth century. In 1170, the cathedral was the scene of the assassination of Archbishop Thomas Becket. A Frenchman, William of Sens, designed the Gothic choir as Becket's shrine. The shrine became the focus of one of the most important pilgrimages in medieval Europe until it was destroyed and its wealth confiscated by order of Henry VIII in 1538. The pilgrimage to Canterbury was memorialized in the 1390s by Geoffrey Chaucer's *Canterbury Tales*. The great fifteenth-century tower of the cathedral is 235 feet (72 meters) high. The cathedral also holds the tombs of Henry IV (d.1413) and Edward the Black Prince (d.1376). In June 1942, a German air raid destroyed several surrounding buildings, including a relatively new library, but did not harm the cathedral itself.

Additional Information: http://www.lonelyplanet.com.au/dest/eur/eng.htm; Keates, Jonathan and Hornak, Angelo. *Canterbury Cathedral*. London: Scala/Philip Wilson, 1980.

Canton Viaduct (Canton, Massachusetts; 1835)

Captain William Gibbs McNeil built the Canton Viaduct to carry the Boston and Providence Railroad over the East Branch of the Neponset River in eastern Massachusetts south of Boston. The **viaduct** is made of granite and is 615 feet (188 meters) long and 22 feet (6.7 meters) wide; it rises 70 feet (21 meters) above the river. The insides of the tall arches are almost totally bricked in; small arches at the bottom allow the river to flow through.

Additional Information: DeLony, Eric. *Landmark American Bridges*. Boston: Bullfinch Press, 1993.

Cape Cod Canal (Cape Cod, Massachusetts; 1909-1914)

The Cape Cod Canal runs for 8 miles (13 kilometers) across the neck of Cape Cod, a long, sandy peninsula that curls for 65 miles (105 kilometers) into the Atlantic from southeastern Massachusetts. The canal connects Cape Cod Bay with Buzzards Bay to the southwest, and runs past the towns of Sagamore, Bourne, Gray Gables, and Buzzards Bay. In 1927, the federal government enlarged and improved the canal.

Additional Information: http://www.virtualcapecod.com/towntext/bourke.html.

Cape Creek Bridge (Florence, Oregon; 1932)

Conde B. McCullough built the Cape Creek Bridge to carry the Oregon Coast Highway (U.S. Route 101) over Cape Creek on the Pacific coast of Oregon. This attractive bridge is unusual because of its two-tier arrangement. Because the bridge crosses an extremely steep valley, McCullough chose to span the river with a 220-foot (67-meter) arch with the two-tiered approaches on either side.

Additional Information: DeLony, Eric. *Landmark American Bridges*. Boston: Bullfinch Press, 1993.

Cape Hatteras Lighthouse (Cape Hatteras National Seashore, North Carolina; 1870)

The massive 205-foot (63-meter) Hatteras Light boasts a revolving first-order **Fresnel lens**, the brightest available. Painted in a "barber pole" design with black-and-white stripes that require 165 gallons of paint on each repainting, the lighthouse is a massive presence at the "graveyard of the Atlantic," the confluence of the Gulf Stream with the Labrador Current in the waters off Cape Hatteras, North Carolina. The current light replaced a 95-foot (29-meter) light that was built in 1803; the old light had acquired the reputation of being the "worst light in the world." Because of beach erosion, plans have been made to move the structure back from the shore, although the tower is still open to tourists.

Additional Information: http://zuma.lib.utk.edu/lights/banks1.html.

Capilano Suspension Bridge (Vancouver, British Columbia, Canada; 1889)

This 350-foot-long (107-meter) hemp and cedar plank footbridge across Capilano Creek in southwest British Columbia was built as a tourist attraction, a role it still fills as the central attraction at Capilano Suspension Bridge and Park. The original owner of this **suspension bridge** was George Grant Mackay, a Vancouver park commissioner. Native Americans call the structure the "laughing bridge." Three Native American carvers on site have produced three totem poles near the bridge; the surrounding park contains life-sized red cedar statues carved during the Great Depression of the 1930s by two Danes who traded their wood carving services for food. Because the carvers had never seen a local Canadian Indian, the statues depict images of Plains Indians the carvers had seen in movies and magazines of the time. Approximately 750,000 people visit the bridge each year.

Additional Information: http://www.bendtech.com/attractions/ capilano/index.html; http://worldtel.com/vancouver/ susbridg.html; http://capbridge.com/fastfax.htm; http:// www.capbridge.com/history.htm.

Capitol Building. *See* United States Capitol Building.

Carondelet Canal (New Orleans, Louisiana; 1794-1795)

Only 2 miles (3.2 kilometers) long and 15 feet (4.6 meters) wide, the Carondelet Canal, an extension of the Bayou St. John, was of great economic importance to New Orleans, allowing the city direct access to the Gulf of Mexico. The canal was built in the 1790s, while New Orleans was under Spanish control, and was named for Governor Francisco Luis Hector, baron de Carondelet, who opened it in 1795. The canal's original purpose was to carry trade from Lake Ponchartrain southward to New Orleans. In 1805, with New Orleans in American hands, the canal was dredged, widened, and improved. In 1807, Congress authorized an extension of the canal to the Mississippi River, but the work was never undertaken and Canal Street now runs along the proposed route of the extension. The Carondelet was one of the earliest true canals to be built in what is now the United States.

Additional Information: Hadfield, Charles. *World Canals.* Newton Abbot, England: David & Charles, 1986; Spangenburg, Ray and Moser, Diane K. *The Story of America's Canals.* New York: Facts on File, 1992.

Capital Building. *See* United States Capitol Building.

Carquinez Straits Bridge (Crockett, California; 1923-1927)

This **cantilever bridge** over the Carquinez Straits, which connect the northern end of San Francisco Bay with Suison Bay, was designed by David B. Steinman; the structure is 3,350 feet (1,022 meters) long, with two main spans of 1,100 feet (336 meters) each. At the time of construction, two of the bridge's piers were the deepest in the United States. The piers and their supporting **caissons** had to be protected against **scour** and also against the incursions of the teredo (a sea animal). De-

The Carquinez Straits Bridge near Crockett in northern California was designed by David B. Steinman. Photo by Bob Colin, courtesy of the State of California Department of Transportation.

signed to withstand earthquakes; the bridge has expansion stops and giant hydraulic buffers at the joints. To avoid a disaster similar to the 1907 collapse of the QUEBEC BRIDGE, which buckled while under construction, throwing some 70 workmen to their deaths, the spans of the Carquinez Straits Bridge were lifted with balance weights on the cantilever arms. Although placement of the spans took a day, lifting took only 35 minutes in the first case, and 30 minutes in the second. The bridge opened on 21 May 1927, the last link in the Pacific Coastal Highway that runs from Canada to Mexico.

Additional Information: Condit, Carl W. *American Building Art: The Twentieth Century.* New York: Oxford University Press, 1961; Steinman, David B. *Famous Bridges of the World.* London: Dover Publications, 1953, 1961.

Carrollton Viaduct (Baltimore, Maryland; 1829)

The Carrollton Viaduct, built by Benjamin Henry Latrobe, Jr., was the first major engineered structure to be built for an American railroad line. The **viaduct** carried the first line of the BALTIMORE AND OHIO RAILROAD over Gwynns Falls southwest of Baltimore. A two-span masonry **arch bridge**, and the first masonry viaduct to be constructed in the United States, the Carrollton Viaduct remains in full service today. The construction process was unique for the day, making use of 12,000 pieces of stone. The structure has a span of 100 feet (30.5 meters)—an 80-foot (24-meter) main span and the smaller side span. Gwynns Falls passed through the larger arch, and a wagon road passed under the side span. The total length of the viaduct is 312 feet (95 meters) and its height above the stream is 51.75 feet (almost 16 meters).

Additional Information: Condit, Carl W. *American Building Art: The Twentieth Century.* New York: Oxford University Press, 1961; Schodek, Daniel L. *Landmarks in American Civil Engineering.* Cambridge, MA: MIT Press, 1987.

Casselman's Bridge (Grantsville, Maryland; 1818)
Notable for its impressive stonework, this masonry **arch bridge** across the Casselman River in western Maryland was one of the first bridges built on the western section of the CUMBERLAND ROAD. When constructed, the bridge was the largest stone arch in the country. The bridge's 354-foot (108-meter) length is exceptionally high, necessitating steep approaches to the bridge. Now closed to traffic, the bridge is part of a county park and can be accessed only on foot.

Additional Information: DeLony, Eric. *Landmark American Bridges*. Boston: Bullfinch Press, 1993; Jackson, Donald C. *Great American Bridges and Dams*. Washington, DC: The Preservation Press, 1988.

Castel Sant'Angelo, also called the Hadrianeum or Sepulcrum Antoninorum (Rome, Italy; A.D. 135-139.)
Originally built in the late 130s A.D. as a mausoleum for the Roman Emperor Hadrian, and serving until the early third century A.D. as the burial place of the Antonine emperors, the Castel Sant'Angelo was converted into a fortress in the fifth century. The structure stands on the right bank of the Tiber, where it guards the PONTE SANT'ANGELO, one of the principal bridges of ancient Rome. The fort is a circle within a square, with each corner of the square protected by an individually designed barbican (a defensive tower). The central circle is a high cylinder containing chapels, hallways, apartments, prison cells, and a courtyard. The fortress took its name from a vision seen by Pope Gregory the Great in 590. While conducting a penitential procession to ask God to end a plague then ravaging Rome, the pope saw the archangel Michael sheath his sword over the fortress, thus signifying the end of the pestilence. A marble statue of the archangel now tops the fortress. In the Middle Ages, the fortress served the popes as a place of refuge; the structure was connected by a protected passage to the papal Lateran Palace, and so could be easily and quickly reached by the pontiff in times of invasion or disorder in Rome. In 1527, for instance, Clement VII took refuge in Castel Sant'Angelo while the mutinous troops of Emperor Charles V sacked Rome. The castle was used as a prison until 1901 when it was restored and became a museum of military history.

Additional Information: D'Onofrio, Cesare. *Castel S. Angelo: Images and History*. English Edition. Rome: Roma Società Editrice, 1984.

Catacombs (outside Rome, Italy; first century A.D.)
The catacombs are subterranean burial places and worship sites used by the early Christian community in Rome, and elsewhere in the Roman Empire, from the first to the fifth centuries A.D. Catacombs have been found at Naples, Paris, Syracuse, Alexandria, and in Asia Minor (modern Turkey) and North Africa, but the most important and extensive catacombs are outside Rome. The pagan Greeks and Romans preferred cremation, but Christians buried their dead and so openly prepared underground burial sites outside the city, municipal ordinances forbidding interment within the walls. The Roman catacombs lie from 22 to 65 feet (6.7 to 20 meters) beneath the surface and occupy a space of more than 600 acres (243 **hectares**) in several levels, one on top of the other. Constructed exclusively in areas where the soft subsurface rock possessed a suitable granular structure, the catacombs consisted mainly of narrow passages about 3 feet (0.9 meters) wide. Lining the walls of the passages are the *loculi*, or recesses, excavated from the rock to a sufficient length to contain a burial. The *loculi*, arranged one above another in tiers, were sealed with slabs of marble or terra cotta that bore painted or incised inscriptions. Some passages open to separate chambers called *cubicula*, which were usually 12 feet (3.6 meters) square, but might be circular or polygonal. These chambers were privately owned family burial vaults or the tomb of a martyr. In the *cubicula*, bodies were held in sarcophagi set in arched niches that lined the room. In times of persecution, which began in the 60s A.D. under Emperor Nero, these chambers were used as underground churches and as places of refuge. The use of candles in the Roman liturgy owes something to the need for light of worshipers in dark catacombs. As the chambers were expanded and new levels were added, the catacombs became a labyrinthine complex of passages, chambers, and galleries. The walls and ceilings were usually plastered and decorated with symbols and frescos; the catacomb walls hold the beginnings of Christian art. Burials continued in the catacombs even after the empire recognized Christianity in A.D. 313. Many people desired to be buried near the early martyrs, and many others wished to make pilgrimages to the site. The Germanic invasions of Italy led to the plundering of the catacombs and the robbing of graves for saints' relics. By the tenth century, most of the bodies had been transferred to churches, and the catacombs, filled with debris, were largely forgotten. In 1578, the catacombs were rediscovered, and their preservation and maintenance have been under the control of the papacy ever since .

Additional Information: Demby, William. *The Catacombs*. Boston: Northeastern University Press, 1991.

Cathedral of Saint Basil the Blessed. *See* Saint Basil's Cathedral.

Cathedral of Santa Maria del Fiore, also known as Il Duomo (Florence, Italy; mid-thirteenth century)
Medieval cathedrals typically took many years to complete. In the case of Il Duomo, the Gothic cathedral of Florence in north central Italy, the long construction

period resulted in a fortuitously attractive whole, rather than a hodgepodge of different styles. In 1294, Arnolfo di Cambio was selected to design and build a new church to replace the church of Santa Reparta; work on the new structure began two years later. When di Cambio

Statues atop the Duomo overlook Florence.

died in 1302, work ceased until 1334, when the famed painter, Giotto, was chosen as di Cambio's successor. Until his death in 1337, Giotto spent most of his energies on the campanile (the nearby bell tower). Several other important artists worked on the cathedral: Filippo Brunelleschi built the cupola of the cathedral between 1420 and 1434 and Lorenzo Ghiberti worked on the famed bronze doors of the cathedral baptistry for over 20 years after 1403. Brunelleschi was awarded the commission to build the dome for the cathedral after winning a design competition. His octagonal ribbed dome is one of the most celebrated and innovative domes in European architectural history. Ghiberti's baptistry doors are perhaps even more celebrated. The great north doors contain 20 panels depicting scenes from the life of Christ in relief; four panels depict fathers of the early church and four others show the evangelists. A molding with busts of prophets and sibyls frames the reliefs.

The eastern doors, which Michaelangelo declared fit to grace the portals of heaven, depict 10 scenes from the Old Testament.

Additional Information: Holler, Anne. *Florencewalks*. rev. ed. New York: Henry Holt and Company, 1993.

Causey Arch Bridge (County Durham, England; 1727)

The world's first railway **viaduct**, the Causey Arch bridge was constructed by Ralph Wood in northeastern England as part of the Tanfield Wagon Way, a mining system railroad begun in 1671. The Tanfield was the first railway to be built under an Act of Parliament. An engineering monument, the railway still operates, albeit under limitations and restrictions.

Additional Information: Ellis, Hamilton. *The Pictorial Encyclopedia of Railways*. Prague, Czechoslovakia: Paul Hamlyn, 1968.

Centennial Bridge (Cottage Grove, Oregon; 1987)

The Centennial Bridge across the Coast Fork Willamette River may be the newest covered bridge in the United States. A **Howe truss**, this 84-foot-long (26-meter) bridge was constructed to commemorate the centennial of the founding of Cottage Grove. The new bridge was built on trusses from two earlier covered bridges that stood at the same site. Built to handle only foot traffic, the Centennial Bridge has never carried vehicles or trains.

Source: http:william-king.www.drexel.edu/top/bridge/CBCent.html.

Center Street Bridge (Cleveland, Ohio; 1901)

Built by James Ritchie and James T. Pardee, the Center Street Bridge carries Cleveland's Center Street over the Cuyahoga River. The structure is a fascinating example of a "bobtail" bridge, which combines both a horizontally rotating section and a section that lifts up from a hinge.

Additional Information: DeLony, Eric. *Landmark American Bridges*. Boston: Bullfinch Press, 1993.

Cento Dam (Cesena, Italy; c.1450)

Builder Domenico Malatesta was a member of the ruling family of Cesena, which is in northern Italy near Milan. The dam crosses the River Savio 2.5 miles (4 kilometers) upriver from Cesena. Still in use, the dam is 234 feet (71 meters) long at its crest and approximately 45 feet (14 meters) thick at its base. A parapet wall that has been added to the crest of the dam raises the height another 4.5 feet (1.4 meters). The sloping air face has a thick concrete facing that was added well after the dam's construction. Because the dam has no spillway, the Savio flows right over the dam in times of high water. The way the dam was constructed is unusual, but has proven successful. The dam is made of bricks, set in lime mortar, and bound together with wooden bars and poles. The dam's **abutments** are made of brick, and on one side of the abutment is a massive brick wall that both supports the dam and guides overflow away from the river bank. A canal, also made of brick, carries water to two small hydroelectric plants; originally, the water was delivered to several water mills downstream.

Additional Information: Smith, Norman. *A History of Dams.* Secaucus, NJ: Citadel Press, 1972.

Central Plaza Building (Wanchai, Hong Kong, China; 1992)

Designed by Ng Chun Man and Associates, the beautiful Central Plaza Building stands in Wanchai, east of Hong Kong's central district. The roughly triangular tower is 374 meters (1,227 feet) high. The base of the building is 30 meters (98 feet) high and set under a podium block. At street level, the building is bordered by a large, open landscaped garden and a public area covering 8,400 square meters. The building has 58 floors of office space and 5 floors for equipment (e.g., elevator rooms, heating and air conditioning systems). Floodlit at night, the building appears to glow in a number of different neon colors, especially gold.

Additional Information: Hong Kong Architecture at http://bcwww.cityu.edu.hk/b-s/cenplaz.html.

Century Freeway, now the Glenn Anderson Freeway (Los Angeles, California; 1981-1993)

This 17.3-mile (28-kilometer) stretch of eight-lane road runs from Los Angeles International Airport to Norwalk. As of 1994, the road was the costliest stretch of highway in the United States; it cost over $2.2 billion (approximately $127 million per mile). Roughly 46 percent of the cost represents money spent to support various community and social programs in the areas the road traverses to alleviate the social effects of construction. The highway is one of the last portions of the 42,796-mile (68,859-kilometer) national interstate highway system begun in the 1950s; less than 114 miles (183 kilometers) of the interstate system are left to complete. Running down the center of the Century Freeway is a rail line, the 23-mile (37-kilometer), 10-station Glenn Anderson Freeway-Transitway (named after a local politician). At its interchange with the San Diego Freeway (I-405), the Century covers more than 100 acres (40.5 **hectares**) and is more than seven stories high. This particular interchange has five levels, 7 miles (11.3 kilometers) of ramps, 11 bridges, and 2 miles (3.2 kilometers) of tunnels. The Century Freeway also has electronic sensors in its lanes, which monitor and control the flow of traffic into intersecting roads by automatically regulating traffic lights.

Additional Information: Reinhold, Robert. "Opening New Freeway, Los Angeles Ends Era." *New York Times*, October 14,

1993, A16; http://www.tmn.com/iop/introduction.html; http://www.tmn.com/iop/decree.html.

Chain Bridge (Georgetown area of Washington, DC; 1807)

James Finley built this **suspension bridge** across the Potomac River. Based on a patent held by Finley, the bridge design used cables made of iron chains. The use of chains caused concern—if one link failed, the whole bridge would fail. This concern proved unfounded. Although the bridge has been repaired several times, it is still in use today and is the main route from Washington, DC, to McLean, Virginia.

Additional Information: Kirby, Richard Shelton et al. *Engineering in History*. New York: Dover Publications, 1990.

Chain of Rocks Canal (St. Louis, Missouri; 1953)

In constructing this canal, the U.S. Army Corps of Engineers removed a chain of rocks immediately north of St. Louis and eliminated the final obstacle to full navigation on the Mississippi River. The canal has an average depth of 32 feet (9.8 meters), and is 300 feet (91.5 meters) wide on the bottom and 550 feet (168 meters) wide on top. At the time it opened, the canal's main lock was the largest in the Western Hemisphere; 110 feet (33.6 meters) in width, the lock can lift or lower a 1,200 foot-long (366-meter) train of barges.

Additional Information: Kirby, Richard Shelton et al. *Engineering in History*. New York: Dover Publications, 1990.

Champs Élysées (Paris, France; eighteenth century)

The Champs Élysées is one of the great avenues running along the right (north) bank of the River Seine in Paris. The avenue runs from the Place de la Concord to the ARC DE TRIOMPHE; it is one of 12 great avenues to form the Place de l'Étoile, the circular plaza in which stands the Arc. The Champs Élysées is celebrated for its tree-lined beauty; its shops, theaters, and cafes; and its fountains. The avenue was built and named by Louis XV in the eighteenth century, and ran through open country until the mid-nineteenth century, when Napoleon III rebuilt much of Paris.

Additional Information: Saalman, Howard. *Haussmann: Paris Transformed*. New York: G. Braziller, 1971.

Channel Tunnel, also known as the "Chunnel" (England-France; 1995)

Although the construction of a link between the island of Britain and mainland Europe involved highly creative resolutions to engineering challenges, obstacles to the tunnel's construction were more political than technical. Costing $15 billion, the Chunnel is Europe's largest public works project and one of the major engineering achievements of the twentieth century. The Chunnel runs underneath the 21-mile-wide (34-kilo-

A view of the Champs Élysées, one of the main avenues of Paris, France.

meter) English Channel, the narrow strait that separates the island of Britain from the continent of Europe. Much of the tunnel digging was automated, and special tunnel boring machines that trace their heritages back to Brunel's Shield were developed for the project. In the French terminal, a part of one of Kawasaki Japan's tunnel boring machines has been made into an art display and is shown as a "Symbol of Mankind and Machine." The idea of constructing a link between Britain and the rest of Europe goes back to Napoleonic times at the beginning of the nineteenth century. The idea met strong resistance from the English, who were reluctant to give up the island isolation that had often protected them from invasion and had allowed them to stand apart from the turmoils of Europe. However, when the decision was made to proceed with the tunnel, Great Britain threw itself willingly into the project. The 31-mile-long (50-kilometer) Chunnel consists of three interconnected tubes—two one-way rail lines and a service tunnel. Only trains are allowed to go through the Chunnel, although cars can be accommodated. The Eurostar is a high-speed train that runs the tunnel from London to Paris, and then on to Brussels in Belgium. Traveling at a maximum speed of 186 miles per hour (almost 300 kilometers per hour), the Eurostar can carry up to 794 passengers in various classes of seating. The train departs up to 15 times per day in each direction. A freight shuttle fire on 18 November 1996 limited passenger service until repairs were completed in May 1997.

Additional Information: http://www.doc.gov/ JTP.Mosaic.Materials/JTP/JTLB/JTP.JTLB.JT LB23/ 4.Chunnel.html; http://www.starnetinc.com/eurorail/ eurostar.Chunnel.html; Fairweather, Virginia. "The Channel Tunnel: Larger Than Life, and Late." *Civil Engineering* 64:5. (May, 1994), pp. 42-46.

Chapel Bridge of Lucerne, also known as the Kappelbrucke (Lucerne, Switzerland; 1333)

The Chapel Bridge, which crosses the Reuss River, is the oldest and longest covered bridge in Europe. Its name comes from the nearby Chapel of St. Peter, which stands at one entrance to the bridge. Since its original construction in the fourteenth century, the 657-foot (200-meter) bridge has been restored or reconstructed at least 10 times. A large part of the bridge was destroyed by fire in August 1994. The fire is believed to have been caused by a cigarette carelessly left burning in a rowboat moored to one of the bridge's supports. In the future, boats will not be allowed to moor under the bridge, although smoking on the bridge is still allowed. The bridge's most recent reconstruction, at a cost of more than $2 million, was completed in mid-1995 and was financed in part by the revenue from a special postage stamp. The roof of the bridge is gabled, and in the seventeenth century was adorned with 158 painted panels showing events in Swiss history, and in the life of the patron saints of the cities of Lucerne, Leger, and Maurice. About two-thirds of the paintings were destroyed or damaged by the fire, and the reconstructed bridge will boast copies of the lost art. At the center of the bridge is a water tower that was also damaged by the fire and was rebuilt. The tower had served as a prison from the fourteenth to the nineteenth centuries.

Additional Information: "Travel Advisory: Lucerne's Rebuilt Chapel Bridge Open Again," *New York Times*, May 29, 1994, V:3.

Chapel of Saint Peter ad Vincula. *See* Tower of London.

Charles Bridge. *See* Karlsbrucke Bridge.

Chartres Cathedral (Chartres, France; 1194-1230)

The Cathedral of Notre Dame in the city of Chartres in north central France is considered one of the world's great masterpieces of Gothic architecture. Because the cathedral was built relatively quickly—during a period of some 30 years in the early thirteenth century—its design is almost pure Gothic, resulting in an architectural unity that is almost unique. The stylistic consistency of the cathedral is marred only by the north spire, which was added around 1507 when the Gothic style had changed from its original conception. A fire in 1194 destroyed the original eleventh-century cathedral that stood on the site; only the crypt, the base of the towers, and the west facade of the original structure remain. Because this earlier church had been a place of pilgrimage dedicated to Mary, Chartres received a "tunic" supposedly worn by Mary during Jesus' birth. This garment is now in the cathedral treasury. The present cathedral is known for its beautiful stained-glass windows, its exquisite Renaissance choir screen, and its hundreds of intricately carved religious sculptures. The huge, circular rose window at the west end, facing the square at the front of the cathedral, is the most well known of the church's 160 medieval stained glass windows. Chartres has two bell towers, one rising 378 feet (115 meters) and the other 350 feet (107 meters). The English occupied Chartres for 15 years during the Hundred Years' War, and the town came under attack by French Protestants during the wars of religion in the sixteenth century. In 1594, Henri IV, a Protestant who turned Catholic to attain the throne, was crowned in Chartres Cathedral. The town was severely damaged during World War II, but the cathedral survived largely unhurt.

Additional Information: Branner, Robert. *Chartres Cathedral.* New York: Norton, 1969; http://member.aol.com/detechmendy/capital.html; http://www.designbase.com/was/038.htm.

Cheesman Dam (Denver, Colorado; 1905)

Named after its designer, Walter Scott Cheesman, the Cheesman Dam was, at its completion, the world's highest gravity-arch stone dam and the first gravity-arch dam in the United States. The dam has an arch radius of 400 feet (122 meters). The structure, which stands on the South Platte River 48 miles southwest of Denver, is an essential contributor to the water needs of the Denver area. Without the water it contributed, Denver's growth would have been impossible. Because of the tremendous need for water in the Denver area, Cheesman, a local businessman, designed the dam and first attempted to build it in 1897, but the incomplete structure was destroyed in 1900 by the water from a torrential rainfall. Construction on a new dam began in August 1900. Completed in less than five years, the dam is considered an amazing feat of engineering. The dam stands 221 feet (67 meters) above the stream bed and is 1,100 feet (335.5 meters) long. It is 176 feet (54 meters) thick at its base and 18 feet (5.5 meters) thick at the top, which also accommodates a 14-foot-wide (4.3-meter) roadway. Although native granite was used, all other material, such as nearly 42 million pounds (over 92 million kilograms) of cement, had to be brought in across rough, mountain terrain. The dam creates a reservoir that holds 79,100 acre-feet of water and has a surface area of 875 acres (354 **hectares**) and a shoreline of 18 miles (29 kilometers) in circumference. The land around the reservoir and dam is federal forest, protecting the water from environmental danger. Almost 100 years after its construction, the dam is still vital to the water supply of the Denver area. In 1973, the dam was designated a National Civil Engineering Historic Landmark.

Additional Information: Schodek, Daniel L. *Landmarks in American Civil Engineering.* Cambridge, MA: MIT Press, 1987; http://www.water.denver.co.gov/cheesman.htm.

Chehalis River Riverside Bridge (Chehalis, Washington; 1939)

The Chehalis River Riverside Bridge carries State Route 6 across the Chehalis River near the Lewis County town of Chehalis in southwestern Washington. A riveted Warren truss with verticals, this strong bridge is testament to the durability of the Warren truss design.

Source: http://wsdot.wa.gov/eesc/environmental/Bridge-WA-111.htm.

Chemin de Fer Métropolitain. *See* Paris Metro.

Chesapeake and Delaware Canal (Maryland-Delaware; 1803-1829)

Known popularly at the time of construction as the "Deep Cut," the Chesapeake and Delaware Canal was a major engineering feat. Cutting mostly through granite rock, builder John Randle, Jr.'s construction crew of 2,500 men, a huge crew for the time, worked only with hand tools. Black powder was used for blasting. The cut was as big as 90 feet (28 meters) deep and 60 feet (18.3 meters) wide in some places. The 13-mile-long (21-kilometer) canal connected the northern end of Chesapeake Bay with Delaware Bay at the mouth of the Delaware River. The canal opened on 17 October 1829. Built at the cost of more than $2 million, a huge sum for the time, the Chesapeake and Delaware was the most expensive canal project in the country, as well as one of the most important. Almost always in financial trouble, the canal's assets were bought and rebuilt by the federal government in 1919; the rebuilt canal could accommodate larger, ocean-going ships. Still in operation, the canal remains valuable for Chesapeake Bay shipping.

Additional Information: Shaw, Ronald E. *Canals for a Nation.* Lexington: University Press of Kentucky, 1990; Spangenburg, Ray and Moser, Diane K. *The Story of America's Canals.* New York: Facts on File, 1992.

Chesapeake and Ohio (C&O) Canal (Washington, DC, to Cumberland, Maryland; 1828-1850)

The Maryland legislature chartered the Chesapeake and Ohio Canal Company and the Baltimore and Ohio Railroad Company on the same day in 1825. The canal and the railroad met at Point of Rocks in the Potomac Valley northwest of Washington, DC. From here, the two ran on to Harper's Ferry, Virginia (now West Virginia), further up the Potomac. Controversy over the proper right-of-way for the canal after this point stalled work on the project for four years until the courts could decide the matter. The canal was planned to run to the Ohio River, but actually stops at Cumberland, the eastern terminus of the CUMBERLAND ROAD, in western Maryland. Competition from railroads prevented the C&O Canal from going any further. The canal runs 184 miles (296 kilometers) along the Potomac River. Flooding in the 1880s caused heavy damage to the canal, which was sold to the BALTIMORE AND OHIO RAILROAD in 1899. The canal was closed permanently in 1924 following another massive flood. The federal government acquired the property in 1938, intending to use the canal as the foundation for a parkway. However, Supreme Court Justice William Douglas, supported by the editors of the *Washington Post*, opposed the road project. Mustering public support, the project's opponents convinced the government to maintain the canal under the auspices of the National Park Service. The canal is now officially designated the Chesapeake and Ohio Canal National Historical Park.

Additional Information: Shaw, Ronald E. *Canals for a Nation.* Lexington: University of Kentucky Press, 1990; http://www.fred.net/kathy/canal.html.

Chesapeake Bay Bridge-Tunnel (Norfolk, Virginia; 1964)

The Chesapeake Bay Bridge-Tunnel is over 17 miles (28 kilometers) long. The bridge-tunnel combination crosses the mouth of the Chesapeake Bay, connecting Norfolk, Virginia, with the Eastern Shore of Virginia and Maryland. The bridge sections are of precast **prestressed-concrete** trestle construction, and cover almost 15 miles (24 kilometers) of the bay crossing. The project required four wide, high-level navigation channels. Two of the channels are spanned by bridges, but the other two are crossed by sunken tunnels, each about 1.2 miles (2 kilometers) long. Out in mid-bay, the transitions from trestle bridge to tunnel and back again were achieved by the construction of four man-made islands built of sand, stone, and concrete, each 1,476 feet (450 meters) long and 230 feet (70 meters) wide. The tunnel sections are double-skinned steel tubes, 295 feet (90 meters) long, towed from the Texas Gulf coast as empty, buoyant shells, then sunk into position in a dredged trench by filling their hollow walls with concrete.

Additional Information: Overman, Michael. *Roads, Bridges and Tunnels.* Garden City, NY: Doubleday & Company, Inc. 1968.

Chicago Water Supply System (Chicago, Illinois; 1869)

Designed by Ellis S. Chesbrough, the Chicago water system takes advantage of nearby Lake Michigan for its fresh water. Chesbrough's somewhat revolutionary plan involved driving a tunnel toward the waters of Lake Michigan. Despite doubt and debate, Chesbrough's plan was adopted. Construction began in May 1864, and proceeded for three years with few surprises. The innovative creation of side chambers in the tunnel allowed specially trained mules to haul out clay being dug and bring in supplies and materials. With the addition of a pumping tower, the system began to supply the city with water in 1866. The tower and the pumping station still stand as landmarks. The Chicago water system is one of the major achievements of nineteenth-century civil engineering.

Additional Information: Schodek, Daniel L. *Landmarks in American Civil Engineering.* Cambridge, MA: MIT Press, 1987.

Chicago's Underground Tunnels (Chicago, Illinois; 1898-1904)

Most cities have unknown and forgotten tunnels, but Chicago's tunnels are unique; they were used until the 1940s for commerce, and are still used today by utilities. Chicago's tunnels were built to carry freight and coal from the river to various sites inside the city, and for delivery of office mail from one building to another. Air shafts were even built to nearby theaters to deliver them constant 55-degree air in an age without air conditioning. Small rail lines were built in the tunnels to efficiently move loads. Elliptical in shape, the tunnels typically are about 7.5 feet (2.3 meters) high and about 6 feet (1.8 meters) wide. Beginning in 1973, when the original builders went out of business, the tunnels were used for utility lines. Even the telephone and electrical utilities have no real idea of how many tunnels there are, or where they all go. Few Chicago residents even knew the tunnels existed until April 1992, when the Chicago River broke through a tunnel entrance, flooding most of the tunnels and threatening the utility lines they housed.

Additional Information: Moffat, Bruce. *Forty Feet Below: The Story of Chicago's Freight Tunnels.* Pasadena, CA: Interurban Press, 1982; Terry, Don. "Chicago's Well-Kept Secret: Tunnels." *New York Times,* April 15, 1992, D:27.

Chicoasén Dam (Grijalva River, Mexico; 1980)

This **rockfill** structure is one of the world's highest dams, with a height above its lowest formation of 261 meters (856 feet). It crosses the Grijalva River in the state of Chiapas in southern Mexico.

China, Great Wall of. *See* Great Wall of China.

Chirkey Dam, also spelled Chirkei (Russia; under construction since 1971)

When completed, Chirkey will be a massive arch dam, 764 feet (233 meters) high and 1,109 feet (338 meters) long. The dam itself will have a volume of 1,604,000 cubic yards; its reservoir will have a capacity of 2,250,000 **acre-feet**.

Additional Information: Smith, Norman. *A History of Dams.* Secaucus, NJ: Citadel Press, 1972.

Choate Bridge (Ipswich, Masschusetts; 1764)

Named after its builder, John Choate, the Choate Bridge, the second oldest existing stone bridge in the United States, carries U.S. Route 1A over the Ipswich River at the town of Ipswich in northeastern Massachusetts. The bridge has a total length of 81.5 feet (25 meters). Its two elliptical arches each have a 30-foot (9.2-meter) span and a 9-foot (2.7-meter) rise.

Additional Information: DeLony, Eric. *Landmark American Bridges.* Boston: Bullfinch Press, 1993.

Chongqing Office Tower (Chongqing, China; expected completion in 1999)

When completed, this tower will be the tallest building in the world at 460 meters (1,509 feet). Its supremacy is likely to be short-lived, however; it will be equaled in height shortly after the turn of the century by the MIL-LENNIUM TOWER in Tokyo.

Additional Information: "Edifice Complex," August, 1995, *Asia, Inc.*, in Asia, Inc. Online: http://www.asia-inc.com/archive/0895edifice.html.

"Chunnel." *See* Channel Tunnel.

Church of the Bleeding Saviour (St. Petersburg, Russia; 1883-1907)

The Church of the Bleeding Saviour is built on the site where a terrorist from an organization called "People's Will" mortally wounded Czar Alexander II with a bomb on 1 March 1881. The czar's son, Alexander III, began a memorial church in 1883, but it was not completed until 1907 under Nicholas II, Alexander II's grandson. The church was intentionally modelled on SAINT BASIL'S CATHEDRAL in Moscow. After the Russian Revolution in 1917, the church was turned into a warehouse. Soviet authorities had planned to make the church into a museum to honor People's Will, but the Russian Republic currently plans to dedicate the church and a museum to Alexander II, although neither building has yet opened. The church has several other names, all used by the local population, including Resurrection of Christ, Assumption, Church of the Redeemer, and Savior of the Spilled Blood.

Additional Information: http://www.spb.su/fresh/sights/bleeding.html.

Cimitière du Père Lachaise (Paris, France; late eighteenth century)

Covering 44 **hectares** (109 acres), this cemetery is perhaps the most famous in the world. It contains more than 100,000 burial places, and over 1 million burials. The tree-lined Avenue Transersale runs straight through the center of the cemetery. Among the famous people interred here are the English writer Oscar Wilde, the Polish composer Frederic Chopin, and the American rock singer Jim Morrison. Originally a park in a Paris suburb named after Cardinal Lachaise, one of the ministers of Louis XIV in the seventeenth century, the property was purchased in the late eighteenth century by Napoleon for use as a cemetery during the French Revolution. The expansion of Paris in the twentieth century has brought the Cimitière du Père Lachaise within the city limits.

Additional Information: http://www.io.org/cemetery/Lachaise/lachaise.intro.html; Harley, Adam. "Cimitière du Père Lachaise" at http://www.users.globalnet.co.uk/~lilth/tcimit.htm.

Cincinnati Suspension Bridge (Cincinnati, Ohio; 1866)

John Roebling, the designer of the BROOKLYN BRIDGE, designed and built the Cincinnati Suspension Bridge, which crosses the Ohio River to connect Cincinnati with Covington, Kentucky. When completed, the structure was the longest spanning bridge in the world at 1,057 feet (322 meters). Washington Roebling, John's son, supervised cable spinning and other construction activities on the Cincinnati bridge, which served the Roeblings as a prototype for the later Brooklyn Bridge. Because of its importance, the Cincinnati Suspension Bridge was one of the few private civil works projects that continued during the Civil War. Because of the heavy traffic, the **suspension bridge** was stiffened and new cables were added in later years. The original **finials** on the towers were removed in 1899, but restored in the 1980s and 1990s.

Additional Information: DeLony, Eric. *Landmark American Bridges.* Boston: Bullfinch Press, 1993.

City Tunnel No. 3, also known as the Third New York City Water Tunnel (New York City; completion expected 2002)

Although this tunnel is meant to supplement New York City's two existing water tunnels, it will also allow the first two tunnels to be drained, inspected, and repaired, if necessary. Tunnel # 1 was put into service in 1917; the second tunnel was placed in service in 1937. The second stage of the new tunnel is 16 miles (26 kilometers) long—11 miles (18 kilometers) in Queens and Brooklyn and 5 miles (8 kilometers) in Manhattan—and will cost about $1.5 billion. Work on the first stage of the tunnel began in 1970; the current, third stage was begun in 1986 and is expected to be finished in 2002. The nearly completed first stage cost $1.2 billion, and runs 13 miles (21 kilometers) through the Bronx, Manhattan, and Queens. Overall, the new tunnel will be 55 miles (89 kilometers) long, with a 10-foot (3.1-meter) by 24-foot (7.3-meter) diameter. To date, 20 lives have been lost in construction accidents. Construction of a fourth stage of the project is scheduled for the twenty-first century; it will take the tunnel through the eastern areas of Queens and the Bronx.

Additional Information: Chiles, James R. "'Remember, Jimmy, Stay Away from the Bottom of the Shaft!' (Water Tunnel Number Three, New York, New York)," *Smithsonian* 25, (July 1994) p. 60.

Cleft Ridge Span (Prospect Park, Brooklyn, New York; 1872)

The Cleft Ridge Span, built by Calvert Vaux, was the first concrete **arch bridge** in the United States. It has a 20-foot (6.1-meter) span and a 60-foot-wide (18.3-meter) pedestrian underpass. The bridge was constructed of artificial, pre-cast stone ("beton Coignet"), which could handle the bridge's elaborate Gothic-style decorative workings more cheaply than sculpted stone. Vaux's masonry bridges, many of which are in New York City parks, are notable for their attractive and delicate decorations.

Additional Information: Jackson, Donald C. *Great American Bridges and Dams*. Washington, DC: The Preservation Press, 1988; Jacobs, David and Neville, Antony E. *Bridges, Canals and Tunnels*. New York: American Heritage Publishing Company, 1968.

Cleopatra's Needles (originally Heliopolis, Egypt; c. 1500 B.C.)

In the late nineteenth century, the government of Egypt divided a pair of obelisks, giving one to the United States and the other to Great Britain. The American obelisk now stands in Central Park in New York City; the British obelisk stands along the Thames embankment in London. Both obelisks are known as Cleopatra's Needle, though neither one has any historic connection with Cleopatra, the first-century B.C. queen of Egypt. The two monuments are much older, originating in Heliopolis where they were dedicated by Pharaoh Thutmose III about 1500 B.C. Each obelisk bears incriptions to Thutmose and to the later pharaoh Ramses II, who died about 1237 B.C. Carved from red granite, the obelisks stand over 69 feet (21 meters) high and weigh 180 tons (162 metric tons). The methods originally used by the ancient Egyptians to quarry and raise these large monuments are still not fully understood.

Additional Information: Hayward, R.A. *Cleopatra's Needles*. Buxton, England: Moorland Publishing Company, 1978.

Clifton Suspension Bridge (Bristol, England; 1864)

William Vick, a Bristol merchant, bequeathed money to the city for the building of a bridge in 1753. In 1829,

This ancient Egyptian obelisk in New York's Central Park is one of two, each known as Cleopatra's Needle; its mate stands along the Thames in London, England. Photo courtesy of the Library of Congress, Detroit Publishing Company Photographic Collection.

the bridge society which had controlled and invested Vick's bequest held a bridge design competition and purchased land for the structure. Isambard Kingdom Brunel submitted four designs for a timber bridge, but none were chosen. Brunel then resubmitted to a second competition and won. Construction of the bridge across the Avon Gorge was delayed until 1836 by the Bristol riots of 1831. In 1837, the contractor went bankrupt, but some work continued on the structure until 1843. When Brunel died in 1859, the piers and the tower were still incomplete. After Brunel's death, his colleagues at the Institution of Civil Engineers formed a company to complete the work. Brunel had distrusted the use of iron in bridges, but the new bridge company made the timber deck Brunel had proposed of **wrought-iron**. Chains from another Brunel bridge, which was being demolished at the time, were also used, along with other modifications. The 245-foot (75-meter) bridge, although not what Brunel had planned, is a practical interpretation of what he would have built had he been able to complete his plan. The Clifton Bridge is the last surviving echo of Brunel's work, for all the timber bridges Brunel constructed are now gone, the last having been disassembled in 1934.

Additional Information: Brown, David J. *Bridges*. New York: Macmillan, 1993; Hopkins, H.J. *A Span of Bridges: An Illustrated History*. New York: Praeger, 1970; http://pyt.avonibp.co.uk/Gallery/Clifton/Bridge.html.

CN Tower (Canadian National Railroad Tower)

CN Tower (Canadian National Railroad Tower)
(Toronto, Canada; 1976)

One of the tallest free-standing towers in the world, the 533.3-meter-high (1,749-foot) CN Tower is an amaz-

The CN (Canadian National Railroad) Tower rises from the Toronto skyline.

ing attraction both for tourists and Toronto residents. The foundation of the tower, which required 7,650 cubic meters (9,945 cubic yards) of concrete, is specially protected from winter cold and carefully designed to slope inward in height. From top to bottom, the tower varies from true vertical no more than 2.7 centimeters (1.05 inches). Among the many attractions of the tower are glass elevators moving to the observation deck at a speed of 6 meters (20 feet) per second, a glass floor in the outside observation deck that gives visitors a feeling of what it was like for the construction team to work at such a height, a motion-simulator movie theater, and CYBERMIND, Canada's first virtual reality center.

Additional Information: http://ppc.westview.nybe.north-york.on.ca/Ernest/CNTower.html; http://www.banfdn.com/cntower.htm (for construction details).

Coalbrookdale Bridge (west central England; 1779)

The Coalbrookdale Bridge, the first significant iron bridge, was built by Abraham Darby III and John Wilkinson to replace a ferry across the Severn River in western England. The bridge, a semicircular arch of five parallel **cast-iron** ribs, has a 100-foot (30.5-meter) span, with cross braces supporting a level roadway 24 feet (7.3 meters) wide and 55 feet (17 meters) above the river. Each rib was cut in halves 70 feet (21 meters) long, weighing about 38 tons (34 metric tons). The ribs were floated from a nearby foundry on barges, raised with block and tackle, and then joined in the center with cast-iron bolts. The total weight of the bridge is about 378 tons (340 metric tons). The heavy weight of the supporting banks eventually tilted the **abutments** inward, causing the arch to have a slight point at the center. The bridge is still standing and is now a national monument.

Additional Information: Kirby, Richard Shelton et al. *Engineering in History*. New York: Dover Publications, 1990.

Coedty Dam. *See* Eigau Dam/Coedty Dam.

Coleman, George P. Memorial Bridge. *See* George P. Coleman Memorial Bridge.

Coleroon Dams (Coleroon River, Tanjore, India; second century A.D.)

The original dam on the Coleroon River in southern India was a masonry dam set in clay, 1,080 feet (55 meters) long, 15 to 18 feet (4.6 to 5.5 meters) high, and 40 to 60 feet (12.2 to 18.3 meters) thick. By the early nineteenth century, the dam and its canals were heavily silted and a new dam was built by Arthur Cotton in 1836. That dam was not well constructed; after several failures and leaks, it was rebuilt again at the end of the nineteenth century. The new structure was the first dam to draw attention to the important problem of piping, a phenomenon in which a dam's foundation is undermined by water percolating under it.

Additional Information: Smith, Norman. *A History of Dams.* Secaucus, NJ: Citadel Press, 1972.

College Park Airport (College Park, Maryland; 1909)

The College Park Airport is the world's oldest continually operated airport, a national historic site, and the location of many aviation firsts. The first woman passenger to fly in an airplane embarked here in 1909, the first army aviation school was established here in 1911, the first mile-high flight was made here in 1912, the first controlled helicopter flight was made here in 1924, and the first use of radio aids to support instrument landings was conducted here in 1927. Still fully operational, the airport has an asphalt runway that is 2,610 feet (796 meters) long. An airport museum contains a history of early aviation and photographs and aviation artifacts.

Additional Information: Doug Nebert in http://www.wp.com/avianet/cgs.html.

Colosseum (Rome, Italy; A.D. 80)

The Colosseum is a giant amphitheatre built in Rome in the late first century A.D. by the Flavian emperors Vespasian and Titus, and so known originally as the Flavian Amphitheatre. Begun about A.D. 70 by Vespasian, the structure was dedicated in A.D. 80 by Titus with an elaborate ceremony that included 100 days of games. In A.D. 82, Emperor Domitian completed the structure by adding the uppermost story. Most earlier Greek and Roman amphitheatres were built into hillsides for extra support, but the Colosseum is a completely freestanding structure of stone and concrete. The oval Colosseum measures 620 feet (189 meters) by 513 feet (157 meters), and could seat 50,000 spectators. The Romans used the Colosseum to stage hand-to-hand cambats between gladiators, contests between humans and animals, and mock naval engagements, for which the amphitheatre was flooded. In medieval times the structure was damaged by lightning and earthquakes, and more severely in recent times by pollution, traffic, and vandalism. Today, only the stone shell of the orginal structure remains; all the marble seats and decorative material have disappeared.

Additional Information: Pearson, John. *Arena: The Story of the Colosseum*. New York: McGraw-Hill, 1973.

Colossus Bridge at Philadelphia

(Philadelphia, Pennsylvania; 1812)

The Colossus Bridge, built by Lewis Wernwag, was a covered arch and truss bridge with a clear span of 340 feet (104 meters); it was perhaps the longest bridge ever built of wood or stone. It crossed the Schuykill River west of Philadelphia in an area that became Fairmount Park. It used five parallel arch ribs, braced and cross-braced, and also added iron rods in every panel to adjust joints. Wood pieces were fastened with iron bolts and links without **mortising** so that joints could be adjusted if they came loose. The arch rose a low 20 feet (6.1 meters), and the roadway had less than half that height. The bridge was destroyed by fire in 1840 and replaced by the FAIRMOUNT BRIDGE.

Additional Information: Kirby, Richard Shelton et al. *Engineering in History*. New York: Dover Publications, 1990.

Colossus of Rhodes (harbor entrance of the island of Rhodes, Greece; 294-282 B.C.)

Chares of Lindos sculpted the gigantic statue of the Greek sun god Helios that stood at the harbor entrance of Rhodes, an island in the eastern Mediterranean. The Colossus was one of the SEVEN WONDERS OF THE ANCIENT WORLD. The statue was paid for with money raised from the sale of military equipment left on the island when Greek invaders from the mainland made peace and lifted a siege against three city-states on Rhodes. The inscription on the statue read:

> To you, O Sun, the people of Dorian Rhodes set up this bronze statue reaching to Olympus when they had pacified the waves of war and crowned their city with the spoils taken from the enemy. Not only over the seas but also on land did they kindle the lovely torch of freedom.

Construction of the statue to commemorate the peace took 12 years. The 102-foot (31-meter) statue was constructed near the entrance of Mandraki, one of

The Colosseum is a huge freestanding amphitheatre built in Rome in the late first century A.D.

several harbors into the city of Rhodes. Standing on a base of white marble, the feet and ankles of the statue had been placed first, and the rest built up from there. An earth ramp was built to reach the higher portions of the statue. Cast in bronze, the statue was reinforced with an inner iron and stone framework. When an earthquake hit the city in 226 B.C., the statue broke at the knee. The Roman writer Pliny recorded that "few people can make their arms meet round the thumb" as the statue lay on the ground. Ptolemy III of Egypt offered to fund the statue's restoration, but his offer was refused after the city fathers consulted with an oracle. The statue's ruins lay on the island until A.D. 654, when a Jewish trader from Syria purchased the remains; legend says 900 camels were required to transport the fragments of the statue.

Additional Information: Ashmawy, Alaa K., ed., "The Seven Wonders of the Ancient World," at http://pharos.bu.edu/Egypt/Wonders/colossus.html.

Commodore Barry Bridge (Bridgeport, New Jersey, to Chester, Pennsylvania; 1974)

This much-needed bridge across the Delaware River replaced a ferry service that had operated in the area for 20 years. The Commodore Barry Bridge is one of the longest cantilevered highway bridges in the world. Five lanes wide, it has a main span of 1,622 feet (495 meters) and reaches a high point of 418 feet (128 meters) above the water. With its approaches, the bridge has a total length of 4 miles (6.4 kilometers).

Additional Information: Bishop, Gordon. *Gems of New Jersey*. Englewood Cliffs, NJ: Prentice-Hall, 1985.

Confederation Bridge, also known as the Northumberland Strait Crossing (Prince Edward Island to New Brunswick, Canada; 1997)

Canada's largest concrete bridge, the Confederation Bridge is a multi-span concrete girder structure that runs 12 kilometers (7.4 miles) across the narrowest point of

A main girder and marine jetty of the Confederation Bridge under construction in 1995. Upon its completion in 1997, the bridge connected Prince Edward Island with New Brunswick. Photo courtesy of Marlene Boily, Prince Edward Island, Canada.

the Northumberland Strait. The bridge's 44 main spans total 11,000 meters (36,080 feet) in length. The typical span is 250 meters (820 feet) long. The bridge has a width of 11 meters (36 feet), allowing for one lane of traffic and one emergency shoulder in each direction. A specially durable hydraulic cement was used in the bridge's construction to deal with the challenge of frequent ice in the strait below. The support piers are placed in water up to 115 feet (35 meters) deep, and are protected by conical ice shields. Portions of the bridge were floated to their position, and then set in place by the Svanen, the world's largest floating catamaran crane.

Additional Information: http://www.peinet.pe.ca/SCI/br_d.html.

Contra Dam (near Lugano, Switzerland; 1965)

This arch dam is one of the world's highest dams at 754 feet (230 meters). The dam is 1,246 feet (380 meters) long and has a volume of 863,000 cubic yards (660,000 cubic meters) and a reservoir capacity of 69,000 **acre-feet**. In the 1996 movie *Golden Eye*, a bungee jump is staged from the top of the Contra Dam.

Additional Information: Smith, Norman. *A History of Dams*. Secaucus, NJ: Citadel Press, 1972.

Coombs Dam. *See* Peak Forest Canal.

Coos Bay Bridge. *See* McCullough Memorial Bridge.

Coquet Dam (County Durham, northern England; 1776)

Coquet Dam, on the Coquet River in County Durham, is one of many still-existing English dams built in the eighteenth century by John Smeaton. The dam is curved to a radius of 170 feet (52 meters) at the air face, is 8 feet (2.4 meters) thick at its base, and stands 8 feet high. The dam is built of a rubble-masonry core faced with large masonry blocks; the water face is sealed with earth covered with rubble-masonry and protected against **scour** by flagstones laid on their edges. The **abutments** have diversion openings that were used during construction so the foundation could be laid dry. Smeaton's instructions for use of the diversion openings are the earliest known directions specifying how to lay a dam foundation on dry ground.

Additional Information: Smith, Norman. *A History of Dams*. Secaucus, NJ: Citadel Press, 1972.

Corinth Canal (Corinth, Greece; 1882-1893)

Work on a canal across the Greek isthmus of Corinth began under the Roman Emperor Nero in A.D. 67, but was abandoned two years later when Nero committed suicide after being deposed by the army. In 1879, a Hungarian named Istvan Turr negotiated the construction of a canal at Corinth with the king of Greece. French contractors began work on the canal in 1882 under the direction of another Hungarian named Bela Gerster. Work on the canal stopped when the canal company that had been formed to complete the project collapsed. In 1890, a newly formed Greek company took over the project and resumed construction. The canal opened on 6 August 1893; it was damaged during World War II, but has been repaired and improved and is still in use today. The 4-mile-long (6.4-kilometer) canal connects the Gulf of Corinth on the west with the Saronic Gulf to the east. The Corinth Canal is 8 meters (26 feet) deep and varies in width from 22 to 25 meters (72 to 82 feet).

Additional Information: Hadfield, Charles. *World Canals*. Newton Abbot, England: David & Charles, 1986; Hawkes, Nigel. *Structures: The Way Things Are Built*. New York: Macmillan, 1990.

Cornell Dam. *See* New Croton Dam.

Cornish-Dingleton Hill Bridge #22 (Cornish, New Hampshire; 1882)

James Tasker, who is known to have built at least 11 covered bridges in the New England area, built this bridge across Mill Brook in western New Hampshire. Tasker assembled the 77-foot-long (24-meter) bridge in a school yard and then moved it to its final site. A multiple kingpost truss design, the bridge cost only $812; it has an extraneous beam above the opening to warn approaching users of loads that are too high for the bridge.

Additional Information: http://vintagedb.com/guides/covered2.html.

Cornish-Windsor Covered Bridge (Cornish, New Hampshire, to Windsor, Vermont; c.1860)

James Tasker and Bela Fletcher built this two-span covered bridge across the Connecticut River to connect Cornish, New Hampshire, with Windsor, Vermont. The bridge, which today carries State Route 44 over the river, measures 460 feet (140 meters) between **abutments** and is the longest American covered bridge still in existence.

Additional Information: DeLony, Eric. *Landmark American Bridges*. Boston: Bullfinch Press, 1993; Schodek, Daniel L. *Landmarks in American Civil Engineering*. Cambridge, MA: MIT Press, 1987.

Cortland Street Drawbridge (Chicago, Illinois; 1902)

The Cortland Street Drawbridge, built by Edward Willman and John Ericson, was the first "**Chicago-type**" trunnion-**bascule bridge** opened to traffic in the United States. Although the bascule is no longer used, the bridge could both raise and swivel through a maximum angle of 76 degrees. With various modifications, 50 out of the 53 movable bridges in Chicago are of this type.

Additional Information: Jackson, Donald C. *Great American Bridges and Dams*. Washington, DC: The Preservation Press, 1988; Schodek, Daniel L. *Landmarks in American Civil Engineering*. Cambridge, MA: MIT Press, 1987.

Couzon Dam (Rive-de-Gier, France; 1788-1811)

Couzon Dam stands 4 miles (6.4 kilometers) south of Rive-de-Gier on the River Couzon in southeastern France. The central masonry core of the dam is 656 feet (200 meters) long and 108 feet (33 meters) high. The crest is 16 feet (4.9 meters) thick and the base is 22.5 feet (6.9 meters) thick. The downstream end of the dam is a 17-foot-thick (5.2-meter) masonry wall. The upstream earth bank reaches 170 feet (52 meters) into the reservoir and ends in a thin masonry wall 35 feet (10.7 meters) high. As long as the reservoir is more than half full, the upstream bank is completely submerged. Two tunnels run through the deepest portion of the embankment. The lower tunnel follows the course

of the River Couzon and is a **scouring** gallery (used to desilt the reservoir). The upper outlet tunnel has valves at the water-face end. The dam also has two spillways with a combined length of 131 feet (40 meters). Built originally to supply water to the now unused Canal de Givors, the dam is currently used to supply water for Rive-de-Gier. When first built, the dam had some leaks and suffered some sliding, but it was rebuilt in 1895-96 and now successfully holds back 1,180 **acre-feet** of water.

Additional Information: Smith, Norman. *A History of Dams*. Secaucus, NJ: Citadel Press, 1972.

Craigellachie Bridge (Craigellachie, Scotland; 1815)

Thomas Telford built this spectacular iron **arch bridge** across the River Spey in northern Scotland. The Craigellachie Bridge inspired the English poet Robert Southey to write the following: "As I went along the road by the side of the water I could see no bridge; at last I came in sight of something like a spider's web in the air. . .and, oh, it is the finest thing that ever was made by God or man!"

Additional Information: Brown, David J. *Bridges*. New York: Macmillan, 1993.

Craponne Canal (Rhône Valley to Provence, France; 1554)

Built by Italian-born Adam de Craponne, the Craponne Canal is France's first large man-made waterway. Over 100 miles (161 kilometers) long, with several branches, the canal was designed to carry water from the Durance River eastward to Provence, a region of southeastern France on the Mediterranean. The dam at Cadenet, once part of the waterway, is now gone, but the canal still carries water into Provence.

Additional Information: Smith, Norman. *A History of Dams*. Secaucus, NJ: Citadel Press, 1972.

Crazy Horse Memorial (outside Custer, South Dakota; scheduled completion 2048)

Begun in 1948 and scheduled to be dedicated in 1998, artist and designer Korczak Ziolkowski's monument to Crazy Horse is expected to be completed by 2048. Ziolkowski took up the challenge of Sioux Chief Henry Standing Bear, who had seen the gradual construction of the MOUNT RUSHMORE project, to create something to show that "the red man has great heroes, too." Unlike Rushmore, the Crazy Horse monument is being dynamited rather than carved out of a mountain. Although Ziolkowski died in 1982, his wife, Ruth, and many of his children continue to work on the massive statue of the famed Indian chief riding a horse. Beginning with small donations, and always refusing government assistance, the Ziolkowskis have completed the head for the dedication in 1998. Funding now comes from private donations, some small and some corporate. Other monies come from admission fees; 1.2 million people visited the monument in 1996. When finished, the statue will be about 700 feet (213.5 meters) tall, larger than any of the Mount Rushmore presidents. At the bottom of the monument will be a round, "hogan-shaped" museum and a university and medical training center for American Indians. Some exhibits and community college courses are already being given in a building near where the base of the statue will be.

Additional Information: Brooke, James. "Crazy Horse Is Rising in the Black Hills Again." *New York Times*, June 29, 1997, A:16.

Croton Aqueduct (New York City; 1837-1842)

The New York City Water Supply System was an engineering model for similar city water systems throughout the United States. The 45-mile-long (72-kilometer) Croton Aqueduct, which ran from an upstate dam on the Croton River to reservoirs in New York City, was the most significant feature of the system. The most significant reservoir in the city was located at Fifth Avenue and 41st Street, the current site of the New York Public Library. Apart from wells, earlier city water delivery systems were unsatisfactory. Because there were no fire hydrants, fire fighters had to drill holes in underground supply pipes and fill in the holes after the fire was out. The process gave rise to the term "fireplug." Despite the tremendous cost involved, city voters in 1835 passed a proposal to bring water from the Croton by a ratio of nearly 3 to 1. Most of the work was done under the direction of John Jervis, an experienced canal builder. Disaster struck the project in 1841 when the unfinished dam was unable to contain flood waters and the earth collapsed near the dam. The resulting flood caused three deaths and destroyed bridges, mills, and houses downstream. Jervis rebuilt the dam, this time using a "**stilling basin**" to protect the dam itself. After completing various other portions of the project, Jervis brought the first water to New York City in the summer of 1842. Although the system was augmented and replaced in the 1880s by the NEW CROTON AQUEDUCT, the original structure was the model for the type of system necessary to bring fresh water to a modern, bustling city.

Additional Information: Kirby, Richard Shelton et al. *Engineering in History*. New York: Dover Publications, 1990; Schodek, Daniel L. *Landmarks in American Civil Engineering*. Cambridge, MA: MIT Press, 1987.

Croton Dam (New York City; 1837-1842)

John B. Jervis built this dam across the Croton River as part of a scheme to bring water to New York City along the 45-mile (72-kilometer) CROTON AQUEDUCT. The Croton system was one of the earliest and most successful schemes to bring water to New York City. Although the Croton Dam was originally intended to be an earthen

dam, various problems, including the destruction of the partially built dam by a flood in 1841, led Jervis to create a masonry and concrete dam based on a wooden foundation. The dam was 430 feet (131 meters) long and 50 feet (15 meters) high. The water face was protected by an earth bank with a slope of 1 to 5, leading to a base thickness of 275 feet (84 meters). On top, the earth bank was faced with stone to protect against overflow. Downstream, another small timber dam flooded the space between it and the Croton Dam and provided a force against the water pushing against the water face of the Croton Dam. The Croton Dam served New York City's water supply until the NEW CROTON DAM was built downstream between 1892 and 1906. The older dam was flooded and now lies underneath the reservoir of the new dam.

Additional Information: Kirby, Richard Shelton et al. *Engineering in History*. New York: Dover Publications, 1990; Smith, Norman. *A History of Dams*. Secaucus, NJ: Citadel Press, 1972.

Crozet Tunnel, also called the Blue Ridge or Rockfish Tunnel (Rockfish Gap, Virginia; 1850-1858)

This railroad tunnel is named after its builder, the French-born civil engineer Claudius (Claude) Crozet. The Crozet Tunnel was a major achievement of nineteenth-century civil engineering. The 4,273-foot (1,303-meter) tunnel is the largest of four railroad tunnels that were built through the Blue Ridge Mountains to facilitate the development of commerce between the Ohio River basin and southern seaports. A replacement tunnel with a wider opening to accommodate more modern railroads was constructed in 1944. Construction of the tunnel was challenging for both technical and social reasons. A labor riot in 1850 among groups of Irish workers from different areas of Ireland became a significant historical and sociological event known as the "Irish Rebellion." Technically, drilling through the mountains was an almost impossible feat. After 20 months of digging, Crozet reported that he had only progressed 755 feet (231 meters). Because the rock was so difficult to work, the intermediate vertical shafts were impossible to dig. Work had to progress only through the two tunnel openings. Crozet solved the problem of ventilation without vertical shafts by creating a system of pipes and valves to clear the air in the tunnel, which was especially fouled after explosions from black powder were set off to break rock. Draining the new tunnel of water required the use of a record-breaking 2,000-foot-long (610-meter) **cast-iron** pipe as a siphon to draw off about 60 gallons (227 liters) of water a minute. When the tunnel was finally holed through, another year's work was required to widen it sufficiently for track to be laid.

Additional Information: Schodek, Daniel L. *Landmarks in American Civil Engineering*. Cambridge, MA: MIT Press, 1987.

Cumberland Road, also known as the National Road or National Pike (Cumberland, Maryland to Wheeling, Virginia [now West Virginia]; 1818)

The Cumberland Road, the first stretch of the National Road, was, at the time of its construction, the most ambitious road-building project undertaken in the United States. It became, from the time of its opening, the great highway of western migration, leading pioneer families from the original states of the Atlantic seaboard to Kentucky, Ohio, Missouri, and beyond. Agitation for a road to the west began in the 1790s, and was supported by Presidents George Washington and Thomas Jefferson. Congress approved the route and appointed a committee to plan details of construction in 1806. Contracts were negotiated in 1811, but the War of 1812 delayed construction until 1815. The Cumberland Road was made of crushed stone, and, when opened for traffic in 1818, ran from Cumberland in western Maryland to Wheeling, on the Ohio River in western Virginia (now West Virginia). The road followed in part the Indian trail over which George Washington and General Braddock had passed in the 1750s. Through the efforts of Henry Clay of Kentucky, the Cumberland Road was extended in the late 1820s and became known as the National Road. It ran westward through Ohio to Vandalia, Illinois, and then to St. Louis, Missouri. The older Cumberland portion of the road was, by the 1830s, in need of repair, so Congress turned control of the road over to the states through which the road passed. The states financed upkeep of the road by levying tolls on road users. Modern U.S. Route 40 follows closely the original National Road.

Additional Information: Lay, M.G. *Ways of the World*. New Brunswick, NJ: Rutgers University Press, 1992; Rose, Albert C. *Historic American Roads*. New York: Crown Publishers, 1976.

D&H Canal. *See* Delaware and Hudson Canal.

Dai-Shimuzu Tunnel (Japan; 1979)
The 13-mile-long (21-kilometer) Dai-Shimuzu Tunnel is one of the longest railroad tunnels in the world. Cutting through Tanigawa Mountain, the Dai-Shimuzu was the first tunnel to surpass in length the 1908 SIMPLON TUNNEL between Switzerland and Italy.

Additional Information: Epstein, Sam and Epstein, Beryl. *Tunnels.* Boston: Little Brown and Company, 1985.

Dale Dyke Dam (Sheffield, England; 1858)
Built to supply water to Sheffield in north central England, the Dale Dyke Dam was 95 feet (29 meters) high and 1,250 feet (381 meters) long. It was 12 feet (3.7 meters) thick at the crest and 500 feet (152.5 meters) thick at its base. The core wall was made of puddled clay that was 4 feet (1.2 meters) thick at the top, 16 feet (4.9 meters) thick at the base of the embankment, and extended as much as 60 feet (18.3 meters) into the foundations. The dam had a spillway that was 24 feet (7.3 meters) wide and 11 feet (3.4 meters) deep. The dam collapsed on 11 March 1864 when the outlets were closed to raise the water level in the Bradfield reservoir to its maximum. A crack in the air face was noticed by a passerby, but too late for preventive action; at 11:30 p.m., the dam collapsed, pouring approximately 200 million gallons (730 **acre-feet**) of water into the area below the dam. Moving at a speed of 18 miles per hour (29 kilometers per hour), the water bore down on Sheffield, 7 miles (11 kilometers) away, destroying everything in its path. Parts of the city were flooded to a depth of 9 feet (2.7 meters); after the water subsided, parts of the city were left covered with wood, mud, sand, and stones. The flood killed 250 people, destroyed 798 houses, and seriously damaged 4,000 other houses. The disaster was most likely caused by faulty positioning of the outlet pipes, undermining of the core wall, and a poorly designed dam crest.

Additional Information: Sandstrom, Gosta E. *Man the Builder.* New York: McGraw-Hill, 1970; Smith, Norman. *A History of Dams.* Secaucus, NJ: Citadel Press, 1972; http://pine.shu.ac.uk/~engrg/victorian/DDDam/Flood1.htm.

Dam of the Pagans. *See* Sadd el-Kafara.

Damietta Dam. *See* Rosetta Dam/Damietta Dam.

Danube Canal Bridge (Vienna, Austria; 1830)
Ignaz von Mitis designed and built this narrow **suspension bridge** across the Danube using steel **eyebars**, the first-ever use of steel in bridge building. The steel sections of the bridge were half the weight that **wrought-iron** would have been. Unfortunately, the cost of steel made its further use in other bridges prohibitive for another 60 years. Von Mitis was disappointed that his bridge vibrated and swung in the wind. The 334-foot (102-meter) span was dismantled in 1860 for a stronger railroad suspension bridge that was itself replaced in 1885.

Additional Information: DeLony, Eric. *Landmark American Bridges.* Boston: Bullfinch Press, 1993; Kirby, Richard Shelton et al. *Engineering in History.* New York: Dover Publications, 1990.

Dashiqiao Bridge. *See* An Ji Bridge.

Delaware and Hudson (D&H) Canal (Honesdale, Pennsylvania, to Kingston, New York; 1828)

Benjamin Wright, the former chief engineer of the ERIE CANAL in central New York, designed this 108-mile (174-kilometer) canal to transport anthracite coal from Honesdale in northeastern Pennsylvania to Kingston, a town on the Hudson River about half-way between New York City and Albany. The canal followed the Delaware River to Port Jervis, just over the Pennsylvania border in southeastern New York, and then turned north to connect with the Hudson at Kingston. A round trip usually took 8 to 10 days. Wright's waterway was 4 feet (1.2 meters) deep, 32 feet (9.8 meters) wide, and contained 108 locks, 137 bridges, and 26 basins, dams, and reservoirs. The private company formed to finance the $1.2-million cost of the canal, the Delaware & Hudson Canal Company, was the first million-dollar private company in the United States. Five of the **aqueducts** built for the canal were constructed by John Roebling, who later designed the BROOKLYN BRIDGE. The canal was removed from service in 1898 when the D&H Company became a railroad company.

Additional Information: Shaw, Ronald E. *Canals for a Nation.* Lexington: University of Kentucky Press, 1990; http://www.mhroc.org/kingston/kgndah.html.

Delaware Aqueduct (Lackawaxen, Pennsylvania, to Minisink Ford, New York; 1849)

John Roebling, the designer of the BROOKLYN BRIDGE, built this **aqueduct** on the Delaware River where the river serves as the boundary between northeastern Pennsylvania and southeastern New York. The aqueduct is the oldest cable-**suspension bridge** in the United States still standing with its original elements. It is also the only survivor of several suspension aqueducts built by Roebling for the DELAWARE AND HUDSON CANAL between 1847 and 1850. Roebling used the structure to perfect the suspension bridge technique that he later used for the Brooklyn Bridge; some historians consider the construction of the Delaware Aqueduct to be an engineering event of equal significance to the building of the Brooklyn Bridge. The Delaware Aqueduct has towpaths on either side of the center flume (the heavy, wooden water channel). The trunk was about 6 feet (1.8 meters) deep, and 19 feet (5.8 meters) wide at the water line. The sides of the trunk were made of two pieces of untreated white-pine planking, placed with diagonal grains, and caulked. The trunk was able to carry its own **dead weight**, leaving the cables to carry only the weight of the water. The floor was hung from the bridge suspenders, as in a conventional suspension bridge. Because of the construction of the trunk, the weight of the water inside created a cantilever effect, relieving some of the strain from the cables and supporting beams. Each cable is made of 2,150 wires that comprise seven strands. Each strand was formed by carrying the wires across

from anchorage to anchorage, making the strands continuous wires that were then compacted into a cylindrical form. Bought by the National Park Service in 1980, the bridge has been rehabilitated and the original superstructure restored. A concrete slab will make up for the weight of the water that originally flowed through the aqueduct.

Additional Information: DeLony, Eric. *Landmark American Bridges.* Boston: Bullfinch Press, 1993; Schodek, Daniel L. *Landmarks in American Civil Engineering.* Cambridge, MA: MIT Press, 1987.

Delaware Memorial Bridge (New Jersey to Wilmington, Delaware; 1951, 1968)

These two attractive bridges across the Delaware River are the longest twin spans in the world, with overall lengths of 3,650 feet (1,113 meters)—a main span of

The last of four anchorage saddles being put in place for the Delaware Memorial Bridge. Photo courtesy of the Bethlehem Steel Corporation.

2,150 feet (656 meters) and two side spans of 750 feet (229 meters) each. Including the approach spans, the project is 24,465 feet (7,462 meters) long. A major concern before construction was clearance under the bridge for boats and ships leaving the Philadelphia Navy Yard; as a result, vertical clearance under the spans reaches a maximum height of 175 feet (53 meters). The first structure was begun in 1948, and opened to traffic in 1951. Because of increased traffic, the second span, 250 feet (76 meters) north of the first structure, was begun in

1964 and opened in 1968. Current traffic across the two bridges reaches nearly 80,000 vehicles per day.

Additional Information: Condit, Carl W. *American Building Art: The Twentieth Century*. New York: Oxford University Press, 1961; The Delaware River and Bay Authority, New Castle, Delaware.

Delaware River Bridge. *See* Benjamin Franklin Bridge.

Denver International Airport (Denver, Colorado; 1989-1995)

Critics of this hotly debated project charged that it was not needed as a replacement for nearby Stapleton Airport and was merely a job creation venture encouraged

Denver International Airport, Denver, Colorado. The roof is made of a Teflon-coated fiberglass material that has been formed into 34 peaks representing Colorado's Rocky Mountains. Photo courtesy of Airport Public Affairs, Denver International Airport.

by the federal and local governments. Despite the on-going discussion about its usefulness, the project was an engineering masterpiece; it was the second largest public works project in the world, smaller only than the CHANNEL TUNNEL that links Britain with Europe. Twice the size of Manhattan, the airport consumed the work of about 10,000 workers at the height of construction activity from 1992 to 1995. Preparations for construction were gargantuan; about 110 million cubic yards (over 84 million cubic meters) of earth had to be moved, enough to fill a 10 by 30-foot (3 by 9-meter) ditch running from Denver to New York City. The airport has electric-powered subway lines capable of carrying 6,000 people an hour and 12,000 covered parking spaces. Although amazingly efficient and fast, a problem with the

automated baggage delivery system delayed the airport's opening for about a year until engineers figured out how to stop the system from mangling the baggage it handled. Twelve runways allow an incredibly large amount of traffic into and out of the airport. Spaced at least 4,300 feet (1,312 meters) apart, the runways meet Federal Aviation Authority (FAA) requirements for use during bad weather conditions. A landmark in its own right, the FAA tower at the airport is 327 feet (100 meters) high, the tallest FAA tower in North America. Overall, the airport can accommodate up to 600 flights and 100,000 passengers per day through its 135 gates. Air conditioning the various concourses and terminals involves 17 mechanical rooms, 1.3 million pounds (585,000 kilograms) of duct work, and 39 miles (63 kilometers) of piping. The system also uses more than 18 miles (29 kilometers) of steel and copper pipe for the cooling/heating system, which delivers air to parked aircraft as well as to the airport itself. The airport boasts the largest jet fuel distribution system in the world. Built to meet strict environmental protection standards, the system has 35 miles (56 kilometers) of pipe with 7,000 welded joints to carry fuel directly to aircraft.

Additional Information: Caile, Bill. "Mechanical Systems Match Complexity of Denver Airport." *Engineered Systems* 12, (August 1, 1995), p. 74; Lindgren, Kristina, "Denver Airport: A Beauty with a Few Flaws." *Los Angeles Times*, March 10, 1995, Business: 7; Searles, Denis. "Debate Flies Over Denver's Big, New Airport. . . . ." *Los Angeles Times*, July 19, 1992, A:2; C.A. Merriman in http://infodenver.denver.co.us/-aviation/factrvia.html.

Detroit-Windsor Tunnel (Detroit, Michigan, to Windsor, Ontario; 1930)

At 5,160 feet (1,574 meters), this tunnel under the Detroit River is one of the longest underwater vehicular tunnels in North America, and the third great subaqueous tunnel for vehicles built in the United States. The tunnel provides the long desired Detroit-Windsor link between the U.S. and Canada. In 1871, workers attempted a 15-foot (4.6-meter) bore masonry tunnel, but work stopped at a length of 135 feet (41 meters) when sulphurous gas made workers sick. A second attempt was made in 1879, but was abandoned when the workers hit limestone and the projected costs of the project soared out of sight. The third attempt in the 1920s used all three tunneling techniques—cut and cover, use of a shield, and creation of a trench and sunken tubes. With a diameter of 32 feet (9.8 meters), the tunnel required 65 miles (105 kilometers) of arc welding, the first major use of arc welding in tunneling history. The approach ramps to the Detroit entrance used a **helical** (spiral) design, another major innovation.

Additional Information: Schodek, Daniel L. *Landmarks in American Civil Engineering*. Cambridge, MA: MIT Press, 1987.

Devil's Bridge (Italy to Switzerland; early thirteenth century)

Little is known about the original construction of this bridge in the mountainous St. Gotthard Pass area between Italy and Switzerland. The bridge was built across the Schollengen Gorge over the Reuss River. The approach to the bridge was a wooden walkway, 80 yards (73 meters) in length, fastened with chains to the rock mountainside. The bridge acquired its name because it seemed that only the devil could have built it. The wooden bridge was replaced with a stone arch in 1595. The stone structure was completely destroyed in 1799 during the wars of the French Revolution.

Additional Information: Hawkes, Nigel. *Structures: The Way Things Are Built*. New York: Macmillan, 1990.

Dismal Swamp Canal (Elizabeth City, North Carolina, to Norfolk, Virginia; 1787-1829)

George Washington was among those who advocated a canal to connect the Chesapeake Bay with the Albemarle Sound in northeast North Carolina. The Dismal Swamp Canal is so named because it runs through the Dismal Swamp in North Carolina to connect with the Pasquotank River before it empties into Albemarle Sound. The canal is 22 miles (35 kilometers) long and 40 feet (12.2 meters) wide. Funded by the Dismal Swamp Canal Company, Congress, and the state of Virginia, the canal cost $800,000 to construct. Construction began in 1787 and the canal opened to traffic in 1794. It was enlarged in 1807 and fully reopened in 1812. Further additions and improvements were made in 1828 and 1829. The canal was sold to the U.S. government in 1929 and is now maintained and operated by the Army Corps of Engineers. The Dismal Swamp Canal is the oldest continually operating man-made canal in the country. Still in use, the canal is now part of the Atlantic Intracoastal Waterway, a continuous series of rivers, canals, and bays on the Atlantic and Gulf Coasts.

Additional Information: Shaw, Ronald E. *Canals for a Nation*. Lexington: University of Kentucky Press, 1990; Spangenburg,

Ray and Moser, Diane K. *The Story of America's Canals*. New York: Facts on File, 1992; http://www.albemarle-nc.com/camden/history/canal.htm.

Drift Creek Bridge (Lincoln City, Oregon; 1914)

Drift Creek Bridge, which carries Highway 101 across Drift Creek near Lincoln City in western Oregon, is the state's oldest covered bridge. Closed to traffic, it is now situated in a small roadside park. A picture-perfect **Howe truss**, this red-painted bridge is 66 feet (20 meters) long.

Source: http://william-king.www.drexel.edu/top/bridge/CBDrift.html.

Druid Lake Dam (Baltimore, Maryland; 1871)

Like most major American cities in the late nineteenth century, Baltimore was continuously seeking fresh, clean water for its growing population. Although the city had relatively new reservoirs nearby, additional water was still needed. Robert W. Martin conceived the idea of building a reservoir and dam within the city both to alleviate the water supply problem and provide a park for city residents. Druid Lake is the country's first **earthfill** dam. With this type of dam, water is held back by soil and natural rock that is covered with a protective layer of clay. Martin built his dam from clay excavated from the dam site; he also used natural stone and reused excavated earth. Completed in January 1871, the dam held back the waters of a 429-million-gallon (1.6-billion-liter) reservoir. The dam is 119 feet (36 meters) long at its peak, and 640 feet (195 meters) wide at its base, tapering to a 60-foot (18.3-meter) width at its top. The area was developed as a park and recreational area, although the increased mobility brought by the automobile made the park less important than Martin had originally envisioned.

Additional Information: Schodek, Daniel L. *Landmarks in American Civil Engineering*. Cambridge, MA: MIT Press, 1987.

Duisburg Autobahn Bridge (Duisburg, Germany; 1963)

Built in the 1960s for one of the newer highways in the German AUTOBAHN ROAD SYSTEM, this multispan bridge over the Ruhr River at Duisburg in northwestern Germany presented a design problem; the bridge seemed likely to sink into the soft soil left by old coal workings. If constructed with conventional building techniques, the bridge was expected to sink up to 2 meters (6.6 feet) during the first eight years of its existence, after which the ground would stabilize. To deal with the potential sinking problem, the pier designs for the bridge included a system of hydraulic jacks and packing to keep spans at their original level during subsidence. Horizontal jacks were also employed to correct any lateral settlement of the piers.

Additional Information: Overman, Michael. *Roads, Bridges and Tunnels*. Garden City, NY: Doubleday & Company, 1968.

Dulles Greenway (Leesburg, Virginia, to Dulles International Airport, Virginia; 1993-1995)

Designed and built by the firm of Dewberry & Davis in Fairfax, Virginia, this 14.1-mile (23-kilometer) toll road in northeastern Virginia west of Washington, D.C., is the first privately built road to be constructed in the United States since 1918. With limited government financing available, privately built toll roads may be the model for future road construction projects. The Virginia General Assembly granted authority to build the road in 1993. It was completed six months ahead of schedule at a cost of $326 million. The project involved the construction of seven major interchanges and 36 bridges, including a 662-foot-long (202-meter) continuous steel-**girder** bridge over Goose Creek in Virginia. Work on the road also involved the largest wetlands-mitigation project ever undertaken in Virginia.

Additional Information: Austin, Teresa. "U.S. Toll Project Update." *Civil Engineering* 64:2, (February 1994), p. 58; Fowler, John P., II and Thompson, Kurt R. "The Toll Road That Wouldn't Die," *Civil Engineering* 65:4 (April 1995), pp. 48-51; Janofsky, Michael. "New Toll Road Offers Glimpse at Future." *New York Times*, Sept. 29, 1995, A:16; http://www.his.com/~cwealth/greenway/facts.html.

Dunlap's Creek Bridge (Brownsville, Pennsylvania; 1839)

Dunlap's Creek Bridge, built by Captain Richard Delafield, carries Old U.S. Route 40 over Dunlap's Creek in southwestern Pennsylvania. The structure was the first major all-iron bridge constructed in the United States. Many earlier bridges had been built at the same crossing since 1794, including a chain bridge built by James Finley. These earlier structures all succumbed to foul weather or bad water conditions, such as **scour**. Designed for the National Road (see CUMBERLAND ROAD entry), the bridge has elliptical arches made of iron castings that are resistant to corrosion. With an 80-foot (24-meter) span and a rise of 8 feet (2.4 meters), the bridge has five identical arch ribs made of hollow, **cast-iron** tubes. Although the bridge has been widened and sidewalks have been added, its original design has not been modified much.

Additional Information: DeLony, Eric. *Landmark American Bridges*. Boston: Bullfinch Press, 1993; Rose, Albert C. *Historic American Roads*. New York: Crown Publishers, 1976; Schodek, Daniel L. *Landmarks in American Civil Engineering*. Cambridge, MA: MIT Press, 1987;

Duomo. *See* Cathedral of Santa Maria del Fiore.

Dusseldorf-Neuss Bridge (Dusseldorf to Neuss, Germany; 1951)

The fighting during World War II destroyed many bridges across Europe, and especially bridges across the Rhine River on the western approaches to Germany. The Dusseldorf-Neuss Bridge was one of many new bridges built after the war. This **box girder** across the Rhine in western Germany connects the cities of Dusseldorf and Neuss. The bridge has two 338-foot (103-meter) side spans on either side of a central span of 676 feet (206 meters).

Additional Information: Brown, David J. *Bridges*. New York: Macmillan, 1993.

Dwight D. Eisenhower Interstate and Defense Highway System (nationwide; 1954)

Construction of this federally funded network of highways connecting the major cities of the United States was begun in the 1950s during the presidency of Dwight D. Eisenhower. Congressional authorization of the system was motivated in part by Eisenhower's observations during World War II of the effectiveness of Germany's AUTOBAHN ROAD SYSTEM. In some cases, existing highways were upgraded to qualify for participation in the system; in other cases, new roads (and bridges and tunnels) were built with large amounts of federal assistance. Generally complete, the interstate system today includes such roads as I-168 in western Maryland, which involved making a gigantic cut in the side of a mountain near Hancock, Maryland. The cut exposed the geological record of 350 million years. The system also includes U.S. INTERSTATE ROUTE 70, which runs through the Eisenhower Memorial Tunnel in Colorado, and Interstate 310 in Louisiana, which involved the construction west of New Orleans of the Hale Boggs Memorial Bridge, the first major steel **cable-stayed bridge** in the United States.

Additional Information: Weingroff, Richard, "Engineering Marvels" in Public Roads On-Line (summer, 1996) at http://www.tfhrc.gov/pubrds/summer96/p96su28.htm.

E

Eads Bridge (St. Louis, Missouri; 1874)

The Eads Bridge, built by James Buchanan Eads, carries Washington Street in St. Louis across the Mississippi River. The center arch is 520 feet (159 meters) long; the two side arches are 502 feet (153 meters) each. When completed, the arches were significantly longer than any ever built before. Eads used the new (to the

A pre-1920 view of Eads Bridge across the Mississippi at St. Louis, Missouri. Photo courtesy of the Library of Congress, World's Transportation Commission Photograph Collection.

United States) method of **cantilever** construction in which the three arches were built out from the piers until the steel work met in the middle. No centering or **falsework** was used. Each span has four truss-stiffened arches with parallel chords, 12 feet (3.7 meters) apart. Made of **wrought-iron** tubes, the chords are 16 inches (41 centimeters) in diameter, with chrome steel staves inside. Because of the depth of the river, pneumatic **caissons**, rather than **cofferdam**s, were used to build the piers of the bridge. The caissons enabled Eads to perform the deepest underwater work that had ever been done in the world up to that time; the use of pneumatic caissons, which had already been employed in Europe, led to an understanding of and a cure for "caisson disease," or the **bends**, in the United States. As the bridge was nearing completion, Eads took a loan of $500,000 from European bankers; the loan was guaranteed with the proviso that the first arch be completed by 19 September 1873. On 16 September, workers found that the final tubes did not fit; excessively warm weather had made the metal expand; the pieces were too long to fit together. Workers tried packing 15 tons (13.5 metric tons) of ice around the metal, but temperatures of 98 degrees kept the tubes five-eighths of an inch too long. Another 45 tons (40.5 metric tons) of ice was applied all day, but at sundown the ribs were still too long. When wired in Europe over the newly laid Atlantic cable, Eads explained that he had manufactured some adjustable members to get around just such a problem. These members were successfully inserted into the gap at 10:00 p.m. on 17 September. The simple and dramatic telegram sent to Eads the next day merely said, "Arch safely closed." The bridge is a double-decker, the lower deck for railroads, the upper for vehicles and pedestrians. A successful public test of the bridge's strength, with 14 heavy locomotives filled with people, was made on 2 July 1874. The bridge was essential in making St. Louis the most important city on the Mississippi and a gateway to the western United States. Still standing and carrying traffic today, the bridge became in 1898 the first ever pictured on a United States postage stamp. Hailed immediately as a major engineering achievement, the bridge was the only one ever built by Eads.

Additional Information: DeLony, Eric. *Landmark American Bridges*. Boston: Bullfinch Press, 1993; Schodek, Daniel L. *Landmarks in American Civil Engineering*. Cambridge, MA: MIT Press, 1987; Steinman, David B. *Famous Bridges of the World*. London: Dover Publications, 1953, 1961.

East River Gas Tunnel (connecting Queens and the Bronx, New York City; 1910–1916)

The East River Gas Tunnel was designed to move coal-gas for consumers between the boroughs of Queens and the Bronx in New York City. Planned to be 18 feet (5.5 meters) high by 16.75 feet (5.1 meters) wide, the 4,662-foot-long (1,422-meter) tunnel had to be built 250 feet (76 meters) below the water bed to avoid meeting the water from the East River. The two **headings**, one from each shore, met without major incident in the summer of 1913. However, when additional rock at the bottom of the tunnel was removed, holes drilled in the side of the Astoria heading on the Queens end began to gush water. Soon after, the same problem also appeared on the Bronx end. The pressure of the water was so intense workers had to leave their tools and run for their lives. Bulkhead doors had been installed in the tunnels, but only on the Bronx end did workers have the time to close the doors; on the Astoria end, the tunnel was completely flooded. Engineers concluded that the water would be impossible to pump out because it came from the river. The eventual solution was to drill through the closed bulkhead on the Bronx end and inject more than 1,000 tons (900 metric tons) of cement into the flooded heading. After pumping began, some of the new concrete gave way, and another 200 tons (180 metric tons) of concrete were needed before the tunnel could be pumped dry. The tunnel was finally completed in 1916.

Additional Information: Beaver, Patrick. *A History of Tunnels*. Secaucus, NJ: Citadel Press, 1973.

Eddystone Lighthouse (on a reef off Plymouth, England; 1759)

John Smeaton built this lighthouse 14 miles (23 kilometers) out in the English Channel off the southwestern coast of England near Plymouth. The lighthouse replaced an earlier light that had been built in 1698, but burned down in December 1755. From an engineering point of view, the new Eddystone Lighthouse is the most important lighthouse built in modern times because its construction required the development of new techniques and materials that heavily influenced all subsequent lighthouse construction. Smeaton, justly renowned for a multitude of engineering projects, de-

signed the tower in the shape of an oak trunk to spread the base over the widest possible area and give the structure the strength to withstand fierce Atlantic gales. Smeaton had each stone used in the tower cut to dovetail with the stone next to it; each **course** of dovetailed stone was pegged to the course above and below to strengthen the entire fabric and make the tower a solid monolith. Smeaton had to find a cement that would harden underwater to make this light last significantly longer than its two wooden predecessors. By examining carefully the properties of lime, Smeaton successfully developed a hydraulic cement to bond the masonry. To make the cement, he collected limestone from all over England, and tested numerous mixtures of clay, lime, and other ingredients. From his tests, he concluded that a cement capable of setting under water had to consist of limestone mixed with large amounts of clay. With this knowledge, Smeaton was able to devise an "eminently hydraulic lime." Construction of the lighthouse began on 12 June 1757 and was completed on 16 October 1759. The lighthouse remained in operation until 1876.

Additional Information: McNeil, Ian, ed. *An Encyclopedia of the History of Technology*. London: Routledge, 1990; Sandstrom, Gosta E. *Man the Builder*. New York: McGraw-Hill, 1970.

Edinburgh Castle (Edinburgh, Scotland; eleventh century)

The precipitous black basalt rock upon which Edinburgh Castle stands rises 250 feet (76 meters) above the city of Edinburgh, the capital of Scotland since the fifteenth century. The castle itself, standing 443 feet (135 meters) above sea level, dominates the city. Edinburgh is located in southeastern Scotland on the south coast of the Firth of Forth, the waters of which can be seen from the castle battlements. The castle height, a place of refuge and settlement since about 1000 B.C., looks down upon a deep depression that once held the waters of North Loch (lake), which was drained in the eighteenth century. The site was held in the seventh century by King Edwin of Northumbria, who may have given his name to the later town and castle. The castle is first mentioned in the eleventh century as the home of Malcolm III, the supplanter of Macbeth. Malcolm's English queen, Margaret, gave her name to the tiny chapel of Saint Margaret, on the rock's highest point; the chapel is the oldest surviving building on Castle Rock, dating from the twelfth century. Most of the current battlements of the castle were added in the eighteenth century. Palace Yard, or Crown Square, occupies most of the summit of Castle Rock. The town of Edinburgh grew up around the castle, which was an important royal residence throughout the medieval period. The castle was besieged on numerous occasions, especially by English armies, and was rebuilt by numerous Scottish kings. James IV, for instance, built a magnificent banqueting hall in the early sixteenth century. The royal apartments, now open to the public, contain the tiny bedchamber in which Mary, Queen of Scots, gave birth in 1566 to her son, the future James VI of Scotland and James I of England. The castle also houses the Scottish Crown jewels and is the military headquarters of the Scottish Division of the British Army. The Royal Mile, which begins outside the castle, runs east down the crest of the rock to the palace of Holyroodhouse. The palace occupies part of the site on which the abbey of Holyrood was built in 1128 and rebuilt in 1220. The current palace comprises a sixteenth-century wing and a seventeenth-century quadrangle. The palace is the monarch's official residence in Scotland.

Additional Information: MacIvor, Iain. *Edinburgh Castle*. London: B.T. Batsford/Historic Scotland, 1993; http://www.ibmpcug.co.uk.

Eiffel Tower (Paris, France; 1887-1889)

The famed engineer Alexandre Gustave Eiffel submitted the design for the Eiffel Tower to a design competition for a structure to symbolize the 1889 International Exhibition of Paris, an event similar to a twentieth-cen-

The Eiffel Tower in Paris, France, was built for the 1889 International Exhibition of Paris.

tury "world's fair." Although Eiffel's design was unanimously selected from the more than 700 proposals submitted, several French notables, including architect Charles Garnier and writers Emile Zola and Guy de Maupassant, opposed the tower as too unattractive to serve as a symbol of the Exhibition. Nonetheless, Eiffel began construction of the tower in 1887 on the Champ-de-Mars, one of the main avenues of Paris. The 300-meter-high (984-foot) iron tower was the tallest building in the world until 1930; its height can vary by as much as 15 centimeters (almost 6 inches), depending on the temperature. The tower rises in four columns from four masonry piers; the columns unite at a height of 189 meters (620 feet) to form one shaft. The tower contains 2.5 million rivets, is covered with 40 tons (36 metric tons) of paint, and is made of 15,000 iron pieces. Tourists can climb 1,652 steps to the top to see a restoration of Eiffel's office and a spectacular view of Paris and its environs. The top level and two other obeservation levels lower down can also be reached by elevators. The tower was nearly torn down in 1909, but the new and growing radio industry wanted to make use of the tower's height as a platform for an antenna. The tower's telegraph antenna, which added 20.75 meters (68 feet) to the structure's height, was the deciding factor in preserving the building. French radio and television have since made use of this antenna, and the destruction of the Eiffel Tower is now unthinkable. The top of the tower also houses a meteorological station.

Additional Information: Kirby, Richard Shelton et al. *Engineering in History*. New York: Dover Publications, 1990; Sandstrom, Gosta E. *Man the Builder*. New York: McGraw-Hill, 1970; http://www.paris.org/Monuments/Eiffel/html

Eigau Dam/Coedty Dam (Dolgarrog, Wales; 1908, 1924)

The Eigau Dam was built in 1908 on the slopes of Carnedd Llewelyn, and was raised to a height of 35 feet (10.7 meters) by 1911. With a crest length of 3,253 feet (992 meters), the dam's reservoir impounded 3,670 **acre-feet** of water that was used to supply hydroelectric power to an aluminum plant and the nearby town of Dolgarrog. In 1924, to supply even more water storage, a second dam, the Coedty, was built 2.5 miles (4 kilometers) below and to the west of the Eigau. The Coedty had a height of 36 feet (11 meters) and a length of 860 feet (262 meters); it impounded 252 **acre-feet** of water. On 2 November 1925, the Eigau Dam failed. Several factors contributed to the collapse, including poor construction and materials. The base of the dam was very shallow. Water percolating under the dam broke free, washing away a channel 70 feet (21 meters) wide by 10 feet (3 meters) deep and releasing an estimated 50 million cubic feet (over 1.4 million cubic meters) or 1,150 acre-feet of water per hour. Rushing toward the Coedty

Dam, the water washed out the spillway along with half the embankment. A 200-foot-wide (61-meter) breach that tapered to 60 feet (18.3 meters) at the bottom released almost all the water in the reservoir at once. The water swamped the downstream village of Dolgarrog, killing 16 people. Fortunately, many of the town's citizens were at a movie on high ground, or the death rate would have been significantly higher.

Additional Information: Smith, Norman. *A History of Dams.* Secaucus, NJ: Citadel Press, 1972.

Eisenhower, Dwight D. Interstate and Defense Highway System. *See* Dwight D. Eisenhower Interstate and Defense Highway System.

Eisenhower Memorial Tunnel. *See* U.S. Interstate Route 70.

Elephant Butte Dam (Truth or Consequences, New Mexico; 1916)

Engineers Arthur P. Davis, Louis C. Hill, and E.H. Baldwin built this concrete gravity structure in Sierra County in southwestern New Mexico, 4 miles (6.4 kilometers) east of Truth or Consequences. The dam was to be the centerpiece of the Rio Grande Project, an early project of the U.S. Bureau of Reclamation. The reservoir associated with the dam is one of the largest in the world; it was the first dam reservoir to send water flowing across international boundaries. The dam is 301 feet (92 meters) high, with a hydraulic height of 193 feet (59 meters). From a narrow top, the dam reaches a base width of 228 feet (70 meters). Covering a surface area of 36,600 feet (11,163 meters), the reservoir has a capacity of over 2.2 million **acre-feet**. Because of expected silting, the reservoir was intentionally constructed on a large scale; Caballo Dam was built 25 miles downstream in 1938 to compensate for storage capacity lost at Elephant Butte because of silting. Under a 1906 treaty with Mexico, the Elephant Butte system provides that country with water through the American Diversion Dam and Canal system 2 miles (3.2 kilometers) northwest of El Paso, Texas. The Elephant Butte Powerplant, which went into operation in 1940, has a capacity of 27,945 kilowatts.

Additional Information: Schodek, Daniel L. *Landmarks in American Civil Engineering.* Cambridge, MA: MIT Press, 1987; http://www.usbr.gov/power/elephant.html.

Elizabeth Tunnel. *See* First Owens River—Los Angeles Aqueduct.

Empire State Building (New York City; 1930-1931)

Located in New York City at Fifth Avenue between 33rd and 34th Streets, the Empire State Building was created specifically to be the tallest building in the world. The building was planned during the real estate boom and high economic optimism of the 1920s; by the time it was complete in 1931, the country was in the grip of the Great Depression. Because of the hard times, leasing the building's 2 million square feet of space proved difficult, and the structure was sometimes referred to as the "Empty State Building," a name it subsequently overcame by becoming one of the world's busiest and most desirable office buildings. The building has steadily decreasing tiers, rising to 102 stories and a total height of 1,472 feet (449 meters), including its television mast. Dominated by touches of **Art Deco** style, the 86th floor observatory allows sights of Manhattan from all sides and, on a clear day, for 50 miles (80 kilometers). A special elevator goes to the base of the television mast near the top of the building. Developers had toyed with the idea of making the building a mooring point for dirigibles, with an actual connection on the 102nd floor. For decades after it opened, the Empire State Building, with its murals of the Seven Wonders of the World and its own post office in its lobby, represented the essence of the skyscraper ideal.

Additional Information: Condit, Carl W. *American Building Art: The Twentieth Century.* New York: Oxford University Press, 1961; Kirby, Richard Shelton et al. *Engineering in History.* New York: Dover Publications, 1990; http://www.hotwired.com/rough/usa/mid.atlantic/ny/nyc/city/midtown.html; http://www.stern.nyu/~hlee10/empstat.htm; The New York Insider: http://theinsider.com/topthing/2EMPIRE.htm.

Engineer's Castle, also known as Mikhailovsky Castle (St. Petersburg, Russia; 1801)

Built as a fortress at the beginning of the nineteenth century by Czar Paul I, the Engineer's Castle was inspired by a dream in which the Archangel Michael (hence the other name of the structure) instructed Paul to build the fortress for his own safety. Each of the four facades of the building is different, and Paul even installed a moat around the building. Soon after the castle's completion, Paul was overthrown and assassinated by his palace guards. Paul's son, Alexander I, approved of the coup, although he had tried to prevent his father's murder. As a result of the assassination, no members of the imperial Romanov family would live at the castle, so it was converted into a school for military engineering. Paul's great-grandson, Alexander III, commissioned a statue of Paul I that now stands just south of the castle.

Additional Information: http://www.spb.su/fresh/sights/engineer.html.

Englewood Dam. *See* Miami Conservancy District.

Entwistle Dam (Bolton, England; 1837)

At a height of 108 feet (33 meters), Entwistle Dam in northwestern England near Bolton was the first dam in England to exceed 100 feet (30.5 meters) in height. The dam impounds almost 3,000 **acre-feet** of water over a length of 360 feet (110 meters). Originally built to power mills, the dam is now used for water supply.

Additional Information: Smith, Norman. *A History of Dams.* Secaucus, NJ: Citadel Press, 1972.

Erie Canal (Buffalo, New York, to Albany, New York; 1825)

The Erie Canal, one of the earliest barge canals in the United States, was built by Benjamin Wright, but was made politically possible by New York Governor DeWitt

A view (c.1906) of the Erie Canal at Rochester, New York. Photo courtesy of the Library of Congress, World's Transportation Commission Photograph Collection.

Clinton. Built between 1818 and 1825, the canal connected the Great Lakes with the Hudson River and the Atlantic Ocean. Dubbed by the governor's opponents as "Clinton"s Folly" or "The Governor's Ditch," the canal, the longest in the world at the time, was soon recognized as a major achievement for New York State. The canal brought many economic benefits to the state, and was a major factor in making New York City an important international port. When completed, the canal was 4 feet (1.2 meters) deep, 40 feet (12.2 meters) wide on top, 28 feet (8.6 meters) wide at the bottom, and 363 miles (584 kilometers) long. Economic uncertainty dictated the original shallowness of the channel (4 feet was a minimum for barges); the canal was deepened in 1835 when it became apparent that the project was a success. Eighty-three locks lifted barges a vertical distance of 627 feet (191 meters), and 18 major **aqueduct**s carried the canal over streams and rivers. On ei-

ther side of the canal were paths for people and for the horses and mules used to tow the canal boats. Large portions of the original canal eventually became part of the NEW YORK STATE BARGE CANAL, but other large sections were abandoned and can still be seen running near or through many New York villages and towns.

Additional Information: Russell, Solveig Paulson. *The Big Ditch Waterways.* New York: Parents' Magazine Press, 1977; Schodek, Daniel L. *Landmarks in American Civil Engineering.* Cambridge, MA: MIT Press, 1987; Shaw, Ronald E. *Canals for a Nation.* Lexington: University Press of Kentucky, 1990.

Erskine Bridge (near Glasgow, Scotland; 1971)

The Erskine Bridge crosses the River Clyde near Glasgow in southwestern Scotland. It was the first British **cable-stayed bridge** over 1,000 feet (305 meters) in length.

Additional Information: Brown, David J. *Bridges.* New York: Macmillan, 1993.

Ertan Dam (western China; under construction)

Ertan is one of the world's highest dams. It rises 245 meters (804 feet) above its lowest formation and has a generating capacity of 3,300 megawatts. The Ertan Dam is located over 1,000 kilometers (620 miles) northwest of Canton on the Yalong Jiang River.

Source: Scott K. Nelson, The National Performance of Dams Program (NPDP), Stanford, California.

Escorial (near Madrid, Spain; 1584)

The Escorial is a large complex of buildings and courtyards located about 30 miles (48 kilometers) northwest of Madrid in the Guadarrama Mountains in central Spain. The Escorial comprises a domed church, a college, a monastery, and a royal palace, all surrounded by a great wall. Construction was begun in 1563 by the architect Juan Bautista de Toledo, who had earlier worked in Italy. After Toledo's death in 1567, the structure was completed by Juan de Herrera. Philip II ordered construction of the monastery of San Lorenzo del Escorial on the site as a burial place for Spanish kings, beginning with his own father, Charles V. Philip and most of his successors on the Spanish throne are today buried in the monastery, which is one of the largest religious establishments in the world, measuring 675 feet by 528 feet (206 meters by 161 meters). The monastery was designed by Toledo as a great three-part rectangle, with the center portion occupied by a large church. South of the church are five cloisters that include the royal palace, the offices of royal ministers and servants, and the Courtyard of the Four Evangelists, a monastic cloister that is particularly impressive in scale and character. North of the church are the living quarters of the monks. Herrera revised Toledo's original plans in 1572, redesigned the church, and completed the Escorial by 1584. The whole structure is massive, austere, and orderly, both outside and inside. The sheer size of the structure combines with the yellow-gray granite walls to give the exterior an especially forbidding appearance. The severe Herreran style of the Escorial became the prevailing architectural style in Spain for the next half century. The El Escorial Library, founded by Philip II, houses a collection of more than 4,700 rare and important manuscripts, many of them illuminated, and more than 40,000 printed books.

Additional Information: Kubler, George. *Building the Escorial*. Princeton, NJ: Princeton University Press, 1981.

Essex-Merrimack Bridge (Newburyport, Massachusetts; 1810)

Designed by James Finley and built by John Templeman, the Essex-Merrimack Bridge over the Merrimack River at Newburyport in northeastern Massachusetts has a central span of 244 feet (74 meters). Finley supported 10 chains, each one 515 feet (157 meters) long, on two timber towers; the chains supported two 15-foot-wide (4.6-meter) roadways. The world's oldest major **suspension bridge**, the Essex-Merrimack Bridge was the first suspension bridge to have a deck that was stiff and

A view of the Essex-Merrimack chain bridge before its total reconstruction with wire cables in 1913. Photo courtesy of the Library of Congress, World's Transportation Commission Photograph Collection.

strong enough to be crossed by heavy vehicles. However, under the weight of snow and a heavily laden ox wagon, one of the chains broke in 1827, and from that time on the bridge underwent continual repairs. In 1913,

a major reconstruction replaced the timber towers with **reinforced concrete** ones, the chains with parallel wire cables, and the two separate roadways with a single, 27-foot (8.2-meter) roadway. Efforts were made to keep the bridge looking as much like the original as possible, even when the towers were replaced.

Additional Information: Hopkins, H.J. *A Span of Bridges: An Illustrated History*. New York: Praeger, 1970; Lay, M.G. *Ways of the World*. New Brunswick, NJ: Rutgers University Press, 1992.

Euphrates Tunnel (below River Euphrates, Babylonia; c.3000 B.C.)

No trace of this 3,045-foot-long (929-meter) tunnel now exists, but according to legend, the tunnel was built to connect the two sides of the city of Babylon (south of Baghdad in modern Iraq), with tunnel entrances built inside the walls of each half of the city. To build the tunnel, a temporary dam was constructed; the tunnel was built using a "cut and cover" technique, and the dam was then opened to allow the river to flow into its original channel. The tunnel was supposedly lined with brick and waterproofed with asphalt. No other subaqueous tunnel was attempted until Marc Brunel built the THAMES TUNNEL in London between 1824 and 1842.

Additional Information: Epstein, Sam and Epstein, Beryl. *Tunnels*. Boston: Little Brown & Company, 1985; Sandstrom, Gosta E. *Tunnels*. New York: Holt, Rinehart and Winston, 1963.

Exeter Canal (Exeter, England; 1566)

Built by John Trew, the Exeter Canal was the first canal in England to use pound-locks. It has vertically rising gates and runs 3 miles (4.8 kilometers) along the bank of the River Exe, connecting Exeter, in Devon in southwestern England, with the sea. Water for the locks was supplied by "Trew's **Weir**," a masonry dam near the head of the canal. The project involved Trew in controversy with the civic authorities over the completion date. Exeter asked Sir Peter Carew and other local Devon gentlemen to arbitrate the dispute. The canal was enlarged between 1698 and 1701. The Exeter Canal may have been the first true canal built in England since Roman times.

Additional Information: Hadfield, Charles. *World Canals*. Newton Abbot, England: David & Charles, 1986; McNeil, Ian, ed. *An Encyclopedia of the History of Technology*. London: Routledge, 1990; Smith, Norman. *A History of Dams*. Secaucus, NJ: Citadel Press, 1972.

Exploratorium. *See* Palace of Fine Arts-Exploratorium.

Extraterrestrial Highway (State Highway 375, Nevada)

This 52-mile (84-kilometer) stretch of lonely, two-lane road in south central Nevada north of Las Vegas crosses one of the most active areas in the United States for Unidentified Flying Object (UFO) sightings. Nearby is base 51 where, according to UFO believers, the Air Force is holding the remains and spacecraft of aliens who have landed in the United States. Perhaps with tongue in cheek, the state of Nevada renamed the road the Extraterrestrial Highway on 18 April 1996. Apart from people searching for beings from outer space, the area is extremely quiet; the major business activity in the area is a diner, recently renamed The AlieInn.

Additional Information: http://www.csicop.org/si/9609/highway.html; http://www.pcaf.com/extrater.htm.

F

Fairmount Bridge, also known as Fairmount Park Bridge (Philadelphia, Pennsylvania; 1841)

Designed and built by Charles Ellet to replace Lewis Wernwag's wooden COLOSSUS BRIDGE across the Schuykill River in Philadelphia, the Fairmount was a wire **suspension bridge** using 10 iron-wire cables that were manufactured on one bank of the river. The Fairmount was the first major wire-suspension bridge in the U.S. Ellet's construction used a 26-foot-wide (8-meter) main span with a length of 357 feet (109 meters). Although the bridge was known for excessive vibration, and had no stiffening **girder**, it lasted until 1874 when it was taken out of service and replaced with a double-deck truss.

Additional Information: Hopkins, H.J. *A Span of Bridges: An Illustrated History*. New York: Praeger, 1970; Schodek, Daniel L. *Landmarks in American Civil Engineering*. Cambridge, MA: MIT Press, 1987.

Fairmount Park Bridge. *See* Fairmount Bridge.

Farmington Canal, also known as the Northampton Canal (New Haven, Connecticut, to Northampton, Massachusetts; 1825-1828)

The first 56 miles (90 kilometers) of the canal north of New Haven, Connecticut, were completed in three years. The canal was later extended to Northampton, Massachusetts, on the Connecticut River. The canal served small New England manufacturers. Canal boats, the largest with a capacity of 25 tons (22.5 metric tons), were horse drawn. In the late 1840s, the canal was abandoned and a large part of the Hartford and New Haven Railroad was built along the towpath.

Additional Information: Kirby, Richard Shelton et al. *Engineering in History*. New York: Dover Publications, 1990.

59th Street Bridge. *See* Queensboro Bridge.

Fink Deck Truss Bridge (Lynchburg, Virginia; 1850s-1870s)

Still in use as a siding to an industrial site, the Fink Deck Truss Bridge is the only known survivor of its type—a bridge in which the roadway lies on top of the upper chords of the supporting trusses. The bridge was placed in its present location in Lynchburg in southwestern Virginia in 1893, but was probably constructed and used elsewhere between the 1850s and the 1870s. The bridge has a span of 52.5 feet (16 meters). The truss consists of vertical and diagonal members of **wrought-iron**, all below the deck. The top chords, which support the floor system, are made of 15-inch-thick (38-centimeter) untreated oak. The bridge is scheduled to be removed from service, but may remain in place, or be moved to another location as part of a public museum or park.

Additional Information: Schodek, Daniel L. *Landmarks in American Civil Engineering*. Cambridge, MA: MIT Press, 1987.

al Firdan Lift Bridge (Suez Canal, Egypt; 1965)

Part of the SUEZ CANAL in northeastern Egypt, the al Firdan Lift Bridge is the longest swinging **drawbridge** ever built. The bridge has a length of 558 feet (170 meters) with two arms that swing about pivots.

Additional Information: *Academic American Encyclopedia*, (electronic version). Danbury, CT: Grolier Electronic Publishing, 1991.

First Owens River—Los Angeles Aqueduct (Los Angeles, California; 1907-1913)

This 240-mile (386-kilometer) water system enabled southern California to continue its spectacular growth in the early twentieth century. Los Angeles had a tradition of providing municipal water to its residents, but by the turn of the century the water table had sunk so low that the city had to bring in water from outside. The most difficult part of the project was the construction of the 26,780-foot (8,168-meter) Elizabeth Tunnel through granite under the San Fernando Mountains. The tunnel alone required temporary towns and the construction of roads and power plants. To supply cement for concrete, Los Angeles actually created the town of Monolith to produce cement from ground deposits in land that Los Angeles had bought. Other sections involved the creation of a 35-mile-long (56-kilometer) canal, and the Haiwee Storage Reservoir, complete with its own **earthfill dam**. Although the project was successful, the ever increasing need for water eventually set the city of Los Angeles against farmers in the Owens Valley. After several cases of sabotage and several court cases, the San Francisqito Dam collapsed in the late 1920s, causing the deaths of 500 persons and destroying tremendous amounts of real estate. To settle the ongoing water issues, Los Angeles agreed to purchase all the land around its **aqueduct**; this practice successfully settled the city's problems, but devastated the Owens Valley and effectively ended any chances for agriculture in the valley.

Additional Information: Reisner, Marc. *Cadillac Desert: The American West and Its Disappearing Water.* New York. Penguin Books, 1987; Schodek, Daniel L. *Landmarks in American Civil Engineering.* Cambridge, MA: MIT Press, 1987; http//www.ladwp.com/aboutwp/facts/supply/supply.htm.

Firth of Forth Bridge, also known as the Forth Rail Bridge or Forth Bridge (near Edinburgh, Scotland; 1882-1890)

Designed by Sir John Fowler and Sir Benjamin Baker, who began work in 1882, the Firth of Forth Bridge is considered by many engineers to be the greatest bridge built in the nineteenth century. When completed in 1890, this 513-meter (1,683-foot) span across the Firth of Forth along the east coast of Scotland was the longest bridge in the world; it was surpassed in length in 1917 by the QUEBEC BRIDGE. The Quebec Bridge, however, has two spans while the Forth Bridge has a single long span. The Forth Bridge was built in extreme weather conditions—rapidly descending fogs, frequent rains, high winds, and frost. The area has an ex-

tremely short working season. The structure has two approach **viaducts** to the main **cantilever bridge**, which crosses the water with three towers. The bridge has room for two railroad tracks. The approach spans are made of stone arches and **girder** spans. Villages for the workers, including stores and workshops, were erected on either side of the river. To transport workers, a paddle steamer capable of carrying 450 persons was built. A fleet of rowboats manned by expert swimmers was stationed at each cantilever tower; these fleets saved at least eight lives. For artificial lighting, Fowler and Baker used newly invented electric arc lamps, which gave great brightness and reduced the danger of fire. Except for some small parts of the bridge (**cast-iron** washers and anchor plates), the structure was made totally of steel, a new material at the time. Steelwork was finished at the site, instead of at a factory. Cantilever arms were built simultaneously on each side of their respective piers, with the suspended arms built out. When the cantilever arms were almost complete, the engineers counted on warm weather to get the two halves to meet perfectly before bolting them together. At one point, a cold wind prevented the bolt holes from meeting equally; to leave the job until the next day would have put unequal stress on the one bolted side, causing distortion and buckling. The engineers had wood shavings soaked in naphtha placed along the problem east chord, extending 20 meters (66 feet) in each direction. When lit, the heat caused enough expansion in the metal for the gap to close and for the bolts to be inserted. Two test trains successfully crossed the bridge, side-by-side, on 21 January 1890. Each train had two locomotives, 50 cars, and one engine at the rear, for a total weight of 900 tons (810 metric tons). The bridge opened to traffic on 4 March 1890.

Additional Information: Lay, M.G. *Ways of the World.* New Brunswick, NJ: Rutgers University Press, 1992; Steinman, David B. *Famous Bridges of the World.* London: Dover Publications, 1953, 1961.

The Firth of Forth Bridge near Edinburgh, Scotland. Photo courtesy of Gilbert Siegel.

Flatiron Building (New York City, 1902)

Originally called the Fuller Building, the name of the company that built the structure and first occupied it, the Flatiron Building is located between Broadway and Fifth Avenue and between 22nd and 23rd Streets. The Flatiron Building was New York City's first skyscraper. Its tapered form was built to take advantage of the new iron-frame construction methods. The steel frame is covered with rusticated limestone. The building is 312 feet tall (95 meters) and has 22 stories. At the time of its construction, it was the tallest building in the world. The building's design incorporates both Gothic and Renaissance influences and the structure resembles a classical column. The expression "23 skidoo" comes from the warnings police officers gave to prevent voyeuristic men from gathering to watch women's skirts blow up in the unpredictable winds created by the building.

Additional Information: http://www.hotwired.com/rough/usa/mid.atlantic/ny/nyc/city/midtown.html.

Flood Rock Demolition (New York City; 1885)

Flood Rock was a 9-acre (3.6-**hectare**) granite island near Long Island Sound, New York. The decision to destroy the island, which was a hazard to navigation, was probably based on economics; it was cheaper to destroy the island than to build and maintain a lighthouse on it. A vertical shaft was sunk from the surface to a depth of 64 feet (19.5 meters) below sea level. Four miles (6.4 kilometers) of galleries (smaller tunnels to the side) went out from the vertical shaft in all directions. The gallery tunnels were from 10 to 24 feet (3.05 to 7.3 meters) high, and were supported by 467 pillars, each 15 feet (4.6 meters) square. The drilling of these tunnels required the excavation of 80,000 cubic yards (61,200 cubic meters) of rock. More than 13,000 holes were drilled to hold explosives—the holes were about 3 inches (7.6 centimeters) in diameter and approximately 9 feet (2.7 meters) long. Charged with 110 tons of explosives, the resulting explosion, which was set off electronically, destroyed the remaining 275,000 cubic yards (210,375 cubic meters) of the island. The explosion itself lifted the sea surrounding the island to a height of over 100 feet (30.5 meters).

Additional Information: Beaver, Patrick. *A History of Tunnels.* Secaucus, NJ: Citadel Press, 1973.

Florianopolis Bridge (Florianapolis, Brazil; 1920-1926)

The 1,114-foot-long (340-meter) Florianopolis Bridge is one of the longest **eyebar** suspension spans in the world. The bridge connects Florianopolis on the island of Santa Catarina with San Jose on the mainland of southeasten Brazil. Designer David B. Steinman, one of the twentieth century's preeminent engineers, incorporated several innovations into his design—eyebars

The Flatiron Building in New York City shown shortly before its completion in 1902. Photo courtesy of the Library of Congress, Detroit Publishing Company Photographic Collection.

instead of wire for cables, a new form of stiffening truss, and high-**tension**, heat-treated steel, a new material that is two to three times as strong as ordinary steel. Proud that the bridge was built without any loss of life, Steinman described the building process in fascinating detail in 1953: "All the work was done from the top down. No **falsework** was used; instead, the eyebars were hung from temporary rope cables, and the members of the stiffening truss were likewise put in place from these overhead ropes. Even the temporary footbridges or scaffolds ordinarily used were eliminated" (Steinman, p. 59). The Florianopolis Bridge was also the first bridge to use rocker towers, an arrangement where the cables are placed on top of the towers on a moveable base that automatically adjusts for expansion or contraction according to the temperature. The bridge was the dream of a governor of Santa Catarina. As the governor lay on his deathbed only hours before construction on the bridge was to begin in 1920, the citizens of the island gathered outside his window and built a scale model of the bridge to assure him that it would be completed.

Additional Information: Brown, David J. *Bridges.* New York: Macmillan, 1993; Steinman, David B. *Famous Bridges of the World.* London: Dover Publications, 1953, 1961.

Fontana Dam (Fontana Village, North Carolina; 1945)

The highest dam in the TENNESSEE VALLEY AUTHORITY system, Fontana is 480 feet (146 meters) tall and 2,385 feet (727 meters) long. This **concrete gravity dam** across the Little Tennessee River is located in a particularly beautiful setting in extreme western North Carolina; historian Carl Condit has called Fontana "a perfect symbol of man and nature in harmony."

Additional Information: Jackson, Donald C. *Great American Bridges and Dams*. Washington, DC: The Preservation Press, 1988.

Forbidden City (Beijing, China; 1406-1420)

Begun in 1406, this gigantic set of buildings was the home and the seat of government for the Ming and Qing emperors of China until the monarchy was over-

The Forbidden City, the former center of imperial government, in Beijing, China.

thrown at the start of the twentieth century. Glowing with the imperial colors of red and golden yellow, the palace complex contains over 9,000 rooms and covers an area of more than 720,000 square meters (almost 18 acres). The Forbidden City is surrounded by a wall that is over 10 meters (33 feet) high. Now a museum open to the public, the Forbidden City contains displays of historic relics, paintings, carvings, and other artifacts of Chinese imperial history.

Additional Information: http://www.ihep.ac.cn/tour/bj.html.

Fort Laramie Bowstring Arch-Truss Bridge
(Fort Laramie National Historic Site, Wyoming; 1875)

The Fort Laramie Bowstring Arch-Truss Bridge is a three-span, 400-foot-long (122-meter) bridge across the North Platt River in southeastern Wyoming. Designed by Zenas King, the bridge was made of **wrought-iron**,

which is cheaper and stronger than **cast-iron**. Because the bridge was better able to handle both tensile and compressive forces than cast-iron bridges, wrought-iron soon became the preferred material for bridges in the United States.

Additional Information: DeLony, Eric. *Landmark American Bridges*. Boston: Bullfinch Press, 1993.

Fort Madison Railroad Bridge (Fort Madison, Iowa; 1927)

With a main span of 525 feet (160 meters), the Fort Madison Railroad Bridge across the Mississippi River near Fort Madison in southeastern Iowa is one of the longest **swing bridges** in the world. When the bridge opens, the navigation channel is a maximum 246 feet (75 meters) wide, less than half of the span of the bridge. The bridge pivots at a central point.

Additional Information: Overman, Michael. *Roads, Bridges, and Tunnels*. Garden City, NY: Doubleday & Company, 1968; Steinman, David B. *Famous Bridges of the World*. London: Dover Publications, 1953, 1961.

Fort Peck Dam (Fort Peck, Montana; 1940)

This dam across the Missouri River in northeastern Montana is one of the most impressive earth dams in the United States. It is a mountain, a half mile thick, made by human hands. The dam stands 250 feet (76 meters) high and 21,126 feet (6,443 meters) long. The volume of the dam is 125,026,000 cubic yards (95,645,000 cubic meters), with an additional reservoir capacity of 19,400,000 acre feet (14,841,000 cubic meters). Creation of the dam enabled navigation to Kansas City, and controlled flooding along the lower Missouri River. The dam also provides a significant amount of hydroelectric power to the region. Fort Peck, the oldest dam on the Missouri River, was built by the U.S. Army Corps of Engineers.

Additional Information: Farb, Peter. *The Story of Dams*. Irvington-on-Hudson, NY: Harvey House, 1961; Jackson, Donald C. *Great American Bridges and Dams*. Washington, DC: The Preservation Press, 1988; Sandak, Cass R. *Dams*. New York: Franklin Watts, 1983; Smith, Norman. *A History of Dams*. Secaucus, NJ: Citadel Press, 1972; http://www.lonelyplanet.com.au/dest/ind/cal.htm#pix01.

Forth Bridge. *See* Firth of Forth Bridge.

Forth Rail Bridge. *See* Firth of Forth Bridge.

Frankford Avenue Bridge, also called the Pennypack Bridge (Philadelphia, Pennsylvania; 1697)

The three-span Frankford Avenue Bridge, which crosses Pennypack Creek, is the oldest stone **arch bridge** still in use in the United States. It was the first major stone arch that is known to have been built in the United States. Originally narrow, the bridge's roadway was widened and a sidewalk was added in 1893.

Additional Information: DeLony, Eric. *Landmark American Bridges.* Boston: Bullfinch Press, 1993; Schodek, Daniel L. *Landmarks in American Civil Engineering.* Cambridge, MA: MIT Press, 1987.

Frankfort Railroad Bridge (Frankfort, New York; 1840)

Built by Earl Trumbull, the Frankfort Railroad Bridge, a 77-foot (24-meter) span over the ERIE CANAL in central New York, was one of the first iron bridges to adapt wooden truss designs to the use of iron. Because the growing American railroad system required bridges of greater strength, iron began replacing wood in bridge construction in the mid-nineteenth century.

Additional Information: Schodek, Daniel L. *Landmarks in American Civil Engineering.* Cambridge, MA: MIT Press, 1987.

Franklin, Benjamin Bridge. *See* Benjamin Franklin Bridge.

Fraser, Alex Bridge. See Alex Fraser Bridge.

Fred Hartman Bridge (Houston Texas; 1995)

Designed by Greiner, Inc. of Tampa, Florida, and built by Williams Brothers Construction Company and Traylor Brothers of Houston, Texas, the Fred Hartman Bridge crosses the ship channel at Baytown-LaPorte near Houston. A **cable-stayed bridge** with a 2,475-foot (755-meter) main span, the structure has eight lanes to meet the traffic demands of the area. The bridge has two decks running side by side, with a combined width of 156 feet (48 meters). Eight acres (3.2 **hectares**) of decking make the bridge the largest cable-stayed structure in the world when measured by square footage. The towers for the bridge are double diamonds made of composite steel and concrete; they were designed especially to withstand expected high wind forces from hurricanes. The bridge also has a specially designed traffic barrier rail made to prevent a large truck from leaving the bridge deck and overturning; this is the first use of such a rail, which was designed after a truck fell from a Houston freeway onto a lower level. The bridge also allows State Highway 146 to act as an escape route from the area in times of hurricanes; in the past, a low tunnel on the highway had to be closed before a hurricane, thus effectively closing the road. The bridge received a 1996 Merit Award in the Outstanding Civil Engineering Achieve-ment competition held by the American Society of Civil Engineers (ASCE).

Additional Information: Brown, David J. *Bridges.* New York: Macmillan, 1993; Prendergast, John. "Dynamic Duo." *Civil Engineering* 66:7 (July, 1996), pp. 40-43.

Fredericksburg and Spotsylvania National Military Park (Fredericksburg, Virginia; 1865, 1927)

The park comprises three cemeteries: Fredericksburg Confederate Cemetery, Spotsylvania Confederate Cemetery, and Fredericksburg National Cemetery. Three months after the end of the Civil War in April 1865, Congress heeded the urgings of General Daniel Butterfield, commander of the Union Fifth Corps at the battle of Fredericksburg, and authorized a national cemetery to honor federal war dead. The cemetery was placed on Marye's Heights behind Fredericksburg, the scene of terrible slaughter among federal troops during the battle of Fredericksburg on 13 December 1862. The Fredericksburg area witnessed four major battles during the war: Fredericksburg (December 1862), Chancellorsville (May 1863), and the Wilderness and Spotsylvania Court House (May 1864). In 1927, Congress created the park, which encloses almost 9,000 acres (3,645 **hectares**), making it the largest military park in the world. Thousands of soldiers, all but 2,473 unknown, lie buried in the three cemeteries. Although some veterans of later wars are buried there, the majority of graves are of Civil War veterans. The cemeteries were officially closed to further interments in the 1940s; the number of interments totalled 15,300. Among the many monuments and markers conveying the solemnity and sadness of the site are the Butterfield Monument, the Moesch Monument, and the Parker's Battery Memorial. Parker's Virginia Battery had two guns in position on the heights when the Union soldiers broke through the line during the Chancellorsville campaign in 1863. Joseph Anton Moesch was a Swiss immigrant who served as a colonel in the 83rd New York Volunteers, also known as the "Swiss Rifles." Moesch was killed during the battle of Fredericksburg; the monument to him was erected by his comrades years after the war.

Additional Information: http://woodstock.mro.nps.gov/frsp/natcem.htm.

Fredericksburg National Cemetery. *See* Fredericksburg and Spotsylvania National Military Park.

Fréjus Tunnel. *See* Mont Cenis Tunnel.

Friant Dam (Friant, California; 1935-1942)

Like the SHASTA DAM in northern California, Friant Dam is part of the Central Valley Project designed by the Reclamation Service (later the Bureau of Reclamation) to bring water to southern California. The dam is located in central California 20 miles (32 kilometers) north of Fresno, in the foothills of the Sierra Nevada Mountains. The Friant is a 319-foot-high (97-meter) **concrete gravity dam** with a crest length of 3,488 feet (1,064 meters). The crest is 20 feet (6.1 meters) wide and the base is 267 feet (81 meters) wide. It is able to impound 520,500 **acre-feet** of water. The Friant Dam controls the flow of water from the San Joaquin River and makes possible massive irrigation of dry lands to the south. Because of numerous problems, the dam's spillway has been modified several times.

Additional Information: Jackson, Donald C. *Great American Bridges and Dams*. Washington, DC: The Preservation Press, 1988; Reisner, Marc. *Cadillac Desert: The American West and Its Disappearing Water*. New York: Penguin Books, 1987; http://www.usbr.gov/cdams/friant.html.

Fuller Building. *See* Flatiron Building.

Furens Dam (St. Etienne, France; 1866)
Still in perfect condition, this well-known dam across a tributary of the Loire River near St. Etienne in southeastern France is curved to a radius of 830 feet (253 meters) at its crest. The dam is 164 feet (50 meters) high, 16.5 feet (5 meters) thick at its crest, and 161 feet (49 meters) thick at the base. Messieurs Graeff and Delocre, the builders of Furens Dam, were the first engineers to design dams according to mathematical principles, taking into account such factors as internal stress. The design, construction, and successful operation of Furens Dam revolutionized masonry dam building in the nineteenth century.

Additional Information: Smith, Norman. *A History of Dams*. Secaucus, NJ: Citadel Press, 1972.

G

Galloping Gertie. *See* Tacoma Narrows Bridge.

Ganter Bridge (Brig, Switzerland; 1969)
Christian Menn built this bridge on the New Simplon Pass Road near the town of Brig in southern Switzerland near the Italian border. Although a futuristic-looking bridge, the Ganter's form carefully follows its function. Sitting on top of paired narrow concrete **box girder**s, the Ganter is a **prestressed concrete** bridge that curves in a shallow S pattern. The straight main span of 571 feet (174 meters) is met on either side by oppositely curved spans of 417 feet (127 meters). The valley over which the bridge crosses is so deep that one of the piers is 492 feet (150 meters) high; the massiveness of the pier is offset by the thinness of the deck depth. Cable stays for the deck need to be curved, and are encased in concrete walls on the sides of the roadway.

Additional Information: Brown, David J. *Bridges*. New York: Macmillan, 1993.

Garabit Bridge, also known as the Garabit Viaduct (Massif Central region, southeastern France; 1884)
Designed and built by Alexandre Gustave Eiffel, the builder of the EIFFEL TOWER, the Garabit Bridge is an iron railroad **viaduct** that crosses the Truyére River in the Massif Central region of southeastern France. An **arch bridge**, the Garabit rises 400 feet (122 meters) above the river. With several towers on either side, the bridge's 541-foot (165-meter) length supports a railway deck that is 1,850 feet (564 meters) long. The Garabit was the longest arch span in the world when it was built. Like all Eiffel's structures, the Garabit is both strong and impressively minimalist, using the least amount of material possible. Eiffel took full advantage of the mathematics of engineering as it was known in his day, and used no more material than was needed for any of his structures. Eiffel also used the new "**two-hinged arch**" concept to reduce the weight of the structure.

Additional Information: Hawkes, Nigel. *Structures: The Way Things Are Built*. New York: Macmillan, 1990; Kirby, Richard Shelton et al. *Engineering in History*. New York: Dover Publications, 1990; Lay, M.G. *Ways of the World*. New Brunswick, NJ: Rutgers University Press, 1992.

Garabit Viaduct. *See* Garabit Bridge.

Garrison Dam (Riverdale, North Dakota; 1960)
Built by the U.S. Army Corps of Engineers, this earth dam across the Missouri River is 200 feet (61 meters) high and 11,300 feet (3,447 meters) long. The dam's volume is 66,500,000 cubic yards (50,873,000 cubic meters) with a reservoir capacity of 24,500,000 **acre-feet**. Built to provide electric power, flood control, and irrigation for eastern North Dakota, the dam's irrigation capabilities have never been made fully functional. Garrison was the first major dam built by the Corps of Engineers after the creation in the 1950s of the Pick-Sloan Plan, which outlined a water management plan for the Missouri River watershed. The dam acquired a certain notoriety because its construction flooded the historical home of the Arikara Indians.

Additional Information: Jackson, Donald C. *Great American Bridges and Dams*. Washington, DC: The Preservation Press, 1988.

Gateway Arch, also known as the Gateway to the West (St. Louis, Missouri; 1968)

Designed by Eero Saarinen, a famous Finnish-American architect, the Gateway Arch stands before the city of St. Louis on the west bank of the Mississippi River,

The Gateway Arch in St. Louis, Missouri. Photo courtesy of Andrew Shelofsky.

greeting westbound travelers as they cross the river and prepare to enter the city. The arch is 630 feet (192 meters) high and an equal distance across at the bottom; Saarinen borrowed the arch concept from the EADS BRIDGE. The stainless steel arch takes the shape of an inverted catenary curve, which is the natural shape a chain would take if it hung freely between two supports. Each leg of the arch is an equilateral triangle. At ground level, the sides are 54 feet (16.5 meters) long; the triangles taper to 17 feet (5.2 meters) at the top. Double walls of steel are 3 feet (0.9 meters) apart at ground level, and 7.75 inches (20 centimeters) apart above the 400-foot (122-meter) level. Up to a height of 300 feet (91.5 meters), the space between the walls is filled with **reinforced concrete**; above that height, the space holds steel stiffeners. The arch is part of the Jefferson National Expansion Memorial, which commemorates the westward pioneers. Beneath the arch is the Museum of Westward Expansion. The Memorial site also contains the Old Cathedral and the courthouse where the Dred Scott case was tried in the 1850s.

Additional Information: http://nps.gov/jeff/arch-home/default.htm; http://www.st-louis.mo.us/st-louis/arch/structur.html.

Gateway to the West. *See* Gateway Arch.

Gaunless Viaduct (West Auckland, England; 1823)

Designed by George Stephenson, the builder of the LIVERPOOL-MANCHESTER RAILWAY, the Gaunless Viaduct was the first iron **railroad bridge** in the world. This four-span lenticular-like truss carried the Stockton & Darlington Railroad across the River Gaunless in northeastern England for 76 years. In 1899, the **viaduct** was moved to Great Britain's National Railway Museum in York.

Additional Information: Steinman, David. B. *Famous Bridges of the World.* London: Dover Publications, 1953, 1961.

Gebel Aulia Dam (Sudan; 1937)

Built on the White Nile River in Sudan for the benefit of Egypt, the Gebel Aulia Dam stores extra flood waters to supplement the Aswan Dam (see entries for ASWAN DAM [1902] and ASWAN DAM [1970]).

Additional Information: Smith, Norman. *A History of Dams.* Secaucus, NJ: Citadel Press, 1972.

George P. Coleman Memorial Bridge (Yorktown, Virginia; 1952)

This **drawbridge** across the York River in southeastern Virginia has two swing spans, each 500 feet (152.5 meters) long. The toal length of the bridge is 3,750 feet (1,144 meters).

Additional Information: Steinman, David B. *Famous Bridges of the World.* London: Dover Publications, 1953, 1961.

George Washington Bridge (New York to New Jersey; 1931)

Planned by the New York Port Authority, designed by Cas Gilbert, and built by Othmar Ammann, Leon S. Moisseiff, and Allston Dana, the George Washington Bridge carries Interstate 95 across the Hudson River, connecting Manhattan to Fort Lee, New Jersey. At the time of its construction, the bridge was the longest span in the world, almost double the length of the previous record holder, the AMBASSADOR BRIDGE in Detroit. The 3,500-foot-long (1,068-meter) central span was later exceeded in length by the GOLDEN GATE BRIDGE in San Francisco, the VERRAZANO NARROWS BRIDGE between Brooklyn and Staten Island, and the HUMBER BRIDGE in

This 1976 view of the George Washington Bridge shows the flag that was hung on the bridge for the U.S. Bicentennial celebration. Photo courtesy of The Port Authority of New York and New Jersey.

Humberside, England. The east tower, on the New York side, was built on dry land. The west tower, on the New Jersey side, had to be built underwater. The two piers for the west tower are 89 by 98 feet (27 by 30 meters) wide, and extend 80 feet (24 meters) below the water level. The bridge used the largest **cofferdams** ever created for bridge building. Although the George Washington was generally a much safer project than other bridges, one cofferdam collapsed during construction, killing several workers. Each 635-foot (194-meter) tower (measured above water level) is made of 22,000 tons (19,800 metric tons) of riveted steel and supports a 108-ton (97-metric-ton) steel saddle. Each of the cables is one yard in diameter and almost one mile in length. The spinning of 107,000 miles (over 172,000 kilometers) of wire, enough to go around the equator four times, was unprecedented. Using the technique pioneered by John and Washington Roebling, the builders of the BROOKLYN BRIDGE, John A. Roebling's Sons Company did the cable spinning. A work crew of 3,000 to 4,000 men spun the cables in only 209 working days. Originally, the bridge carried six lanes of Interstate Route 95, but that number was increased to eight lanes in 1946 by using a central median that had been left available for that purpose. An entire second level was added to the bridge in 1962, creating the world's first 14-lane **suspension bridge**, an eventuality that had also been planned for in the original design.

Additional Information: Hopkins, H.J. *A Span of Bridges: An Illustrated History*. New York: Praeger, 1970; Kirby, Richard Shelton et al. *Engineering in History*. New York: Dover Publications, 1990; Schodek, Daniel L. *Landmarks in American Civil Engineering*. Cambridge, MA: MIT Press, 1987; Steinman, David B. *Famous Bridges of the World*. London: Dover Publications, 1953, 1961.

Germantown Dam. *See* Miami Conservancy District.

Gladesville Bridge (Sydney, Australia; 1964)

Designed by the firm of G.A. Mansell & Partners, the Gladesville Bridge is one of the longest concrete arches in the world, and the first arch to span 1,000 feet (305 meters). This eight-lane vehicular bridge crosses the Parramatta River at the western branch of Sydney Harbor in southeastern Australia. It is a **voussoir** bridge with four ribs integrated by stressed wires and was built without **falsework**. It has a main span of 1,000 feet (305 meters), with a 72-foot-wide (22-meter) roadway and two footpaths on columns above its low arch. The design was reviewed and approved by the noted engineer Eugene Freyssinet.

Additional Information: Hopkins, H.J. *A Span of Bridges*. New York: Praeger, 1970; Overman, Michael. *Roads, Bridges, and Tunnels*. Garden City, NY: Doubleday & Company, 1968.

Glasgow Bridge (Glasgow, Missouri; 1879)

Built by William Sooy Smith, the Glasgow Bridge crosses the Missouri River near the town of Glasgow in north central Missouri. The Glasgow is the first all-steel truss bridge built in the United States. Serving the Chicago & Alton Railroad, the bridge's five supported sections are each 300-foot-long (91.5-meter) pre-connected Whipple trusses.

Additional Information: Hopkins, H.J. *A Span of Bridges*. New York: Praeger, 1970; Schodek, Daniel L. *Landmarks in American Civil Engineering*. Cambridge, MA: MIT Press, 1987.

Glasgow District Tube (Glasgow, Scotland; 1891-1897)

The Glasgow District Tube, built under the Clyde River in southwestern Scotland in the 1890s to serve the growing industrial city of Glasgow, was the second "tube" railway in the world. The 6.5-mile (10.5-kilometer) loop line of twin tubes passed mostly through wet, sandy soil, except when it ran under the River Clyde, where it passed through extremely wet mud. Four Greathead shields were used for two bores; no problems were encountered until the tunnel reached the river. Ten blowouts occurred in the first 80 feet (24 meters) of under-river construction, requiring the river bed to be repaired with bags of clay. George Talbot, an engineer who took over the project after the original contractor resigned in frustration, used variable air pressure in the tunnel to prevent blowouts. Talbot also required work to continue round the clock, seven days a week, for he had noticed that blowouts usually occurred when work had ceased. Work was still difficult; in one instance, Talbot successfully led a dangerous and heroic effort to rescue some miners trapped by a fire. Other obstacles to be overcome included an abandoned, waterlogged quarry. The tunnel finally went into operation on 21 January 1897.

Additional Information: Beaver, Patrick. *A History of Tunnels*. Secaucus, NJ: Citadel Press, 1973.

Glen Canyon Dam (Page, Arizona; 1966)

The U.S. Bureau of Reclamation built Glen Canyon Dam across the Colorado River in extreme northern Arizona to supply and control water, and to generate electricity for northern Arizona and southern Utah. The huge dam is a tremendous economic benefit to the region. In 1948, after a two-year study by the Bureau, engineers chose the dam site because of the amount of water nearby, the strength of the canyon walls and foundation, and the available supply of building material. The town of Page was created to service the immense workforce needed for the dam. A bridge across the canyon was completed by 1959, and concrete placement began in 1960 and continued for three years. The dam and power plant contain more than 5 million cubic yards (over 3.8 million cubic meters) of concrete. The dam is

3,700 feet (1,128.5 meters) long and 638 feet (195 meters) high at its crest. The power plant is capable of generating 1.3 million kilowatts of electricity. The eight generators are supplied by **penstocks** through which more than 15 million gallons (57 million liters) of water pass each minute. Water to the dam is supplied by LAKE POWELL, an artificial lake created as a reservoir by the dam's construction. The lake extends northeast more than 180 miles (290 kilometers) into southern Utah. By the 1980s, the dam began to generate concerns about its downstream effects on the environment of the Grand Canyon. Before the construction of the dam, annual floods replenished the environment of the Grand Canyon with needed soil nutrients and water. In 1996, engineers released water from Lake Powell into the Grand Canyon to partially rehabilitate the affected areas. The experiment seemed successful, and controlled annual flooding will be scheduled in the future.

Additional Information: http://www.pagehost.com/lakepowell/ gcdam.htm; Davis, Tony. "Managing to Keep Rivers Wild." *Technology Review*, (May/June 1986); Reisner, Marc. *Cadillac Desert: The American West and Its Disappearing Water*. New York: Penguin Books, 1987.

Glenfinnian Railway Viaduct (Invernesshire, Scotland; 1898)

Glenfinnian, in northern Scotland, is Great Britain's first major reinforced-concrete arch. It consists of 21 arches, each with a 50-foot (15-meter) span. Unlike more modern concrete arches, Glenfinnian was clad with a stone facing to make the concrete look like another material because the public distrusted the concept of a concrete arch.

Additional Information: Brown, David J. *Bridges*. New York: Macmillan, 1993.

Glenn Anderson Freeway. *See* Century Freeway.

Glomel Dam. *See* Nantes-Brest Canal.

Goethals Bridge (Elizabeth, New Jersey, to Howland Hook, Staten Island, New York; 1928)

Named after Major General George W. Goethals, builder of the PANAMA CANAL, the Goethals Bridge was one of the first two projects undertaken by the Port Authority of New York and New Jersey, a quasi-governmental agency dedicated to improving transportation between the two states (the other project is the OUTERBRIDGE CROSSING). The Goethals Bridge crosses the Arthur Kill, a narrow river tributary at the mouth of the Hudson. A steel truss **cantilever** design, the four-lane Goethals is 62 feet (19 meters) wide with a center span of 672 feet (205 meters) and a truss span of 1,152 feet (351 meters).

Additional Information: Bishop, Gordon. *Gems of New Jersey*. Englewood Cliffs, NJ: Prentice-Hall, 1985; Condit, Carl W. *American Building Art: The Twentieth Century*. New York: Oxford University Press, 1961.

Golden Gate Bridge (San Francisco, California; 1937)

The Golden Gate Bridge, which carries U.S. Route 101 over the entrance to San Francisco Bay, was designed by Irving Morrow and built by Charles B. Strauss,

The Golden Gate Bridge in San Francisco looking north toward Marin County. Photo by Bob Colin, courtesy of the State of California Department of Transportation.

Charles Ellis, and Leon S. Moisseiff. The Golden Gate was the longest span in the world until the construction of the VERRAZANO NARROWS BRIDGE in New York in 1964. The center span of the Golden Gate is 4,200 feet (1,281 meters) long. The total length of the bridge is 8,981 feet (2,739 meters). Each of the towers is 746 feet (227 meters) high. Each of the two cables is over 36 inches (91 centimeters) in diameter and contains 27,572 galvanized steel wires. The bridge expands in heat and contracts in cold—on hot days, the heat lengthens the cable enough to lower the center of the bridge 16 feet (4.9 meters) and make the bridge up to 6 feet (1.8 meters) longer than on a cold day. In total, the bridge used more than 100,000 tons (90,000 metric tons) of steel, 693,000 cubic yards (530,000 cubic meters) of concrete, and 80,000 miles (over 128,000 kilometers) of wire cable. The bridge opened on 21 May 1937, with a "Pedestrians' Day," during which 200,000 people walked across the bridge. Safety measures taken during the construction of the bridge included the presence of a doctor and nurses on the wharf daily to examine each worker and the hanging of a safety net below the floor of the bridge for the length of the span. The net is credited with saving 19 lives. Thirteen workers died during the construction of the bridge, 12 of them when a scaffold collapsed on 7 February 1937. Other construction mishaps included the inadvertent destruction of a temporary trestle on the south side by a passing vessel, and the destruction of a section of the bridge by a storm. The beauty of the bridge has made it both an historical and a romantic landmark. Unfortunately, that romanticism has a dark side; as of mid-1995, nearly 1,000 people had committed suicide by throwing themselves over the side of the bridge. In April 1996, suicide prevention teams began patrolling the bridge during daytime hours.

Additional Information: DeLony, Eric. *Landmark American Bridges.* Boston: Bullfinch Press, 1993; Goldberg, Carey. "Golden Gate Bridge to Institute Suicide Patrols." *New York Times,* February 25, 1996, A:32; Steinman, David B. *Famous Bridges of the World.* London: Dover Publications, 1953, 1961.

Grand Canal [China], also known as Pei Yun Ho, "north transport river" (Hangchow to Tientsin, China; begun in fifth century B.C.)

Begun in the fifth century B.C., and almost completed when seen by the Italian explorer Marco Polo around 1290, the Grand Canal runs 860 miles (1,384 kilometers) from Hangchow on the Fuchun River north to Tientsin, just southeast of Beijing. The canal used in-

A section of China's Grand Canal between Tong-Chow and Peking (Beijing). Photo courtesy of the Library of Congress, World's Transportation Commission Photograph Collection.

clined planes, rather than locks, to lift and lower barges. Marco Polo described it as "a wide and deep channel dug between stream and stream, between lake and lake, forming as it were, a great river on which large vessels can ply." The earliest sections were dug north and south of the Yangtze River in the ancient kingdom of Wu. Ten centuries later, the Sui emperors, ruling a united China, extended the canal even further. In the thirteenth century, Kublai Khan extended the canal northward to reach Beijing (then Peking). As part of Mao Tse-tung's "Great Leap Forward" in 1958, the canal, which had fallen into

disuse because of siltation, was reconstructed and enlarged to a width of 148 feet (47 meters) and a depth of 11 feet (3.4 meters).

Additional Information: Harrington, Lynn. *The Grand Canal of China.* Chicago: Rand McNally & Company, 1967; Kirby, Richard Shelton et al. *Engineering in History.* New York: Dover Publications, 1990; Sandstrom, Gosta E. *Man the Builder.* New York: McGraw-Hill, 1970.

Grand Canal [Venice] (Venice, Italy; c. seventh century A.D.)

The Grand Canal of Venice runs through the heart of an interconnected network of more than 200 smaller channels and waterways that form the thoroughfares

The Grand Canal of Venice, Italy, is the heart of the Venetian waterway system.

of the city. Venice is situated on an archipelago lying in a crescent-shaped lagoon that runs for more than 32 miles (51 kilometers) along the northeastern coast of Italy at the head of the Adriatic Sea. The lagoon is protected by a line of sandbanks, or *lidi,* broken by three gaps, or *porti,* that allow both the tides and the city's maritime traffic to pass in and out of the lagoon. The Grand Canal describes a large backward S that runs for more than 2 miles (3.2 kilometers) through the center of the old city from the railway station to the Doges' Palace. The canal varies in width from 100 to 225 feet (30.5 to 69 meters). It is lined by the grand palaces once owned by the great merchant families of the Venetian Republic and by the large public warehouses, or *fondaci,* used for foreign trade. Most of the city's 400 remaining gondolas, the sleek keelless boats for which Venice is famous, still ply the Grand Canal. The canal is spanned by three bridges, the most famous being the sixteenth-century RIALTO BRIDGE, which crosses the most dramatic bend in the waterway. The other two bridges, the marble railway bridge and the high-arched wooden Accademia Bridge, were built in the twentieth century. The palaces lining the canal are made usually of brick faced with stone, most commonly the white marble of northern Italy. The water story of the palaces generally served as the merchant's offices and storerooms, while the family lived on the floor above. The main room of the living quarters, or *salone*, overlooked the Grand Canal through a series of five or six large windows. Each palace ran back from the canal to a central courtyard that was frequently planted with trees and ornate gardens. The Doges' Palace at the eastern end of the Grand Ca-

The Bridge of Sighs in Venice once connected the Doges' Palace with the state prisons. Photo courtesy of Mel Wittenstein.

nal was the political center of the Venetian Republic. It served as the official residence of the elected doge and as the meeting place of the councils and ministries of the republic. The narrow canal on the east side of the Doges' Palace is spanned by the Bridge of Sighs, a small covered bridge that once led to the state prisons. Lord Byron immortalized the bridge in *Childe Harold*, part of which was written in Venice.

Additional Information: Franzoi, Umberto. *Canal Grande.* English Edition. New York: Vendome Press, 1993.

Grand Central Station (New York City; 1913)

Designed by William J. Wilgus, Grand Central Station is an incredibly large and beautiful railroad station located in Manhattan at 42nd Street and Lexington Avenue. The station is a **Beaux Arts**-looking structure built on an iron frame. The current steel and glass building is a reconstruction of an earlier terminal. The station was a major achievement of its time and would still be a significant construction project if it were being built today. The concourse of the station is 470 feet by 150 feet (143 by 46 meters), one of the largest and most impressive open spaces under a roof. Its acoustics are notable; when quiet, a conversation across the concourse can be held in whispers. When the station is crowded, as it usually is, the din is almost painful. The barrel vaulted ceiling has 2,500 stars painted on it. Catwalks run around part of the station, crossing some of its 60-foot-high (18.3-meter) windows.

Additional Information: http://www.hotwired.com/rough/usa/

mid.atlantic/ny/nyc/city/midtown.html; Condit, Carl W. *American Building Art: The Twentieth Century.* New York: Oxford University Press, 1961. [Condit's discussion of both the political and engineering challenges involved in the construction of the station is detailed and worthwhile for readers seeking an extended discussion.]

Grand Concourse and Boulevard (New York City; 1910-1914)

Based on a concept by Frederick Olmstead, the Grand Concourse and Boulevard, which runs from 151st Street to Mosholu Parkway in the Bronx, was the first major street to separate parts of a roadway into local and express sections, and to limit access to the express lanes. Olmstead also had the idea of running cross streets either above or below the roadway, rather than intersecting it.

Additional Information: Condit, Carl W. *American Building Art: The Twentieth Century.* New York: Oxford University Press, 1961.

Grand Coulee Dam (near Spokane, Washington; 1942)

This famous **concrete gravity dam** was the first structure in the world to exceed the volume of the GREAT PYRAMID OF GIZA in Egypt. The major public works project in the Pacific Northwest during the Great Depression of the 1930s, the Grand Coulee Dam was authorized by Congress in 1933. The dam is 550 feet (168 meters) high. It crosses the Columbia River in eastern Washington, and is 4,173 feet (1,273 meters) long. Its volume is 10.6 million cubic yards (over 8.1 million cubic meters) and it creates a reservoir 150 miles (241 kilometers) long. It irrigates 1.2 million acres (486,000 **hectares**), generates 2 million kilowatts of electrical power, and makes the Columbia River navigable up to Revelstoke in Canada. The first irrigation water flowed from the dam on 22 May 1952. To publicize the project, the Bureau of Reclamation created an entire 90-acre (36.5-hectare) farm on irrigated land in one day, and deeded the land to a local war veteran nominated by voters. Some of the electric power of the dam is used to pump water to the Grand Coulee, a geological formation created during the last ice age, which serves as a reservoir for the dam.

Additional Information: Jackson, Donald C. *Great American Bridges and Dams.* Washington, DC: The Preservation Press, 1988; Sandstrom, Gosta E. *Man the Builder.* New York: McGraw-Hill, 1970; Smith, Norman. *A History of Dams.* Secaucus, NJ: Citadel Press, 1972.

Grand Dixence Dam (across the Dixence River, Switzerland; 1961, 1962)

One of the world's highest dams, the Grand Dixence is a **gravity dam** that stands 932 feet (284 meters) high and runs to a length of 2,296 feet (700 meters). It has a volume of 7,792,000 cubic yards (5,961,000 cubic meters) and a reservoir capacity of 324,000 **acre-feet**.

Additional Information: Smith, Norman. *A History of Dams.* Secaucus, NJ: Citadel Press, 1972.

Grand Junction Railway (Manchester to Birmingham, England; 1837)

Joseph Locke, who trained under George Stephenson, constructed this trunk line to Stephenson's LIVERPOOL-MANCHESTER RAILWAY. The line ran 80 miles (129 kilometers) connecting Birmingham in west central England with the Liverpool-Manchester Railroad at Manchester in northwestern England. Although he trained under Stephenson, Locke proved more capable at managing a large project than his mentor. The Grand Junction line has a maximum grade of 1:330, hardly significant by modern standards, but a considerable improvement over the problem-filled 1:180 grade Stephenson had demanded for the Liverpool-Manchester line. Finished on time, Locke's railway used switchbacks and followed the contours of the land, saving money by going around obstacles rather than through them with costly and difficult tunnels.

Additional Information: Sandstrom, Gosta E. *Man the Builder.* New York: McGraw-Hill, 1970.

Grand-Mère Bridge (Grand-Mère, Quebec, Canada; 1929)

Designed by the firm of Robinson and Steinman, which pioneered this "rope-strand" type of **suspension bridge**, the Grand-Mère Bridge crosses the St. Maurice River at Grand-Mère in southern Quebec province, about 20 miles (32 kilometers) northwest of Trois-Rivières on the St. Lawrence River. The cables are composed of twisted strands of wire that look like ropes. The main span is 950 feet (290 meters) long.

Additional Information: Steinman, David B. *Famous Bridges of the World.* London: Dover Publications, 1953, 1961.

Grand Palace (Bangkok, Thailand; after 1782)

Begun when Bangkok became the capital of Thailand in 1782, the Grand Palace is actually a group of buildings. Especially notable in the complex is the Temple of the Emerald Buddha. The Buddha was brought from the palace in Thonburi, the former capital, and placed in the Royal Chapel. The Grand Palace occupies an area of one square mile. The original compound was built during the reign of King Rama I (1782–1809). No longer the residence of royalty, the palace is still the site of several important state occasions each year.

Additional Information: http://www.asiatour.com/

thailand/e-03bang/et-ban70.htm; http://www.asiatour.com/thailand/e-03bang/et-ban71.htm.

Grand Pont Suspendu (Fribourg, Switzerland; 1834)

Designed by Joseph Chaley, the Grand Pont Suspendu, an 896-foot-long (273-meter) span across the Sarine River in western Switzerland, was the longest **suspension bridge** ever built at the time of its construction. Its deck rode 164 feet (50 meters) above the water. Because the bridge was so long, the cables could not be assembled on the ground and then raised to the towers. The solution was to make the cables in sections and then splice them together on the bridge. The cables consisted of 1,000 iron wires bound together with iron wrappings. The cables were then led over the towers and fastened to "saddles" that held them to the towers where they were then connected further toward the anchorages. The bridge was demolished in 1923. Chaley, originally a medical officer in the French army, introduced bundled cables and aerial spinning to the building of suspension bridges.

Additional Information: Brown, David J. *Bridges.* New York: Macmillan, 1993; Hopkins, H.J. *A Span of Bridges.* New York: Praeger, 1970; Kirby, Richard Shelton et al. *Engineering in History.* New York: Dover Publications, 1990; Lay, M.G. *Ways of the World.* New Brunswick, NJ: Rutgers University Press, 1992.

Grants Mill Covered Bridge (Hardenbergh, New York; 1902)

In 1992, Robert Vredenburgh and his son Joseph completed reconstruction of Grants Mill Covered Bridge, a structure originally built by Robert Vredenburgh's great-grandfathers, Edgar Marks and Wesley Alton. One of only 21 covered bridges remaining in the state of New York, Grants Mill, like many other covered bridges in

Construction began on the Grand Palace in Bangkok, Thailand, in the late eighteenth century.

the United States, is being restored as part of a growing movement to preserve the superb work of the past. The 66-foot-long (20-meter) bridge was covered to protect traffic and its own structure from the damaging effects of weather. The reconstructed bridge no longer carries traffic. Restoration of the bridge cost about $15,000, with funding coming partly from the town of Hardenbergh and partly from private sources. The town hopes the reconstructed bridge will become a tourist attraction. For Vredenburgh and his son, the project was "a labor of love."

Additional Information: Faber, Harold. "Hardenbergh Journal: One of New York's Covered Bridges Gets a Reprieve." *New York Times*, May 4, 1992, B:5.

Great Apennine Railway Tunnel (between Grizzana and Vernio, Italy; 1934)

At 11.5 miles (18.5 kilometers) long, the Great Apennine Tunnel is less than a mile shorter than the SIMPLON TUNNEL. Unlike the Simplon, the Great Apennine was built for a double track railroad. The tunnel runs through the Apennine Mountains of north central Italy, carrying the rail line that connects Bologna with Florence. The tunnel cost the lives of 97 men who died during 13 years of construction. The sheer size of the project required the use of new and different approaches to tunneling. Thirty miles (48 kilometers) of railroad track were laid just to transport machinery and materials to the **headings.** Tunnel entrances were over 800 feet (244 meters) up on a mountainside, and a 5-mile long (8-kilometer) aerial ropeway had to be built to get supplies to the headings. Towns had to be built at the headings and at the shafts to accommodate the workers and their support people. An oil-burning plant was built to supply electricity, and an aqueduct was needed to bring in fresh water. Methane gas was the major construction difficulty; fires in the tunnel were a regular event. At one point, a fire ignited that could not be put out; the tunnel was walled in on both sides and a bypass tunnel was constructed alongside. Five months passed before the bypass tunnel could be holed through to the original tunnel. Water was another problem; the mountain was honeycombed with underground caverns and streams. Despite the use of pumps, water flow regularly interrupted work, and the drainage system had to be left in place even after the tunnel was completed. Almost 69 million cubic feet of earth had been removed in the construction process, and nearly 2,000 workers had been employed on the project at one time.

Additional Information: Beaver, Patrick. *A History of Tunnels.* Secaucus, NJ: Citadel Press, 1973.

Great Northern Railway Stone Arch Bridge. *See* Stone Arch Bridge.

Great Pyramid of Cheops. *See* Great Pyramid of Giza.

Great Pyramid of Giza, also known as the Great Pyramid of Khufu or the Great Pyramid of Cheops (Giza [formerly Memphis], now a suburb of Cairo, Egypt; twenty-seventh century B.C.).

The only surviving "wonder" from the SEVEN WONDERS OF THE ANCIENT WORLD, the Great Pyramid was built by the Egyptian pharaoh Khufu (or Cheops) around 2650 B.C. as his future tomb. It is one of a group of three pyramids built at Giza. The tallest structure in the world until the nineteenth century, the pyramid was built over a period of some 20 years, and the building process involved the cutting, movement, and placement of more than 2 million blocks of stone. The exacting detail of the pyramid exhibits a careful, advanced knowledge of engineering and mathematics. Although the Egyptians had no knowledge of the compass, each side of the pyra-

The Great Pyramid of Giza outside Cairo, Egypt.

mid is oriented with one of the cardinal points of the compass (north, south, east, west). At the base, each side is 229 meters (751 feet) in length; at any height, the horizontal cross-section is a square with a maximum error in length of less than 0.1 percent. Originally, the pyramid was 145.75 meters (478 feet) high, although 10 meters (33 feet) have been lost from the top since construction. The angle of each side is a precise 54 degrees, 54 minutes. The entrance to the pyramid is on the north side. At the center of the pyramid is the pharaoh's burial chamber, accessible through the "Great Gallery" and a single, ascending corridor. The sarcophagus of the pharaoh, made of black granite, is wider than the entrance to the chamber, and must have been placed in the pyramid during construction, rather than at the time of the pharaoh's death. Near the pyramid, in a museum, is the sun boat, discovered in 1954 near the south side of the pyramid. The sun boat was probably used to transport the body of Khufu to the pyramid, and may have remained at the site for the pharoah's use in the afterlife.

Additional Information: Ashmawy, Alaa K., ed., "The Seven Wonders of the Ancient World," at http://pharos.bu.edu/Egypt/Wonders/pyramid.html

Great Pyramid of Khufu. *See* Great Pyramid of Giza.

Great River Bridge (Burlington, Iowa, to Gulfport, Illinois; 1993)

Designed by Sverdrup Corporation of St. Louis, this five-lane, **cable-stayed bridge** across the Mississippi River is the first cable-stayed bridge in Iowa. More than 2,260 feet (689 meters) long, the bridge is an attractive and worthy replacement for the old MacArthur Bridge, a two-lane **cantilever** constructed in 1917. Among the most important design elements of the new bridge is its successful connection with the old roadway of U.S. Route 34; a different type of bridge would have required reconstruction of approach roads. The new bridge's single H-shaped tower provides an asymmetrical design that significantly increases traffic flow and fits into the existing roadway system.

Additional Information: Petzold, Ernst H. "Cables Over the Mississippi." *Civil Engineering* 64:2 (February, 1994), pp. 62-65.

Great Seto Bridge (Honshu Island to Shikoku Island, Japan; 1988)

Slightly less than 8 miles (13 kilometers) long, Great Seto Bridge is one of the world's longest double-decker bridges. Capable of carrying cars and railroad trains, the bridge comprises six spans: three suspended, two **cable-stayed**, and one conventional truss. Built at a cost of over $8 billion, the bridge connects two of Japan's four main islands.

Additional Information: Hawkes, Nigel. *Structures: The Way Things Are Built*. New York: Macmillan, 1990.

Great Sphinx (Giza [formerly Memphis], now a suburb of Cairo, Egypt; third millennium B.C.)

A massive statue surrounded by temples in the desert of Egypt near the pyramids at Giza, the Sphinx is a recumbent figure with the body of a lion and the head of a man. No one is sure why it was built or specifically what it represents, and no dedicatory stones, tablets, or

carvings exist to tell us. One theory suggests the Sphinx was built in the twenty-seventh century B.C., about the same time as the nearby GREAT PYRAMID OF GIZA; the face is thought to be that of Pharoah Khufu, the builder of

The Sphinx stands near the pyramids at Giza outside the modern city of Cairo, Egypt.

the Great Pyramid. Other smaller sphinxes in Egypt are known to have borne the likenesses of rulers. A more recent theory based on erosion patterns on the body of the Sphinx dates the statue well before the pyramids, as early as 7000 B.C. This theory is highly speculative and controversial. Nearly 73 meters (239 feet) long and 21 meters (69 feet) high, the figure is carved out of the natural rock, although the paws are masonry. A temple to the sun god Harmachis once stood between the paws. Research is ongoing into the origins and internal configuration of the Sphinx; some testing has suggested the figure contains hidden chambers and walkways, although results have not been conclusive.

Additional Information: http://sunship.com/egypt/giza/phototr1.html provides pictures and drawings of several aspects of the monument. **Source:** http://www.idsc.gov.eg/cgi-win/pser3.exe/10.

Great Wall of China (northern China; begun third century B.C.)

The Great Wall is a 1,500-mile-long (2,414-kilometer) series of fortifications winding across northern China from Gansu province in the west to near Qinhuangdao in Hebei province on the east. The wall, which runs along the southern edge of the Mongolian plain, was erected to protect China from northern invaders. The Emperor Shih Hwang-ti (reigned 246-209 B.C.) began the wall, but the structure took its present form largely from reconstructions and extensions carried out during the Ming dynasty (1368-1644). The wall averages 25 feet (7.6 meters) high and runs from 15 to 30 feet (4.6 to 9 meters) wide at the base, sloping to 12 feet

(3.7 meters) at the top. Guard towers and watch stations are placed at regular intervals along the wall. In the east, the wall is constructed of earth and stone faced

The Great Wall runs for more than 1,500 miles across northern China.

with brick, but in the west it is often no more than an earthen mound. The wall proved of little military use; China was successfully invaded several times from the north.

Additional Information: Waldron, Arthur. *The Great Wall of China: From History to Myth*. Cambridge: Cambridge University Press, 1990.

Green Sergeant's Bridge, also known as
Sergeantsville Bridge (Sergeantsville, New Jersey; 1872)

Green Sergeant's Bridge carries Route 64 across the Wickecheoke Creek, a tributary of the Delaware River. Restored in 1961, this white covered bridge is the last surviving public covered bridge in the state. The oak bridge is itself a replacement for an older bridge, the original abutments of which date to 1750. The bridge is 81 feet (25 meters) long, with a single lane 12 feet (3.7 meters) wide.

Additional Information: Bishop, Gordon. *Gems of New Jersey*. Englewood Cliffs, NJ: Prentice-Hall, 1985; Foderaro, Lisa W. "Strolling in a Bucolic New Jersey Corner." *New York Times*, November 24, 1995, C1:1.

Grosbois Dam (north central France; 1831-1837)

Engineers examining this dam were the first to notice that a masonry dam acted elastically, moving back and forth in relation to the amount of water being held back. The Grosbois Dam was built to supply the Canal de Bourgogne in north central France southeast of Paris;

the canal links the rivers Seine and Saône. The dam is 73 feet (22 meters) high and 1,805 feet (551 meters) long with a vertical air face. The water face is stepped in six increments, increasing from 21.5 feet (6.6 meters) at the crest to 52.5 feet (16 meters) at the base. Early on, the dam was found to have leaks and to have slid on its clay base a distance of about 2 inches (5 centimeters). In 1842, seven large buttresses, 26.5 feet (8 meters) long at top, 38.5 feet (11.7 meters) long at the base, and 33 feet (10 meters) in mean width, were built against the air face. Another two buttresses were added in 1854, but even these did not completely stop the sliding. In 1905, 10 years after the collapse of the BOUZEY DAM at Épinal in eastern France, further efforts were made to stop the sliding at Grosbois by constructing a second dam. This earth dam is about 800 feet (244 meters) downstream, causing water to be impounded to a height of 50 feet (15 meters) up the air face of the first dam and pushing against the weight of the water on the other side. This second dam also had some safety difficulties, but was finally made safe in 1948.

Additional Information: Smith, Norman. *A History of Dams*. Secaucus, NJ: Citadel Press, 1972.

Grotta Tunnel (Naples, Italy; c.30 B.C.)

This famous Roman road tunnel was built to ease pedestrian congestion to the resort town of Baia. The tunnel goes through the ridge of the Posilip Peninsula between the southern Italian city of Naples and Bagnoli, one of its current suburbs. One kilometer long, the tunnel has been widened over the years and currently has a diameter of 6.5 meters (21 feet). The tunnel was excavated by heating rocks and then throwing water on the hot rocks to induce a fracture.

Source: Lay, M.G. *Ways of the World*. New Brunswick, NJ: Rutgers University Press, 1992.

Guavio Dam (on the Chivor and Batatas Rivers, Ubala, Colombia; under construction)

Guavio Dam, near Ubala, Colombia, is one of the world's highest dams, rising 243 meters (797 feet) above its lowest formation. The dam provides Colombia with approximately 8,000 megawatts of hydroelectric power.

Additional Information: http://www.eia.doe.gov/emeu/cabs/colombia.html.

Gunnison Tunnel (Montrose, Colorado; 1910)

This irrigation tunnel was designed to divert waters from the Gunnison River through the Vernal Mesa into a 12-mile-long (19-kilometer) canal to irrigate parts of western Colorado. The tunnel was one of the first projects undertaken by the United States Reclamation Service, which was created under the Reclamation Act of 1902. The act was largely responsible for the economic development of the West at the beginning of the twentieth century. Construction work on the tunnel was done through four **headings**—one east, one west, and two from a centrally located vertical shaft. Water first flowed through the tunnel on 6 July 1910. Workers encountered a number of difficulties, including soft ground, pockets of gas, and dangerous seams of pressurized underground water, both hot and cold. Soft ground caused a cave-in that killed six workers in early 1905. Despite these obstacles, the tunnelers made good progress, achieving a record by cutting through 449 feet (137 meters) of granite in one month. The completed tunnel, which is still in operation, is 30,583 feet (9,328 meters) long, 10 feet (3 meters) wide, and 10 feet deep at the sides. About two-thirds of the tunnel is lined with concrete. The tunnel carries water at a rate of about 1,300 cubic feet per second.

Additional Information: Schodek, Daniel L. *Landmarks in American Civil Engineering*. Cambridge, MA: MIT Press, 1987.

H

Hadley Falls Canal (South Hadley Falls, Massachusetts; 1794)

The first canal built in New England, this 2-mile-long (3.2 kilometer) canal on the Connecticut River in western Massachusetts was the first canal in the United States to make use of an inclined plane. The canal was abandoned in the 1820s.

Additional Information: Spangenburg, Ray and Moser, Diane K. *The Story of America's Canals*. New York: Facts on File, 1992.

Hadrianeium. *See* Castel Sant'Angelo.

Hadrian's Wall (northern Britain; A.D. 122)

Hadrian's Wall is a Roman defensive barrier running for 73 miles (118 kilometers) across the length of the island of Britain near the current border between England and Scotland, from Wallsend (Segedunum) on the River Tyne in the east to Bowness on the Solway Firth in the west. The wall was designed to protect the northern frontier of the Roman province of Britain from invaders from the unconquered northern third of the island. The wall was constructed in the 120s A.D. on the orders of the Emperor Hadrian, who visited the province of Britain in A.D. 122. The eastern section of the wall was made of stone and stood 12 Roman feet high (a Roman foot is slightly longer than a standard foot) and 10 Roman feet wide. The western section was a turf rampart that was 20 Roman feet wide at its base. Both sec-

tions were fronted by a ditch. The Romans placed towers along the wall at intervals of a third of a mile, and small forts, called milecastles, at intervals of one mile. The milecastles were gates through the wall surmounted by a tower and flanked by one or two barracks for housing the wall's defensive force. On the western end, larger forts were built along the wall line at intervals of about seven miles, and an earthwork, called the *vallum*, was constructed behind the wall and the line of forts. From the western end of the wall, a system of forts and towers ran south down the Cumbrian coast for another 26 miles (42 kilometers). The purpose of this extension may have been to protect the province from raiders from Ireland. Although built by the three Roman legions sta-

Hadrian's Wall, running near the current border of England and Scotland, was built in the 120s A.D. to defend the northern frontier of the Roman province of Britain.

73

tioned in the province, the wall was manned by auxiliary troops. The wall's purpose was to control movement on the frontier and to provide a base for a mobile field force capable of moving anywhere it was needed in northern Britain. The troops were never meant to fight from the top of the wall, but to meet and defeat invaders in the open field. During the reign of Hadrian's successor, Antoninus Pius (A.D. 138-161), Hadrian's Wall was abandoned in favor of the Antonine Wall, which was built about 75 miles (121 kilometers) north of Hadrian's Wall in modern Scotland. The Antonine Wall was begun in the early 140s A.D. after a victory over the northern peoples allowed the Roman re-occupation of southern Scotland. The wall ran almost 43 miles (69 kilometers) between the Firth of Forth on the east and the Firth of Clyde on the west, the narrowest neck of land in Britain. The Antonine Wall was built entirely of turf on a 14-foot-wide (4.3-meter) stone base and fronted by a ditch 40 feet (12.2 meters) across and 10 feet (3.05 meters) deep. Like the southern wall, the Antonine Wall had towers and forts, though it lacked the backing *vallum*. Constant pressure from un-Romanized peoples on both sides of the Antonine Wall, and even from peoples south of Hadrian's Wall, caused the abandonment of the Antonine Wall and the reoccupation of the southern defensive line by the mid 160s A.D. Hadrian's Wall was rebuilt and modified on several occasions in the next two centuries, and was finally abandoned when the Romans withdrew from Britain about A.D. 410. Many sections of Hadrian's Wall are still visible today.

Additional Information: Salway, Peter. *Roman Britain*. New York: Oxford University Press, 1981; Wacher, John. *Roman Britain*. London: J.M. Dent and Sons, 1978.

Hall's Sheeder Bridge (Chester County, Pennsylvania; 1850)

Designed by Theodore Burr and built by Robert Russell and Jacob Fox, Hall's Sheeder Bridge carries Hollow Road across French Creek in Chester County in southeastern Pennsylvania. This 100-foot-long (30.5-meter) Burr truss is still used by local traffic. To offer additional support, a concrete pier was later built at mid-span.

Additional Information: Jackson, Donald C. *Great American Bridges and Dams*. Washington, DC: The Preservation Press, 1988.

Hampton Court Palace (southwest of London, England; begun 1514)

Until the early sixteenth century, the site of Hampton Court, along the River Thames southwest of London, was occupied by a small manor house belonging, since the twelfth century, to the knightly Order of St. John of Jerusalem. In 1514, Cardinal Thomas Wolsey, archbishop of York and powerful chief minister to Henry VIII, leased the house for his riverside country residence.

Wolsey substantially rebuilt the structure, greatly enlarging it and adding a suite of rooms for the king and queen. Wolsey planned a progression of courts marked by a series of magnificent gatehouses. Made of red brick,

Henry VIII's palace of Hampton Court along the Thames near London. Photo courtesy of Donna Bronski.

Hampton Court became one of the largest houses in Europe at the time, with the Base Court measuring 167 feet (51 meters) by 142 feet (43 meters). When he fell from power in 1529 after his failure to obtain a divorce for the king, Wolsey was forced to surrender Hampton Court to Henry, who further enlarged the palace to accommodate the full Tudor court of some 1,000 persons. The king added new royal apartments, a new kitchen, a tiltyard, a tennis court, and the Great Hall. Henry also added a number of grand devices, such as the huge astronomical clock made by Nicholas Oursian in 1540. The clock, which still survives, tells the hour, day and month, number of days since the start of the year, phases of the moon, and time of high tide at London Bridge (see entry for OLD LONDON BRIDGE). Henry spent a good part of his last years in the palace. Most of Henry VIII's private apartments were demolished in the 1690s to make way for a new **Baroque** palace built by Sir Christopher Wren for William III and Mary II. The Wren portion of the palace comprises a series of ornately furnished and appointed galleries. The still-existing remnants of the Tudor palace include the kitchens, the Great Hall, and the Great Watching Chamber, which originally gave access to the royal apartments. Also remaining are the Chapel Royal and the Picture Gallery; the latter is said to be haunted by the ghost of Catherine Howard, Henry VIII's fifth wife, who was executed in 1542 for adultery. The last monarch to reside in the palace was George II in the mid-eighteenth century, although Queen Victoria made many changes to the palace during the nineteenth century. Hampton Court is surrounded by 60 acres (24 **hectares**) of Tudor, Baroque, and Victorian gardens, which are known for their exotic plants and statuary and for an intricate maze planted in the late seventeenth century. The gardens also contain a still-bearing grape vine planted in 1768 and an old and massive wisteria vine. Hampton Court has been open to the public since 1838.

Additional Information: http://www.buckinghamgate.com.

Hanging Gardens of Babylon (on the Euphrates River, Babylon [modern Iraq]; fifth century B.C.)

The Hanging Gardens are said to have graced the walls of ancient Babylon, an important city on the Euphrates about 50 kilometers (31 miles) south of Baghdad, the present-day capital of Iraq. Nebuchadnezzar II of Babylon supposedly built the gardens in the fifth century B.C. for the pleasure of his wife (or a concubine), but the gardens are not mentioned in Babylonian literature. It is possible that the gardens, considered one of the SEVEN WONDERS OF THE ANCIENT WORLD, never existed; general tales of the wonders of Mesopotamia spread by the Greek soldiers of Alexander the Great after he conquered the region in the 320s B.C. may have gradually taken on a specific place and pattern, leading to stories of the Hanging Gardens. The Greek sources (by writers who never saw Babylon) that mention the gardens describe a "quadrangular" shaped garden with a terrace (rather than the earth) supporting tree roots. Waterfalls and a superb irrigation system supposedly kept the gardens permanently green and beautiful.

Additional Information: Ashmawy, Alaa K., ed., "The Seven Wonders of the Ancient World" at http://pharos.bu.edu/Egypt/Wonder.Home.html.

Hare's Hill Road Bridge (Kimberton, Pennsylvania; 1869)

Built by Thomas Moseley, this 103-foot (31-meter) span across French Creek in southeastern Pennsylvania is the only surviving example of the Moseley **wrought-iron** lattice **girder** bridge. This type of bridge is characterized by an arched upper chord that is formed by a pair of Z-bars riveted at their flanges and their lattice webbing.

Additional Information: DeLony, Eric. *Landmark American Bridges.* Boston: Bullfinch Press, 1993.

Harlech Castle (Harlech, Wales; 1289)

Designed by Master James of St. George, Harlech Castle is one of the greatest strongholds in Britain, a fearsome fortress well matched to its natural geography. During his second campaign in Wales in the 1280s, Edward I of England began building an "iron ring" of castles along the Welsh coastline. Work began on Harlech in 1283, and continued for another six years. At the height of construction, nearly 950 men were employed. Harlech

Edward I of England began construction of Harlech Castle in North Wales in 1283.

is a concentrically designed structure, with one line of defense enclosed by another. Because of the lay of the land, only the east face of the castle was open to possible attack, and the passage through the gate to the castle was defended by seven obstacles. A 200-foot-long (61-meter) stairway runs down to the foot of the castle; originally, the stairway met the sea and was used for moving supplies into the castle. The castle was last used in wartime in the fifteenth century, during the Wars of the Roses, when Lancastrian supporters of the deposed king, Henry VI, held the castle for years against the reigning Yorkist king, Edward IV. Events at the castle from this period inspired the Welsh song "Men of Harlech."

Additional Information: http://www.wp.com/castlewales/harlech.html.

Harpole Bridge (Colfax, Whitman County, Washington; 1922, 1928)

Built in 1922, Harpole Bridge in eastern Washington is an unusual example of a wooden, boxed-in **Howe truss**. In 1928, the Great Northern Railway acquired and upgraded this bridge across the Palouse River to handle freight traffic. The heavy timber structure of the boxed-in truss was set in place as part of the upgrade. The bridge is now converted to vehicular use. It is the only extant structure of its type in Washington State, and one of the few such structures in the United States.

Additional Information: http://wsdot.wa.gov/eesc/environmental/Bridge-WA-133.htm.

Hartland Covered Bridge (Hartland, New Brunswick, Canada; 1899)

At a length of 1,282 feet (391 meters), the Hartland Bridge across the St. John's River in eastern New Brunswick province in Canada is the world's longest covered bridge. Although New Brunswick had 320 covered bridges in 1944, many are no longer in existence in the 1990s.

Additional Information: http://www.mi.net/tge.html.

Hartman, Fred Bridge. *See* Fred Hartman Bridge.

Hatchie Bridge (Covington, Tennessee; 1934)

The Hatchie Bridge, built in 1934 near Covington in western Tennessee, collapsed in 1989. The wooden timber piles of the bridge were undermined by **scouring**; the channel of the Hatchie River near one of the piers was a full 83 feet (25 meters) from where it had been when the bridge was built. On the day the bridge collapsed, the river was 3 feet (0.9 meters) above flood level. The bridge had no redundancy, an important feature of most major structures. A redundancy is a built-in backup arrangement in case of failure of some portion of the structure.

Additional Information: Matthys, Levy and Salvadori, Mario. *Why Buildings Fall Down.* New York: W.W. Norton & Company, 1992.

Haupt Truss Bridge (now at the Railroaders' Memorial Museum, Altoona, Pennsylvania; 1854)

Herman Haupt built this iron bridge combining the **Pratt truss** and the tied arch. The first Haupt truss was built at the Pennsylvania Railroad's shops in 1851 and was the railroad's first all-iron bridge. This bridge was originally located in Vandevander, Pennsylvania, but was moved to Thompsontown in 1889. It remained in service as a vehicular bridge over the main line until 1984, when it was moved to the Altoona museum. Three other Haupt truss bridges survive in Pennsylvania—at Ronks, Ardmore, and Villanova—all are single spans over the main-line tracks. From 1856 to 1862, Haupt, whose reputation as a civil engineer was impeccable, worked on the HOOSAC TUNNEL in Massachusetts.

Additional Information: DeLony, Eric. *Landmark American Bridges.* Boston: Bullfinch Press, 1993.

Haverhill-Bath Bridge #27 (Bath, New Hampshire; 1829)

The Haverhill-Bath Bridge #27 carries State Route 135 across the Ammonoosuc River in northwestern New Hampshire. The oldest covered bridge in continuous use in the United States, this **Town lattice truss** with two additional arches has two spans of 104 feet (32 meters) and 120 feet (37 meters). Overall, the bridge is 256 feet (78 meters) long.

Additional Information: http://vintagedb.com/guides/covered3.html.

Heceta Head Light (Florence, Oregon; 1894)

Boasting a huge first-order **Fresnel lens**, this lighthouse on the central Oregon coast north of Florence is situated 205 feet (63 meters) above sea level; its light is visible for over 21 miles (34 kilometers). Construction of the light required 1,000 barrels of blasting powder to create a flat surface for the structure on the rocky cliffs. Although now close to Route 101, the Pacific Coast Highway, the lighthouse was in a desolate area until the 1930s when the highway was built. Heceta Head takes its name from Captain Don Bruno de Heceta, a Spanish explorer who passed along the Oregon coast around 1775.

Additional Information: http://zuma.lib.utk.edu/lights/heceta.html.

Hell Gate Bridge (New York City; 1917)

The Hell Gate **railroad bridge** crosses the East River at Hell Gate to connect Queens and the Bronx. Built by Gustav Lindenthal, this two-hinge **spandrel**-braced arch was designed to link railroads in the area. The bridge carries four main railroad tracks. It spans 977 feet (298 meters) from center to center of the pins; its 80,000 tons (72,000 metric tons) of steel support a combined **dead** and **live load** of 70,000 pounds (31,500 kilograms) per linear foot. The lower chord of the arch carries the loads, while the upper chord serves as the member of a stiffening truss. The bridge has massive stone **abutments** and portal towers, with extensive approach spans on either side. Setting one of the foundations for this bridge proved to be significantly more difficult than expected. Instead of solid rock, the underwater area where the foundation was to rest was a crevasse of unknown depth, varying from between 5 and 20 meters (16.4 and 66 feet) in width. The crevasse was bridged

The Hell Gate Bridge across the East River between Queens and the Bronx, New York. Photo courtesy of the Library of Congress, World's Transportation Commission Photograph Collection.

underwater with a solid concrete arch, and the crown of this underwater arch holds the abutment. Although not elegant, the bridge's obvious brute strength has a monumental presence that demands respect from the viewer.

Additional Information: Condit, Carl W. *American Building Art: The Twentieth Century.* New York: Oxford University Press, 1961; DeLony, Eric. *Landmark American Bridges.* Boston: Bullfinch Press, 1993; Overman, Michael. *Roads, Bridges and Tunnels.* Garden City, NY: Doubleday & Company, 1968.

Henniker-New England College Bridge #63

(Henniker, New Hampshire; 1972)

Built by Milton and Arnold Graton, this 136-foot-long (41.5-meter) single-span **Town lattice truss** carries Route 114 across the Contoocook River at Henniker in south central New Hampshire. The bridge is over 18 feet (5.5 meters) wide, with a roadway of almost 14.5 feet (4.4 meters). The structure may be the most recently constructed covered bridge in the United States, if not the world. It's existence is a sign of the continuing American romance with covered bridges.

Additional Information: http://vintagedb.com/guides/covered4.html.

Henry Hudson Bridge (New York City; 1933-1936)

Designed and built by David B. Steinman, who proposed the project as a graduation thesis from Columbia University in 1908, the Henry Hudson Bridge crosses the Hudson River at Spuyten Duyvil, where the Hudson and Harlem rivers meet. The main span of the bridge is 800 feet (244 meters) long; the arch is hingeless steel. The structure won an Artistic Bridge Award.

Additional Information: Steinman, David B. *Famous Bridges of the World.* London: Dover Publications, 1953, 1961. Steinman, David B. and Watson, Sara Ruth. *Bridges.* New York: Dover Publications, 1941, 1957.

High Bridge (High Bridge, Kentucky; 1876, 1911)

John Roebling, the designer of the BROOKLYN BRIDGE, began constructing a **railroad bridge** across the Kentucky River at this site in eastern Kentucky in the 1850s. The outbreak of the Civil War in 1861 prevented Roebling from finishing the structure, and an entirely new bridge was designed and built by Louis F.G. Bouscaren and C. Shaler Smith in the 1870s. The first major **cantilever bridge** in the United States, High Bridge was 1,125 feet (343 meters) long. The bridge was completely replaced by Gustav Lindenthal in 1911 because a stronger bridge was needed to handle more modern railroad traffic. Lindenthal's reconstruction remains in use today. The new bridge has a trackbed that is 31 feet (9.5 meters) higher than the original. Although the new bridge is significantly more massive than the original, it retains the configuration of the original structure.

Additional Information: Jackson, Donald C. *Great American Bridges and Dams.* Washington, DC: The Preservation Press, 1988.

High Coast Bridge (Kramfors, Sweden; 1997)

Built by a European consortium headed by Skansa, the High Coast **suspension bridge** is due to open in 1997. Modeled after the GOLDEN GATE BRIDGE in San Francisco, the High Coast Bridge in east central Sweden will have towers that rise 180 meters (590 feet) above the surface of the Angermanälven River. The bridge will be 1,800 meters (5,904 feet) long and 17.8 meters (58 feet) wide. The span between the two towers will be 1,210 meters (3,969 feet). The bridge will be part of the new European highway.

Additional Information: http://swe.connection.se/hoga-kusten/uk/hk.html.

High Dam at Aswan. *See* Aswan Dam (1970).

Highland Light, also known as the Cape Cod Light

(Truro, Massachusetts; 1797)

Originally lighted with 15 reflector lamps, the Highland Light, on Cape Cod in eastern Massachusetts, acquired a first-order **Fresnel lens** in 1857. The lens allowed a single oil lamp to be seen up to 25 miles (40 kilometers) away. The lens was 12 feet (3.7 meters) high, had a 9-foot (2.7-meter) diameter, and weighed 2,000 pounds (900 kilograms). Estimates made at the time of construction put the life of the lighthouse at about 45 years. In the mid-nineteenth century, writer Henry David Thoreau concurred, believing at that time that erosion would soon bring the lighthouse down. At the time of its opening, the lighthouse was 510 feet (156 meters) from the edge of the Truro Bluffs. Today, erosion has reduced that distance to less than 125 feet (38 meters). In 1996, the lighthouse was moved back from the edge.

Additional Information: http://zuma.lib.utk.edu/lights/cod8.html.

Hokuriku Railway Tunnel (Japan; 1962)

Used by the Japanese Railway, this 9-mile-long (14.5-kilometer) tunnel is one of the longest railway tunnels in the world. The Japanese Railway also operates another of the world's great railway tunnels, the Aki Tunnel, which was built on the island of Shikoku in 1975.

Source: Railway Directory & Year Book; http://www.rtri.or.jp/japanrail/JapanRail_E.html; *The World Almanac and Book of Facts 1993*. New York: Pharos Books, 1993, excerpted at http://booksrv2.raleigh.ibm.com:80/cgi-bin/bookmgr/bookmgr.cmd/BOOKS/BUILDNGS/CCONTENTS.

Holland Tunnel (New Jersey to New York; 1927)

Along with the story of the BROOKLYN BRIDGE, the story of the Holland Tunnel, named after its builder, John Holland, is one of the most romantic stories in the re-

This 1928 photograph shows traffic moving westward through the Holland Tunnel. Photo courtesy of The Port Authority of New York and New Jersey.

cent history of civil engineering. The first long tunnel specifically designed for vehicular traffic, the Holland Tunnel runs under the Hudson River, linking New York City and New Jersey. The project ran into two seemingly insurmountable difficulties—digging a tunnel of such length and supplying fresh air for both tunnelers

and travelers. To overcome these problems, John Holland designed an air conditioning system that is still in use today. He began by measuring car exhaust fumes and identifying the components of exhaust; he then tested the effects of exhaust fumes on volunteers. Holland decided to use four buildings above the surface of the tunnel to serve as outlets and inlets for air. Inside the tunnel, he placed 84 fans (42 blowers and 42 exhaust fans); these fans completely change the air inside the entire length of the tunnel every 90 minutes. Holland's new method of ventilation became known as the "vertical transverse flow." Holland died, supposedly of overwork, several months before the tunnel was completed. The work was continued by his assistant, Milton H. Freeman, and then, after Freeman's death, by Ole Singstad, Holland's design engineer. The original name of the tunnel had been the Hudson River Vehicular Tunnel, but it was later changed to honor its builder. The tunnel's separate entrance and exit openings were unique for the time, and the length of the tunnel itself was 8,557 feet (2,610 meters).

Additional Information: Beaver, Patrick. *A History of Tunnels.* Secaucus, NJ: Citadel Press, 1973; Lay, M.G. *Ways of the World.* New Brunswick, NJ: Rutgers University Press, 1992; Schodek, Daniel L. *Landmarks in American Civil Engineering.* Cambridge, MA: MIT Press, 1987.

Holyoke Dam (Holyoke, Massachusetts; 1849)

Built by the Holyoke Water Power Company to power paper mills, this large rubble and timber dam on the Connecticut River was located 30 miles (48 kilometers) upstream from Hartford, Connecticut, near Holyoke in western Massachusetts. A wooden dam had been built earlier to provide some power and to protect the Holyoke Dam; this earlier structure was destroyed in 1848 as soon as the water level reached its height. The Holyoke Dam was 1,017 feet (310 meters) long and 35 feet (10.7 meters) high. Built on a natural ledge across the river bed, it was almost butterfly shaped, with one side deeper than the other. All the timber beams used in the project were at least 12 inches (30 centimeters) square, and the base used 4 million board-feet of timber; 3,000 iron bolts pinned the base of the dam to the rock ledge in the river. Spaces between the beams were filled with rubble and stones, and the crest was faced with iron sheets. Because water had to flow over the crest, the dam required careful maintenance. Some support was added between 1868 and 1870, but continued and worsening leakage required that the dam be replaced in 1899.

Additional Information: Smith, Norman. *A History of Dams.* Secaucus, NJ: Citadel Press, 1972.

Holyroodhouse Palace. *See* Edinburgh Castle.

Hoosac Tunnel (through Hoosac Mountain in northwestern Massachusetts; 1851-1876)

Groundbreaking ceremonies for the Hoosac Tunnel were held in January 1851. Work on the tunnel began under Chief Engineer A.F. Edwards, who had planned on drilling the tunnel with a "full area" boring machine, a 70-ton tool that would cut concentric grooves in which explosive black powder charges could be placed. Unfortunately, the new machine broke down after work had progressed only about 10 feet (3 meters), and Edwards returned to the older method of working on the tunnel face by hand. In 1856, Edwards was replaced by Herman Haupt, a noted civil engineer who had written the highly regarded and authoritative *A General Theory of Bridge Construction*. Despite Haupt's efforts to use various boring machines and drills, he was always forced to return to hand work. Other problems included the need to shore up the western end of the tunnel because workers were encountering loose stone, and the growing risk of collapse. In 1862, during the Civil War, work came to a halt and the project was re-evaluated. Thomas Doane later became chief engineer and, like Haupt, sought innovative ways to do the work. Doane brought in Charles Burleigh's new invention, the Burleigh drill, a device attached to a movable carriage that was able to drive a drill deep into the rock as well as feed and rotate the drill until the hole was deep enough for an explosive to be packed inside. Each carriage drove four to six drills. The Burleigh drill is the prototype of modern pistoning drills. Doane also encouraged the use of the new "trinitroglycerin" in place of black powder for blasting; under the direction of chemist George Mowbray, the tunnelers found that safety could be enhanced if the trinitroglycerin was frozen during transportation to the site. Mowbray also developed new types of fuses and an electric ignition system. More than 1 million pounds (450,000 kilograms) of nitroglycerine were used in the last eight years of construction. In 1867, a fire in a vertical shaft killed 13 men; in all, 200 lives were lost digging the tunnel. The final cost of the tunnel was $17 million. Although the tunnel was an engineering marvel, travelling through the 4.75-mile (7.6-kilometer) tunnel must have been fearsome. In the inaugural train ride through the tunnel on 19 February 1875, passengers emerged covered with soot and choking from locomotive exhaust. The entire ride took 34 minutes. Although the engineer and fireman had wrapped wet cloths around their heads for protection, they fared little better than the passengers. A ventilation fan was installed in the tunnel in 1897, but the problem was not alleviated until 1911 when the tunnel was electrified and trains could be pulled through by an electric locomotive.

Additional Information: Epstein, Sam and Epstein, Beryl. *Tunnels*. Boston: Little Brown & Company, 1985; Schodek, Daniel L. *Landmarks in American Civil Engineering*. Cambridge, MA: MIT Press, 1987.

Hoover Dam, also known as Boulder Dam or the Boulder Canyon project (Boulder City, Nevada; 1936) Designed by George Kaufmann, this famous concrete-gravity dam on the Colorado River was built by the U.S.

A view of the lower face of Hoover Dam, which spans the Colorado on the Arizona-Nevada border.

Bureau of Reclamation. Hoover Dam is located on the Arizona-Nevada border, about 25 miles southeast of Las Vegas, Nevada. The dam is 727 feet (222 meters) high and 1,244 feet (379 meters) long, and has a capacity of 4,400,000 cubic yards (3,366,000 cubic meters). The dam is part of the Boulder Canyon Project, which also includes a hydroelectric power plant with a capacity of about 1.5 million kilowatts. At completion, the dam contained more than 3.25 million cubic feet (over 828,00 cubic meters) of concrete; to remove the heat generated by the drying concrete, pipes carrying cold brine were placed within the concrete. The dam's reservoir, Lake Mead, is one of the world's largest artificial bodies of water with a length of 115 miles (185 kilometers) and a depth of 589 feet (180 meters). Lake Mead is capable of storing 20 million **acre-feet** of water (35 billion cubic meters). The concrete base of the dam contains 4.5 million cubic yards (over 3.4 million cubic meters) of concrete. The magnificent Hoover Dam is still one of the world's highest dams.

Additional Information: Jackson, Donald C. *Great American Bridges and Dams*. Washington, DC: The Preservation Press, 1988; Kirby, Richard Shelton et al. *Engineering in History*. New York: Dover Publications, 1990; Sandstrom, Gosta E. *Man the Builder*. New York: McGraw-Hill, 1970; Smith, Norman. *A History of Dams*. Secaucus, NJ: Citadel Press, 1972.

"Hope" Statue (John Howell Park, Atlanta, Georgia; approved 1996; installation expected 1998)

In 1996, Dr. Jesse Peel, a member of the Atlanta Park Project Board, arranged for the installation of a statue entitled "Hope" in John Howell Park, a public park built in an area of Atlanta left vacant by an aborted plan for the construction of an interstate highway. The statue will represent hope for a cure for AIDS. Created by sculptor Felix de Weldon, whose other accomplishments include the Iwo Jima Memorial in Washington, D.C., "Hope" is a cast bronze work showing a man, a woman, a child, and an infant. Braced by the woman, the man reaches upward toward a molecule. All four figures are wrapped by a representation of the AIDS Memorial Quilt, each panel of which represents someone who has died of AIDS. All parts of the statue, except the molecule, will turn color as the sculpture weathers; the molecule will remain shiny.

Additional Information: McGowan, Cate. "Howell Park to Display Famous Sculptor's 'Hope.'" *Atlanta Magazine* (December 1995); reprinted at http://www.nav.com/atlanta-30306/decfeat/f1.htm.

Horse Mesa Dam (near Horse Mesa, Arizona; 1924-1927)

Part of the SALT RIVER PROJECT, this dam on the Salt River east of Phoenix, Arizona, is named after nearby Horse

Horse Mesa Dam in Arizona. Photo courtesy of Salt River Project.

Mesa, where thieves once hid stolen herds of horses. The dam is 300 feet (91.5 meters) tall and 600 feet

(183 meters) long. It generates electrical power with the aid of three conventional hydroelectric units and a pumped storage hydroelectric unit, which was added in 1972. The conventional units generate 32,000 kilowatts and the storage unit generates 97,000 kilowatts.

Additional Information: http://www.srp.gov/aboutsrp/water/srplakes.html.

Horseshoe Dam (northeast of Phoenix, Arizona; 1946)

Part of the SALT RIVER PROJECT, Horseshoe Dam is located on a horseshoe-shaped curve of the Verde River, about 10 miles (16 kilometers) north of Bartlett Lake

Horseshoe Dam in Arizona. Photo courtesy of Salt River Project.

in central Arizona. Water for the dam comes from the Phelps Dodge Corporation, which pumps water to the dam from the nearby Black River. The dam supplies part of Phoenix's water. An earthen structure 144 feet (44 meters) tall and 1,500 feet (458 meters) long, including its spillway, Horseshoe Dam provides a recreation site at Horseshoe Lake; activities at the lake vary, depending on the season. Typically, the Salt River Project nearly drains the lake during the summer to provide water for Phoenix.

Additional Information: http://www.srp.gov/aboutsrp/water/srplakes.html#hs_dam.

Houses of Parliament. *See* Westminster Palace.

Howrah Bridge (Calcutta, India; 1943)

Designed by the firm of Rendel, Palmer & Tritton, the 1,500-foot (457.5-meter) Howrah Bridge crosses the Hooghly River at Calcutta in eastern India. The bridge comprises a central section of 564 feet (172 meters) and anchor spans of 468 feet (143 meters) each. The anchor piers, built on land, are the largest piers ever sunk up to that time; they use more than 40,000 tons (36,000 metric tons) of concrete each. Excavations for the piers required a space of 180 by 80 feet (55 by 24 meters), and went down more than 100 feet (30.5 meters). When built, the Howrah was the third largest **cantilever** in the world, and the first large cantilever constructed in the first half of the twentieth century. The Howrah is still an impressive structure today, and is one of the world's busiest bridges. The construction of this bridge is an excellent example of how engineers deal with soft foundations. The two main pier foundations are each 177 feet by 79 feet (54 by 24 meters), divided into 21 vertical shafts, each 16 feet (5 meters) square with 6.6-foot (2-meter) walls between. The muck of the soft foundation was excavated through the hollow walls. The foundations were built on the surface, with a cutting edge below them; as the muck was excavated, concrete was added to fill the spaces in the walls and to keep the top of the foundation above the water line. The north pier eventually wound up in clay, 85 feet (26 meters) below ground. The shafts were then pumped dry and filled with concrete. On the south side, the base of the pier sank 1,023 feet (312 meters) until it rested on blue clay. Since clay is not watertight, the subsoil water was removed from the shafts by sealing and pressurizing the shafts with air. Concrete was added from adjacent air locks. The anchor arms were built first, supported by temporary scaffolding and tied down to the anchorage. The suspended span was cantilevered out, half from each side, with each half 9 inches (23 centimeters) back from final position, leaving only an 18-inch (46-centimeter) gap. Nineteen 433-pound (8,813-kilogram) hydraulic jacks were built into the junction between each half span and the cantilever arm that supported it. Sixteen jacks were used to push out the two halves of the suspended span until they met and were bolted together. Each half of the suspended span is 308 feet (94 meters) long and weighs 2 million kilograms (over 4.4 million pounds).

Additional Information: Brown, David J. *Bridges*. New York: Macmillan, 1993; Overman, Michael. *Roads, Bridges and Tunnels*. Garden City, NY: Doubleday & Company, 1968; Steinman, David B. *Famous Bridges of the World*. London: Dover Publications, Inc., 1953, 1961.

Huddersfield Canal (Huddersfield, England; 1811)

Begun in 1794, the Huddersfield Canal, which linked northeastern and northwestern England across the Pennine Mountains, required 17 years to complete. Much of that time was need to dig the related STANDEDGE TUNNEL. Water for the canal is supplied by two reservoirs at Slaithwaite and at Marsden, near the tunnel's eastern end. Both reservoirs are curved earth banks between 55 and 60 feet (17 and 18.3 meters) high, and more than 500 feet (152.5 meters) long. The water faces are covered with masonry blocks to protect against erosion. Puddled clay was used to make the dams on the canal watertight. The combined capacity of the dams and reservoirs is just 450 **acre-feet**, but provides sufficient water for the canal.

Additional Information: Sandstrom, Gosta E. *Tunnels*. New York: Holt, Reinhart and Winston, 1963; Smith, Norman. *A History of Dams*. Secaucus, NJ: Citadel Press, 1972.

Hudson, Henry Bridge. *See* Henry Hudson Bridge.

Hudson-Manhattan Railroad Tunnel (New York City to Jersey City, New Jersey; 1908)

This first tunnel under the Hudson River was begun by DeWitt Clinton Haskins, who formed the Hudson Tunnel Railroad Company, and completed by William McAdoo and Charles Jacobs. Haskins wanted to use a pneumatic **caisson** to keep the tunnel pressurized during construction and to keep water out and tunnel liners in place. The work face would be exposed. Eventually, a thick concrete wall was built at the mouth of the tunnel and workers had access to their work through an air lock. After a series of accidents and leaks, the tunnel was abandoned in 1882; work resumed shortly thereafter, but was again abandoned in 1887. An attempt was made to use a Greathead Shield to excavate the tunnel, but an English financial crisis caused abandonment again. In 1899, the incomplete project was sold, and McAdoo and Jacobs began their work. The north bulkhead at the New York end was reached in March 1904. The southern tunnel took until February 1908. The original New York terminal was at the site of the current WORLD TRADE CENTER. The tunnel is now part of PATH (Port Authority Trans Hudson), which administers all the traffic tubes that pass under the Hudson between Manhattan and New Jersey.

Additional Information: Cudahy, Brian J. *Rails Under the Mighty Hudson*. Brattleboro, VT: S. Greene Press, 1975; Sandstrom, Gosta E. *Tunnels*. New York: Holt, Reinhart and Winston, 1963; Schodek, Daniel L. *Landmarks in American Civil Engineering*. Cambridge, MA: MIT Press, 1987; .

Huffman Dam. *See* Miami Conservancy District.

Humber Bridge (Kingston-upon-Hull, England; 1981)

Designed by Freeman Fox, this span across the Humber Estuary in northeastern England was the largest bridge in the world when completed in 1981. It was surpassed by the AKASHI-KAIKYO BRIDGE in Japan in 1991. Still one of the longest **suspension bridges** in the world, the Humber Bridge runs 4,624 feet (1,410 meters), with side spans of 919 feet (280 meters) and 1,739 feet (530 meters). Major problems were encountered in setting the south pier at Barton in 1,640 feet (500 meters) of water; the pier is placed on twin **caissons** submerged in the water, sand, and clay of the estuary; problems with the construction of the pier delayed the project for almost two years. The Humber is the first major suspension bridge made with concrete towers.

Additional Information: Brown, David J. *Bridges*. New York: Macmillan, 1993; Lay, M.G. *Ways of the World*. New Brunswick, NJ: Rutgers University Press, 1992.

Humpback Covered Bridge (Covington, Virginia; 1857)

This covered bridge across Dunlaps Creek in the mountains of western Virginia has an 8-foot (2.4-meter) vertical rise, which is unique in the United States. The curve is built into the top and bottom chords of the bridge's multiple king-post trussing system. According to legend, Union and Confederate troops negotiated an agreement that prevented the bridge from being destroyed during the Civil War.

Additional Information: DeLony, Eric. *Landmark American Bridges*. Boston: Bullfinch Press, 1993.

I

Ij Road Tunnel (Amsterdam, Holland; 1968)
The twin 7-meter-wide (23-foot) roadways of the Ij Road Tunnel run for 1.6 kilometers (almost one mile) under the Ij Estuary. The roadways run side by side with a central service channel and with separate under-road ducts for both fresh and stale air. Made of **reinforced concrete**, the ducts are made watertight by means of a 0.75-centimeter (0.3-inch) steel skin on the bottom and sides, and a bituminous membrane along the top. The ducts were fabricated in a nearby dry dock built especially for the purpose, floated to the site with temporary concrete bulkheads, and sunk onto a prepared foundation in a submarine trench prepared by dredging. The foundation is a concrete slab supported along its length by piles with a 113-centimeter (44-inch) diameter; the piles were driven down through soft ground to a solid bed found between 65 and 80 meters (213 to 262 feet) below ground level. Between the foundation and the tunnel elements are PTFE (the chemical abbreviation for the material most commonly known as Teflon®) bearings that allow the elements to slide if necessary; these elements are joined together with bellows-type expansion joints that allow for movement caused by temperature changes or minor subsidence of the piles. To allow for expected post-assembly settlement (which can reach as much as 4 centimeters [1.56 inches]), each section was built with two hinges, each with watertight double-skinned steel bellows, the longitudinal steel reinforcement passing unbroken through these joints. The reinforcement was made rigid after settlement by filling the gaps inside the bellows with concrete. Four shorter sections of this tunnel carry on their tops a ventilation building and a railroad embankment. These sections were constructed differently; they were concreted on site within sheet piling, using compressed air to force out the water.

Additional Information: Overman, Michael. *Roads, Bridges and Tunnels*. Garden City, NY: Doubleday & Company, 1968.

Il Duomo. *See* Cathedral of Santa Maria del Fiore.

Inguri Dam (on the Inguri River, Republic of Georgia; 1980)
At a height of 988 feet (301 meters), this arch dam is one of the highest dams in the world. The dam is located in the Abkhazian region of the Republic of Georgia, a former component republic of the Soviet Union that lies in the Transcaucasia region of Eurasia between the Black and Caspian seas. The Inguri Dam is 2,240 feet (683 meters) long, and has a volume of 3,920,000 cubic yards (almost 3,000,000 cubic meters); its reservoir capacity is 1,257,000 **acre-feet**.

Additional Information: Smith, Norman. *A History of Dams*. Secaucus, NJ: Citadel Press, 1972.

Inistearaght Lighthouse, also known as The Tearaght (County Kerry, Ireland; 1870)
This lonely lighthouse in southwestern Ireland is the most westerly situated lighthouse in Europe. Its keepers are supplied by the most westerly running railroad in Europe. The Inistearaght Lighthouse was lighted on 4 May 1870 after five years of rock blasting and construction. The lighthouse and its keepers maintain a

lonely watch over the dangerous coast of western Ireland, especially over such treacherous areas as the infamous Foze Rocks. The Tearaght's beam reaches 27 nautical miles (50 kilometers).

Additional Information: http://zuma.lib.uk.edu/lights/eagle/eagle10-30.html.

Inter-American Highway. *See* Pan American Highway System.

Interstate Highway System. *See* Dwight D. Eisenhower Interstate and Defense Highway System.

Itaipú Dam (on the Paraná River between Paraguay and Brazil; 1991)

The Itaipú Dam is not only the largest dam in South America, it is one of the largest hydroelectric complexes in the world. Its 18 huge turbines can generate up to 12,600 megawatts of power. The dam was part of a joint project of Brazil and Paraguay that resulted in the construction of several dams. Itaipú is the main dam of the Itaipú hydroelectric power plant. The plant puts out enough power to meet the equivalent electricity needs of the state of California. The plant provides approximately 25 percent of Brazil's energy supply, and approximately 78 percent of Paraguay's. With its support dams, the Itaipú is 7,744 meters (25,400 feet) long, and its highest crest elevation is 225 meters (738 feet). The Itaipú Dam used more than 15 times the volume of concrete required for the CHANNEL TUNNEL. Begun in 1975 and completed in 1991, the dam sometimes had as many as 30,000 workers engaged in its construction.

Additional Information: http://pharos.bu.edu/Egypt/Wonders/Modern/itaipu.html.

Ivrea Dam (on the Dora Baltea River, Italy; c. 1470)

This early **diversion dam** is located near the ancient Roman town of Ivrea in the Piedmont region of northwestern Italy, close to the point where the Dora Baltea descends from the Alps to enter the north Italian plain. The dam is about 1,500 feet (457.5 meters) long. It was heavily reengineered in 1651 and was built to send the waters of the Dora Baltea into a canal at the its southern end. The dam raises the water level behind it about 15 feet (4.6 meters) and is still in use today.

Additional Information: Smith, Norman. *A History of Dams.* Secaucus, NJ: Citadel Press, 1972.

Jacobs Creek Bridge (Uniontown, Pennsylvania; 1801)

James Finley designed this 1,000-foot-long (305-meter) span in southwestern Pennsylvania. At the time of its construction in 1801, the Jacobs Creek Bridge was the longest **suspension bridge** ever built, the first suspension bridge in the United States, and the first modern suspension bridge. Drawn **wrought-iron** wire was used for the main cables. Damaged and repaired in the 1820s, the bridge lasted for 50 years.

Additional Information: Schodek, Daniel L. *Landmarks in American Civil Engineering*. Cambridge, MA: MIT Press, 1987.

James River Bridge (Richmond, Virginia; 1838)

The James River Bridge employed a **Town truss** style of construction, a building style devised and patented in the 1820s. In early April 1865, Confederate forces destroyed the bridge when they evacuated Richmond, which had served as the capital of the Confederate States of America during the Civil War. The bridge was 2,900 feet (884.5 meters) long and rested on 18 supporting piers.

Additional Information: Kirby, Richard Shelton et al. *Engineering in History*. New York: Dover Publications, 1990.

Jami Masjid Mosque (Old Delhi, India; mid-seventeenth century)

Shah Jahan, the ruler of the Mogul Empire of northern India from 1628 to 1658, built this mosque, the largest in India and one of the largest in the world, at his capital of Delhi. The emperor, who was known for his building projects, also constructed the TAJ MAHAL in Agra as a mausoleum for a beloved wife and the RED FORT, or Lal Qila, in Delhi to defend the city from invaders. The Jami Masjid has two 40-meter-high (131-foot) minarets constructed of strips of red sandstone and white marble. The mosque also has three gateways and four towers.

Additional Information: http://www.lonelyplanet.com.au/dest/ind/cal.htm.

Johnson, E. Memorial Tunnel. *See* U.S. Interstate Route 70.

Johnstown Dam. *See* South Fork Dam.

Jones Falls Dam. *See* Rideau Canal.

K

Kanmon Railroad Tunnels (Japan; 1936-1944)

The Kanmon Tunnels were the first tunnels built under the ocean floor by using both rock and soft-ground tunneling techniques. The original tunnels run for 3.8 kilometers (2.4 miles); a 7.4-kilometer (4.6-mile) tunnel was constructed in the same area in 1975. The 1975 tunnel, which is operated by Japanese Railway, is one of the longest railway tunnels in the world.

Additional Information: *Academic American Encyclopedia*, (electronic version). Danbury, CT: Grolier Electronic Publishing, 1991; *The World Almanac and Book of Facts 1993*. New York: Pharos Books, 1993, excerpted at http://booksrv2.raleigh.ibm.com:80/cgi-bin/bookmgr/bookmgr.cmd/BOOKS/BUILDNGS/CCONTENTS.

Kappelbrucke. *See* Chapel Bridge of Lucerne.

Kariba Dam (on the Zambesi River between Zambia and Zimbabwe; 1955-59)

In 1959, when the newly completed Kariba Dam began providing hydroelectricity to the surrounding region, Zambia (then Northern Rhodesia) and Zimbabwe (then Southern Rhodesia) were British colonies. Today, the dam and 165-mile-long (266-kilometer) Lake Kariba, which the dam has created, are major tourist attractions for both countries. Kariba is an arch dam, 420 feet (128 meters) high, with an impressive length of 2,025 feet (618 meters). It has a volume of 1.3 million cubic yards (994,500 cubic meters) of concrete, and its related reservoir holds a volume of 130,000,000 **acre-feet** of water.

Additional Information: Smith, Norman. *A History of Dams*. Secaucus, NJ: Citadel Press, 1972.

Karlsbrucke Bridge (Prague, Czech Republic; mid-fourteenth century)

Construction on the Karlsbrucke (the "Charles Bridge") began in 1357 during the reign of King Charles IV of Bohemia and ended in the 1380s after the king's death. Designed by Peter Parler, this bridge across the Vltava River was possibly the most monumental bridge of its time. The bridge required some maintenance and reengineering work in the fifteenth century, including some strengthening after a flood. The Karlsbrucke is unusual because it is built of sandstone, not granite, the more typical building material for stone bridges. It has a total length of 1,692 feet (516 meters), carried on 16 arches that vary in length from 54.5 to 76.25 feet (16.6 to 23.25 meters), and supports a 33-foot-wide (10-meter) deck. The wedge-shaped piers range in thickness from 26 to 28 feet (about 8 to 8.5 meters); they were designed to be especially massive to withstand floating ice in winter. Beginning in 1683, the parapets were decorated with "devotional statuary" and 24 statues now stand on the bridge.

Additional Information: Brown, David J. *Bridges*. New York: Macmillan, 1993.

Karnali River Bridge (across the Karnali River, Nepal; 1993)

The New York firm of Steinman Boynton Gronquist & Birdsall designed the Karnali River Bridge and supervised its construction. Kawasaki Heavy Industries, Ltd., a Japanese firm, built the bridge. This beautiful asymmetric, single-tower, **cable-stayed bridge** is one of the longest structures of its type in the world. Sixty cable stays come from the main tower, and each cable of the main span has a matching cable on the side span. The bridge is the first fixed link between western Nepal and the rest of the country. The bridge and the road leading to it are a part of the proposed Asian Highway, a project prepared by the Asian Development Bank and intended to link all of Southeast Asia with a road and railroad network. The bridge is 500 kilometers (310 miles) from Kathmandu, the capital of Nepal, and 100 kilometers (62 miles) from the airport in Napalganj. These distances and the mountainous terrain made just gearing up for construction an extreme challenge. The first construction material to come to Nepal took seven days to arrive from the airport. Kawasaki built a minivillage at the work site that accommodated as many as 650 people at the height of construction. Kawasaki also built five single-story buildings for supervising staff. Because of flooding concerns, the engineers decided to build the tower for the bridge outside the river channel during the dry season. This decision led to the asymmetrical design; the main span runs 325 meters (1,066 feet), while a side span of 175 meters (574 feet) extends over the monsoon-season floodplain. Monsoon weather greatly limited construction. When the first building season ended, the tower workers had reached only 6 meters (20 feet) below the water level. When the workers returned the next season, they found the entire area had been silted in by flood waters. Kawasaki brought a larger excavator from Japan, but the work, including large amounts of underwater blasting, still went slower than expected. Precast concrete castings for the deck and other parts of the bridge were made on-site, simplifying transportation problems, but also generating the need for a way to protect the curing concrete from cold weather during the winter months.

Additional Information: Gessner, George and Selvaratnam, Selva. "Bridge Over the River Karnali." *Civil Engineering* 66:4 (April 1996), pp. 48-51.

Kensico Dam (Valhalla, New York; 1916)

J. Waldo Smith was engineer on the Kensico Dam, which was built as part of the expansion of New York City's water supply. This **concrete gravity dam** on the Bronx River about 20 miles (32 kilometers) north of New York City is 307 feet (94 meters) high and 1,025 feet (313 meters) long. Water is delivered to the dam by a 75-mile-long (121-kilometer) **aqueduct** from the northern Catskill area of New York State. The dam forms the Kensico Reservoir from which the city of New York draws part of its water supply. Although not needed for any architectural purpose, the downstream side of the dam is faced with carved masonry ornamentation to emphasize the importance of the structure to the city's water supply. The absence of a spillway (not needed because of the way water is supplied to Kensico) gives the dam a calmness that contrasts with its massiveness. Visitors can climb concrete steps to the top of the

A view of the carved masonry ornamentation at the top of the Kensico Dam near Valhalla, New York. Photo courtesy of Mel Wittenstein.

dam where small Roman-type semicircular structures contain the impressive dedicatory plaques.

Additional Information: Jackson, Donald C. *Great American Bridges and Dams*. Washington, DC: The Preservation Press, 1988.

Keokuk Canal and Hydroelectric Power Plant

(on the Mississippi River at Keokuk, Iowa; 1877, 1913)

As a young officer of engineers, Lieutenant Robert E. Lee, the future Confederate general, examined the Des Moines (or Keokuk) Rapids of the Mississippi in 1837 for ways to improve the channel for navigation. As a result, the army over the next few years blasted and excavated the passage along the rapids several times. After 1865, work began on a canal that was opened to navigation in 1877. The 7-mile-long (11.3-kilometer) canal consisted of three locks that each measured 80 by 350 feet (24 by 107 meters). In 1913, the Mississippi River Dam Company built the Keokuk Hydroelectric Power Plant, one of the first hydroelectric plants built in the United States and, at the time, the largest hydroelectric plant in the world. The new dam is 4,649 feet (1,418 meters) long with a head of 34.5 feet (10.5 meters). The powerhouse has room for 30 turbines of 10,000 horsepower. Another lock, #19,110 by 1,200 feet (33.5 by 366 meters), was installed for navigation to replace the canal. Today, the many small dams (and their associated canal sections) between St. Louis and Minneapolis-St. Paul provide navigation along the whole section of the river.

Additional Information: Kirby, Richard Shelton et al. *Engineering in History*. New York: Dover Publications, 1990.

Kielder Dam (Kielder, England)

This earth embankment in Northumberland in northern England is the largest dam in Great Britain. The dam is 1,740 feet (531 meters) long; the embankment contains nearly 7 million cubic yards (9.1 million cubic meters) of earth. The attached reservoir of 2,684 acres (1,087 **hectares**) supplies water through the dam to a good portion of northern England. The dam's reservoir, Kielder Water, may be the largest man-made lake in Europe.

Additional Information: Hawkes, Nigel. *Structures: The Way Things Are Built*. New York: Macmillan, 1990.

Kiev Dam (Ukraine; 1964)

This earth dam over the Dnieper River is one of the largest in the world; it is 62 feet (19 meters) high and 134,188 feet (40,927 meters) long (more than 25 miles). The dam has a volume of 58,030,000 cubic yards (44,393,000 cubic meters), and the reservoir holds a volume of 3,024,000 **acre-feet**.

Additional Information: Smith, Norman. *A History of Dams*. Secaucus, NJ: Citadel Press, 1972.

Kintai Bridge (Iwakuni, Japan; 1673)

This four-span timber-**arch bridge** across the Nishiki River is the oldest extant timber bridge in the world. The Kintai is a **muleback bridge** with circular-segment arches that are less than full semicircles. One arch of the bridge has been replaced every five years since the bridge was constructed in the seventeenth century.

Source: Lay, M.G. *Ways of the World*. New Brunswick, NJ: Rutgers University Press, 1992.

Kinzua Viaduct (Kushequa, Pennsylvania; 1882, 1900)

Octave Chanute constructed the original Kinzua Viaduct in 1882. The bridge was built to carry the Erie Railroad across the Kinzua Creek Valley. The **wrought-iron** viaduct was 302 feet (92 meters) high and 2,052 feet (626 meters) long; its 20 towers were each made of two sloped or "battered" columns resting on masonry piers. The towers stood 99.5 feet (30 meters) apart and had a maximum height of 279 feet (85 meters). The tallest tower was divided into 10 vertical panels with horizontal struts and intersecting diagonal members. Columns were circular, with an outside diameter of about 10.75 inches (almost 28 centimeters). The columns had **cast-iron** caps and pedestals. The connecting spans were lattice **girder**s, 6 feet (1.8 meters) deep. Because the structure swayed, train speed was restricted to five miles per hour. Because of increases in the weight of trains, the bridge was rebuilt in 1900 by Adolphus Bonzano, who used steel rather than wrought-iron; however, the bridge retained its original design elements. The **viaduct**, which was taken out of service in the late 1960s, is today the main feature of Kinzua Bridge State Park.

Additional Information: DeLony, Eric. *Landmark American Bridges*. Boston: Bullfinch Press, 1993; Schodek, Daniel L. *Landmarks in American Civil Engineering*. Cambridge, MA: MIT Press, 1987; Phillips, D. Harvey. "The Kinzua Viaduct and the Man Who Built It" at http://www.bradford-online.com/blsvduct.html.

Kittatinny Tunnel. *See* Pennsylvania Turnpike.

Koyna Dam (Koyna, India; 1962)

Impounding over 2 trillion cubic meters (2.2 million **acre-feet**) of water, the Koyna Dam seemed to induce seismic activity when it was filled. In one instance, the filling process may have caused a 1967 earthquake that killed 200 people at Koynanagar, near Bombay on the west coast of India.

Additional Information: Smith, Norman. *A History of Dams*. Secaucus, NJ: Citadel Press, 1972.

Kra Canal (across the Kra Isthmus, Thailand; proposed project)

This canal, if built, would cut across the narrow Kra Isthmus of southern Thailand, thus reducing the sea voyage between West and East Asia by 3,000 kilometers (1,860 miles) and six days. Although the political climate in the 1990s seems opposed to building the canal, the project has been a dream in Thailand for 200 years, and the possibility of its construction grows greater as the twenty-first century approaches. The canal was first proposed in 1793 by Prince Surasihanaj, brother of Thailand's king; he intended to build a canal to attack the Burmese, Thailand's neighbors to the west. In the 1880s, Ferdinand de Lesseps, the French entrepreneur most responsible for the building of the SUEZ CANAL in Egypt, proposed building the Kra Canal, but his proposal came to nothing because it ran counter to the regional political interests of the British, who controlled India at the time. Recent studies have suggested that spending $20 billion to build the canal would be a positive economic achievement for modern Thailand. The proposed canal would be 102 kilometers (63 miles) long, 20 kilometers (12.4 miles) longer than the PANAMA CANAL. It would require the removal of 2.8 billion cubic meters (over 3.6 billion cubic yards) of earth and rock, 30 times the amount moved in the digging of the Suez Canal. If built, the canal would significantly shift the economic balance of power in Asia, creating in Thailand the kind of economic clout currently exercised by Singapore and Taiwan.

Additional Information: Mellor, William. "Mirage or Vision? Asia's Suez Canal." *Asia, Inc.* December 1995; reprinted at http://198.111.253.144/archive/1295thaisuez.html.

Kremlin (Moscow, Russia; walled in the fifteenth century).

In the Middle Ages, the walled citadels or kremlins (from *kreml*, meaning "fortress") at the center of several Russian cities served as administrative and religious cen-

A night view of the Kremlin in Moscow, Russia.

ters in time of peace and protective enclosures in time of war. A kremlin was almost a city unto itself, for it might contain palaces, government buildings, churches, and markets. The triangular central fortress of Moscow became known simply as the Kremlin. Its crenallated walls were built in the fifteenth century, when the fortress was already the seat of power of the rulers of Moscow. The walls are 2,235 meters (7,331 feet) long and vary in height from 5 to 19 meters (16.4 to 62.3 feet), and in thickness from 3.5 to 5.5 meters (11.5 to 18 feet). The Kremlin occupies 90 acres (36.4 **hectares**) in the historic core of the city. It houses three cathedrals, including the fifteenth-century Uspenski (Assumption) Cathedral, where the czars were crowned, and the Arkhangelski Cathedral, where the czars were buried. Along its walls are several palaces, including the nineteenth-century Grand Palace, which housed the Soviet parliament in the twentieth century. The walls also enclose the 300-foot (91.5-meter) bell tower of Ivan the Great; the tower's golden cupola dominates the Kremlin skyline. Of the Kremlin's 16 other towers, the most spectacular is the Spassky (Saviour's), which rises over the main gate and contains a famous set of chimes. Built in 1491, the Spassky marks the official entrance to the Kremlin; in the sixteenth and seventeenth centuries, the Spassky gate was used for ceremonial processions. An amazingly large clock tops the tower; the clock's mechanism alone weighs 2,160 kilograms (4,763 pounds). Although the fortress occasionally came under siege, it was in enemy hands only once—in 1812 when Napoleon I of France used the fortress as his headquarters during his occupation of Moscow. The Kremlin was the official residence of the czars until Peter the Great moved the capital of Russia to St. Petersburg in 1712. The fortress was the political and administrative center of the Soviet Union from 1918 to 1991. Most of the leaders of the Soviet state, including Vladmir Lenin and Joseph Stalin, lived and worked in the Kremlin. Lenin's body was put on display in a Kremlin tomb after his death in 1924. During the Soviet period, five Kremlin towers supported enormous red stars, each lit by a 3,700- to 5,000-watt bulb; the stars were designed to revolve in the wind. The office of the president of the Russian Republic, as well as many other government offices, are currently located in the Kremlin.

Additional Information: http://www.russia.net/country/moscow/sights/kremlin.html.

Kurushima Oohashi Bridge (connecting the Japanese islands of Shikoku and Honshu; proposed for completion by 1999)

The Kurushima-Oohashi Bridge will connect Onomichi on the main island of Honshu with Imabari on the smaller southern island of Shikoku. The proposed span, including the land portion of the bridge, will be 19 kilometers (12 miles) long. The bridge will be the world's first three-layer **suspension bridge**.

Additional Information: Matsumoto, Shinji in *Nishiseto Highway and Setonaikai Bridges* at http://www.webcity.co.jp/mi/tour/hashiE.html.

Kwai River Bridge. *See* Bridge Over the River Kwai.

L

Lacey V. Morrow Floating Bridge (Seattle, Washington; 1940)

Built of **reinforced concrete**, this pontoon bridge across Lake Washington is unique in its construction. Because of the depth of the lake, builders Charles Andrew and Clark Eldridge found that the usual supports for a bridge would be too difficult and too expensive to construct. The design of the structure uses 25 reinforced-concrete floating pontoons to support the deck of the bridge. Each 350-foot-long (107-meter) pontoon is 60 feet (18 meters) wide and 14 feet (4.3 meters) deep and constructed of a series of watertight compartments to make the flooding of an entire pontoon highly unlikely. The bridge above the pontoons is made of two steel Warren trusses, a steel arch, and three concrete **girder** spans. Although some maintenance reconstruction has been performed since construction, the bridge continues to bear its share of major road traffic on Interstate 40.

Additional Information: Jackson, Donald C. *Great American Bridges and Dams*. Washington, DC: The Preservation Press, 1988.

Lachine Canal (Montreal, Canada; 1821-1825; rebuilt 1843)

Built to provide a detour for shipping around the Lachine Rapids of the St. Lawrence River, the Lachine Canal ran for 3 miles (4.8 kilometers) around Montreal to the city of Lachine, which lies south of Montreal on Montreal Island. The canal connected the Saint Lawrence with Lake Saint Louis. The canal had five locks and created a drop of 42 feet (12.8 meters). It carried a heavy volume of river traffic, and greatly enhanced Montreal's shipping business with the Great Lakes. The canal was abandoned in 1960 after the completion of the SAINT LAWRENCE SEAWAY. In 1995, the city of Montreal attempted to drain and clean the canal, but stopped because the bottom of the canal was filled with toxic material. The side of the canal is now a park and bike path, and there is some speculation that at some time in the future pleasure craft will be allowed to use the canal in the summer.

Additional Information: http://www.cam.org/~fishon1/oldmtl.html

Lago di Ternavasso. *See* Ternavasso Dam.

Lake Fucinus Emissarium (central Italy; A.D. 52)

This drainage tunnel is the best known Roman land reclamation project. Designed to lower the level of Lake Fucinus and open an estimated 38,000 acres (15,390 **hectares**) of new farmland, the project involved a 3.5-mile-long (5.6-kilometer) tunnel that connected Lake Fucinus with the River Liri, which runs southeast of Rome through the Apennine Mountains of central Italy. Construction of the tunnel began in A.D. 41 under Emperor Claudius, who gave direction of the project to an imperial official named Narcissus. The tunnel was built by sinking to grade a series of vertical or sloping shafts, spaced 120 feet (37 meters) apart, to depths of up to 400 feet (122 meters). A clearing of the tunnel in the nineteenth century revealed a height of about 20 feet (6.1 meters), and a width of more than 9 feet (2.7 meters). The tunnel was driven through solid rock and unlined, except in areas of soft ground, where the roof was lined with ashlar, a thin layer of squared and dressed stone. During a decade of construction, some 30,000 workers excavated about 78,000 cubic yards (almost 60,000 cubic meters) of rock by hoisting the fragments to the surface in copper buckets. Claudius arranged a magnificent celebration for the opening of the tunnel in A.D. 51. Galleys manned by 19,000 convicted prisoners staged a mock naval battle on the lake before it was drained. Detachments of the Praetorian Guard lined up along the shore to prevent the escape of any of the prisoner-combatants. After the battle, when the emperor signalled for the **sluices** to be opened, water failed to flow. Narcissus spent the next year deepening the tunnel and adjusting its grade. At the second opening ceremony, which was also preceded by a gladiatorial combat, the rebuilt underground waterway worked too well. The force of the water was too great for the canal to contain. The flood swept away the imperial banquet and turned the second opening celebration into as big a disaster as the first. Lake Fucinus (Lago Fucino, in Italian) survived until 1875 when it was finally drained by Swiss and French engineers working for Prince Allessandro Torlonia. The prince added 38,600 acres (15,633 hectares) to his estates through the project. The nineteenth-century tunnel followed the route of its Roman predecessor, but was 1,590 feet (485 meters) longer. In 1951, 1,900 years after Claudius's first abortive opening ceremony, the Italian government expropriated most of the land reclaimed from the lake in the nineteenth century and divided it up among 8,000 families of settlers.

Additional Information: Sandstrom, Gosta E. *Man the Builder.* New York: McGraw-Hill, 1970.

Lake Maracaibo Bridge (Maracaibo, Venezuela; 1962)

Riccardo Morandi designed and built this **cable-stayed bridge** to carry the PAN AMERICAN HIGHWAY across Lake Maracaibo in northwestern Venezuela. Lake Maracaibo, a large brackish lake extending 180 kilometers (112 miles) inland from the Caribbean Sea, is located in one of the richest oil-producing areas in the world. The Lake Maracaibo Bridge runs just south of the city, near the lake's outlet to the sea. Built of reinforced and **prestressed concrete**, the bridge is 28,470 feet (8,683 meters) long, rises to 148 feet (45 meters) above the water, and is composed of 135 spans.

Additional Information: Brown, David J. *Bridges.* New York: Macmillan, 1993.

Lake Mead. *See* Hoover Dam.

Lake Powell (Page, Arizona, to Hite, Utah; 1966)

The construction of the GLEN CANYON DAM in the 1960s created this lake on the Colorado River. The lake is 186 miles (299 kilometers) long and, with its many flooded canyons, offers more than 2,000 miles (3,218 kilometers) of shoreline. It took 17 years until the lake was completely filled for the first time. Operated by the National Park Service, the lake is almost totally off limits to commercial activities; it is kept available for fishing, hiking, camping, boating, and other outdoor recreations. The lake is named for Major John Wesley Powell, a Civil War veteran, explorer, and head of the first scientific and geological survey of Glen Canyon in 1869. Powell was instrumental in the establishment of the U.S. Geological Survey, which he headed from 1881 to 1894.

Additional Information: http://www.pagehost.com/lakepowell/man-made.htm.

Lal Qila. *See* Red Fort.

Lampy Dam (Revel, France; 1781)

The Lampy Dam was built in the late eighteenth century to add water to the CANAL DU MIDI, which was built in the late seventeenth century to connect the Mediterranean Sea with the Atlantic Ocean. One of Europe's earliest buttress dams, the Lampy has a horizontal sill 32.5 feet (10 meters) down from the crest on the water face; the sill increases the thickness of the dam from 23.5 to 27.5 feet (7.2 to 8.4 meters). At its base, the dam is 37 feet (11.3 meters) thick. Measured from the top, the dam is 385 feet (117 meters) long and 17 feet (5.2 meters) thick; the crest is 53 feet (16 meters) above the base. Although still in use, the dam has had leakages almost since it was built. About 1804, lime was thrown into the adjoining reservoir in an effort to seal cracks in the masonry. Dozens of nuts on the crest are attached to long iron bars descending to the dam's base; the nuts were inserted in the hope that prestressing the bars from the top would strengthen the dam's resistance to the force of the water. The dam crosses the Lampy River about 8 miles (13 kilometers) east of Revel.

Additional Information: Smith, Norman. *A History of Dams.* Secaucus, NJ: Citadel Press, 1972.

Languedoc Canal. *See* Canal du Midi.

LaSalle Street Tunnel (Chicago, Illinois; 1871)

William Bryson designed the LaSalle Street Tunnel, the second tunnel built in Chicago. The tunnel was 1,890 feet (576 meters) long and built at a cost of $566,000. The tunnel remained in use until November 1939, when it was closed during the construction of the nearby Dearborn Street Subway.

Additional Information: http://cpl.lib.uic.edu/004chicago/timeline/tunneltrffc.html.

Laughery Creek Bridge (Aurora, Indiana; 1878)

Built by David A. Hammond and engineered by Job Abbott, the Laughery Creek Bridge carries Old State Route 56 over Laughery Creek in southeastern Indiana. This parallel-chord vehicular truss bridge stands 40 feet (12.2 meters) tall, and extends an unusually long span of 302 feet (92 meters). The bridge is the only known example of a triple-intersection **Pratt truss** in which the diagonals cross three panel points.

Additional Information: DeLony, Eric. *Landmark American Bridges*. Boston: Bullfinch Press, 1993.

Leaning Tower of Pisa (Pisa, Italy; 1174, 1272)

Bonanno Pisano began construction of a tower in Pisa, an ancient city in north central Italy on the River Arno, in 1174. Because the ground was too soft and unstable to support its weight, the tower began to lean shortly

The Tower of Pisa in Pisa, Italy, began to lean shortly after construction began in 1174.

after construction began. All work stopped on the tower after only three floors had been completed. Between the end of construction in 1174 and the completion of the tower in 1272, the structure developed an even more pronounced list. The list grew more marked after 1300. Various attempts, especially by the Italian Ministry of Public Works, have been made to straighten or at least stabilize the tower. These efforts have included tying cables to the tower and pouring concrete around its base. In recent years, the use of concrete has been somewhat successful, causing the structure to list the other way. The tower is about 60 meters (almost 200 feet) high, and lists 4.3 meters (about 14 feet) out of the perpendicular.

Additional Information: Matthys, Levy and Salvadori, Mario. *Why Buildings Fall Down*. New York: W.W. Norton & Company, 1992.

Leeds and Liverpool Canal (northwestern England; built 1770-1816)

The Leeds and Liverpool Canal is the oldest and most northerly waterway to cross the Pennine Mountains of England. The canal has two supply reservoirs, one above

the other, at Foulridge, adjacent to a tunnel that is nearly one mile long. The upper dam is an earth bank, 300 feet (91.5 meters) long and 40 feet (12.2 meters) high. A third reservoir at Barrowford has three of its sides formed by the dam wall, and the fourth side by a hill.

Additional Information: Smith, Norman. *A History of Dams*. Secaucus, NJ: Citadel Press, 1972.

Lehigh Canal (Mauch Chunk to Easton, Pennsylvania; 1830)

Josiah White, the project's developer and first chief engineer, built the Lehigh Canal to carry anthracite coal across eastern Pennsylvania from Mauch Chunk to Easton on the Delaware River. The canal ran 46 miles (74 kilometers) and then continued another 10 miles (16 kilometers) as "slack" (slowly running water). An 1832 extension brought the route south to Philadelphia. Some stone locks on the canal were as much as 100 feet (30.5 meters) long. The canal was 60 feet (18.3 meters) wide at the top but tapered to a width of 40 feet (12.2 meters) at the bottom. At 5 feet (1.5 meters) deep, the Lehigh Canal was one foot deeper than the ERIE CANAL. The Lehigh Canal was an immediate success and carried traffic until well into the twentieth century.

Additional Information: Shaw, Ronald E. *Canals for a Nation*. Lexington: The University Press of Kentucky, 1990; Spangenburg, Ray and Moser, Diane K. *The Story of America's Canals*. New York: Facts on File, 1992.

Leipzig Train Station (Leipzig, Germany; 1915)

The train station of Leipzig, an industrial city on the Pleisse River in east central Germany, is the largest train terminal in Europe. Built jointly by the Royal Saxon State Railroad and the Royal Prussian Railway Directorate, the terminal was constructed at the crossroads of one of Europe's major trading routes. The optimistic spirit that grew out of the political unification of Germany in 1871 inspired the building's size and architectural magnificence. The station boasts a sandstone facade that is approximately 300 yards (271 meters) long. When opened, the station had 13 tracks for Saxon trains and another 13 tracks for Prussian trains. Capable of holding 10,000 people, the terminal is covered with a 100-foot-high (30.5-meter) glass and steel roof that makes the structure seem even more spacious than it is. The waiting room was lit by large glass chandeliers, a glass ceiling, and many stained glass windows. The waiting room also contained displays of wood carving, mosaics, and fine porcelain. A large complex of rooms underneath the railroad tracks was once used to handle luggage and shipping, but served as a bomb shelter during World War II. Although still usable, the terminal is much faded from its original glory and plans for its rehabilitation were announced in April 1996. The building was heavily damaged during World War II and poorly maintained in the post-war years when Leipzig was the second city of communist East Germany. The underground area, if reconstruction plans go forward, will become a vast center for consumers with stores, offices, meeting rooms, a fitness center, and a dance studio.

Additional Information: Kinzer, Stephen. "Leipzig Plans to Restore Once-Regal Train Station." *New York Times*, April 7, 1996, I:10.

Liberty, Statue of. *See* Statue of Liberty.

Lighthouse at Alexandria, also known as Pharos at Alexandria (Alexandria, Egypt; 280 B.C.)

Constructed during the reign of Ptolemy II of Egypt, the lighthouse, one of the SEVEN WONDERS OF THE ANCIENT WORLD, was destroyed by an earthquake in 1385. It was constructed on the peninsula of Pharos in the Mediterranean off Alexandria. The structure was so well known, it gave its name to many other ancient lighthouses. It was probably as tall as the GREAT PYRAMID OF GIZA, but much narrower, with a base of about 86 feet (26 meters) square. A fire light on the top of the lighthouse shone through a lens, magnifying the light so that it could be seen, reportedly, from as far as 40 miles (64 kilometers) away. A giant statue topped the lighthouse. Modern archaeologists have found various masonry blocks in the nearby harbor, along with a multitude of other treasures, including a statue of a sphinx. Many believe these artifacts come from the ruined tower. A 40-ton (36-metric ton) block of granite that may have been part of one of the columns of the lighthouse was hoisted from the sea early in 1996.

Additional Information: Colt, George Howe, "Raising Alexandria." *Life* (April 1996), p. 70; http://www.newton.cam.ac.uk/egypt/news.html; Sandstrom, Gosta E. *Man the Builder*. New York: McGraw-Hill, 1970; Ashmawy, Alaa K., ed. "The Seven Wonders of the Ancient World" at http://pharos.bu.edu/Egypt/Wonder.Home.html.

Limmat River Bridge (Wettingen, Switzerland; 1764-1766)

The brothers Johannes and Ulrich Grubenmann, two Swiss carpenters who built several well-known bridges over the Rhine River, are thought to have built this 200-foot (61-meter) span of covered bridge over the Limmat River at Wettingen near Baden in north central Switzerland. The Limmat River Bridge was the first timber span to use a true arch; its solid curved arch rib was built of timber layers, and the thrust of the arch was held by the end **abutments**. Like the DEVIL'S BRIDGE in St. Gotthard Pass between Switzerland and Italy, and the Grubenmann brothers' Rhine bridges, the Limmat Bridge was destroyed by the French army under Nicolas Oudinot as it retreated from Austria in 1799.

Additional Information: Kirby, Richard Shelton et al. *Engineering in History*. New York: Dover Publications, 1990; Lay, M.G. *Ways of the World*. New Brunswick, NJ: Rutgers University Press, 1992; Steinman, David B. *Famous Bridges of the World*. London: Dover Publications, Inc., 1953, 1961.

Lincoln Highway (Washington, DC, to San Francisco, California; 1912-1923)

U.S. Route 30, the Lincoln Highway, was intended to be the "first American Rock [i.e., gravel] Highway." Supported both by the new automotive industry (primarily Henry Joy, president of the Packard Motor Company) and by local boosters along the route, the highway created a road, albeit a difficult one, that a driver could follow from coast to coast. Because of the federal government's historical reluctance to engage in road building, much of the highway was built with private funds. To promote the project, backers printed and distributed brochures that originated the phrase "See America First." Supporters named the road after Abraham Lincoln because they wanted to echo the former president's commitment to a transcontinental railroad. Although supporters were disappointed the road was not completed in time to bring drivers to the 1915 World's Fair in San Francisco, the 12-foot-wide (3.7-meter) highway became an immediate and obvious success when it was finally finished in 1923. During the same period, other smaller roads were being built in the same manner throughout the country. When the federal government accepted responsibility for the road, it was designated U.S. Route 30, a designation still used today for many stretches of the original route.

Additional Information: Patton, Phil. *Open Road: A Celebration of the American Highway*. New York: Simon and Schuster, 1986. For an excellent history of the Lincoln Highway, see Lin, James *Lincoln Highway History* at http://www.ugcs.caltech.edu/~jlin/lincoln/history/part1.html.

Lincoln Tunnel (New York City to Weehawken, New Jersey; 1937, 1945, 1957)

The Lincoln Tunnel runs under the Hudson River connecting 38th Street in New York City with Weehawken, New Jersey. Begun soon after the completion of the HOLLAND TUNNEL, the 1.6-mile-long (2.6-kilometer) Lincoln Tunnel benefitted greatly from the experiences at its sister tunnel. Its first tube opened in 1937, followed eight years later by a second tube. When the third tube of the tunnel opened in 1957, the Lincoln became the only three-tube underwater vehicular tunnel in the world. Each tube of the tunnel has two traffic lanes, and the center tube lanes are reversible; during rush hours, the heavier traffic can be assigned four lanes.

The New Jersey entrance of the Lincoln Tunnel as it looked in 1987. Photo courtesy of The Port Authority of New York and New Jersey.

Sources: Paaswell, Robert E. "Hudson River Tunnels" and Paulson, Boyd, Jr. "Lincoln Tunnel" in *The World Book Encyclopedia*. Chicago: World Book, Inc., 1995.

Lions' Gate Bridge (Vancouver, British Columbia; 1938)

The Lion's Gate Bridge is one of the longest suspension spans in the world, and the longest **suspension bridge** in western Canada. This structure's main span is 472 meters (1,550 feet) long; its towers reach a height of 111 meters (364 feet) above the water. Including approach spans, the bridge's total length is 1,517 meters (4,978 feet). Two stone lions stand at the southern end of the bridge, which crosses Burrard Inlet. The bridge connects 1,000-acre (405-**hectare**) Stanley Park with North Vancouver. When the bridge was opened in 1939 by King George VI, the span was the longest suspension bridge in the British Empire and the only land route to the north shore of Burrard Inlet.

Additonal Information: http://www.b-t.com/liongate.htm; http://www.bctour.com/bctour/lowmain/stypk.htm.

Liverpool-Manchester Railway (Liverpool to Manchester, England; 1830)

George Stephenson and his 19-year-old son Robert began work in 1825 on the world's first public railway, a line to bring goods from the the port of Liverpool in northwestern England on the Irish Sea eastward to the inland industrial city of Manchester. Although George Stephenson, a recognized builder of engines, was a genius in his craft, engineering and administration were not his strong points. Stephenson severely mismanaged the construction of the 30-mile (48-kilometer) line. An inquiry halfway through construction found that Stephenson had negotiated few contracts for work by subcontractors, and that fees paid for work varied tremendously, depending on the outcome of Stephenson's on-the-spot bargaining. Stephenson also insisted on building his railroad along the shortest route possible. Afraid of handling grades, he insisted on clearing rock, cutting across swamps, and building **viaducts** for a direct route. The railroad eventually required 63 bridges, including the SANKEY VIADUCT across the Sankey Canal, which needed nine arches to support the railroad line 70 feet (21 meters) above the water. The rail line finally opened on 15 September 1830. Stephenson had designed and built the line's locomotive, the Rocket, which had covered a test distance of 1.5 miles (2.4 kilometers) at a speed of 21.5 miles (35 kilometers) per hour. On opening day, eight trains carried 700 guests, including the duke of Wellington and Prime Minister Grey. At Parkwood station, the Rocket accidentally ran over William Huskisson, member of Parliament for Liverpool and a strong supporter of the railroad. Huskisson died the next day. Despite this tragedy and other initial difficulties, the line was an economic success. Manchester mill owners got their imported cotton more quickly and made a handsome return on their investment in the line. Within three years of the line's opening, Parliament had authorized the construction of two trunk lines—the GRAND JUNCTION RAILWAY, connecting Manchester with Birmingham to the south, and the LONDON-BIRMINGHAM RAILWAY, extending the system southeast to the capital.

Additional Information: Sandstrom, Gosta E. *Man the Builder*. New York: McGraw-Hill, 1970.

Loch Thom Reservoir (Greenock, Scotland; 1827)

At the time of its completion, the Loch Thom Reservoir, in west central Scotland near Greenock on the Firth of Clyde, was the largest in Britain. Named after its builder, Robert Thom, Loch Thom was created by an earth dam that measured 1,400 feet (427 meters) long and 66 feet (20 meters) high. Water went to Greenock along a 6-mile (9.7-kilometer) **aqueduct** with an overall fall of 512 feet (156 meters).

Additional Information: Smith, Norman. *A History of Dams.* Secaucus, NJ: Citadel Press, 1972.

Lockington Dam. *See* Miami Conservancy District.

London-Birmingham Railway (Birmingham to London, England; 1836)

Robert Stephenson, the builder of the London-Birmingham Railway, was a much better administrator than his father, George Stephenson, whose lack of administrative ability had severely complicated the construction of the LIVERPOOL-MANCHESTER RAILWAY in 1830. Robert Stephenson divided the construction work on the line into four divisions, each of which was under the direction of a competent engineer. In each division, work was contracted out for 6 miles (9.7 kilometers) of line, with special agreements for **viaducts** and tunnels. Construction of the line eventually involved 29 separate contracts and up to 20,000 men, who laid 112 miles (180 kilometers) of track to Chalk Farm north of London and dug eight tunnels. The 2,400-yard-long (2,196-meter) tunnel through Kilsby Ridge was the most difficult to construct. The workers at Kilsby ran into great quantities of water and quicksand. When John Howell, the contractor for the tunnel, died midway through the project, Stephenson had to take over the work himself. The tunnel took 30 months and three times the expected cost to complete. Other portions of the line were almost as difficult; Stephenson finally finished the project at a cost of £50,000 per mile, more than twice the cost per mile of the Liverpool-Manchester line.

Additional Information: Sandstrom, Gosta E. *Man the Builder.* New York: McGraw-Hill, 1970.

London Bridge. *See* New London Bridge; Old London Bridge.

London Ferris Wheel (Jubilee Gardens, London, England; proposed for 1998)

Plans call for the construction of the largest Ferris wheel in the world in London, on the south bank of the Thames River. With completion scheduled for 1998, the wheel would be part of the celebration of the millennium in England. At a height of 500 feet (152.5 meters), the wheel would provide riders on a clear day with a 60-mile (97-kilometer) view. The current tallest Ferris wheel is in Yokohama, Japan, and is 344.5 feet (105 meters) high. Although approval from city authorities is still pending, the wheel is expected to become reality.

Additional Information: Darnton, John. "Ferris Wheel Will Look Down on Big Ben." *New York Times*, 16 April 1996, I:7.

Los Angeles Aqueduct. *See* First Owens River—Los Angeles Aqueduct.

Lotschberg Railroad Tunnel (Lotschberg, Switzerland; 1906-1913)

Now operated by the Bern-Lotschberg-Simplon Railway, this 9-mile-long (15-kilometer) tunnel is one of the longest railroad tunnels in the world. The tunnel runs through the mountains at an elevation of 3,935 feet (1,200 meters) near Lotschberg in southwestern Switzerland. The tunnel is over 26 feet (8 meters) wide. Completion of the tunnel was delayed seven months by a rock slide, but engineers worked to make up time and were able to open the tunnel in 1913 only five months behind schedule.

Additional Information: *The World Almanac and Book of Facts 1993.* New York: Pharos Books, 1993, excerpted at http://booksrv2.raleigh.ibm.com:80/cgi-bin/bookmgr/bookmgr.cmd/BOOKS/BUILDNGS/CCONTENTS.

Louvre Museum (Paris, France; thirteenth century)

Originally a royal palace and fortress constructed by King Philip II Augustus in 1204, the Louvre is now one of the largest and most prestigious art museums in the

Begun in the thirteenth century as a royal palace, the Louvre is today one of the world's most important art museums.

world. The building was reconstructed after 1541 and converted into a national museum by Napoleon I at the start of the nineteenth century. Napoleon's conquests added greatly to the museum's collections, although many famous pieces were returned to their original owners after the emperor's downfall. Further expansion and renovation was completed by 1852, and the museum now covers approximately 18 **hectares** (over 44 acres). The Louvre collections hold 400,000 works of art, including the *Mona Lisa* by Leonardo da Vinci, the *Venus de Milo,* and the *Nike* or *Victory of Samothrace*. The museum is particularly rich in works by Rembrandt, Rubens, Titian, Leonardo da Vinci, and French masters. The enormous collection, which contains works dating from 7 B.C., is so huge that almost every guidebook about France or Paris ever written cautions the tourist that it is impossible to see it all in one day, and difficult to do so in one week. During World War II, all the pieces that could be removed were stored away for safe keeping; the Louvre reopened in 1947.

Additional Information: http://www.focusmm.com.au/~focus/fr_re_02.html; http://sunsite.unc.edu/wm/paris/hist/louvre.html.

Lover's Leap Bridge (New Milford, Connecticut; 1895)

Built by William O. Douglas, the Lover's Leap Bridge carries Pumpkin Hill Road over the Housatonic River in New Milford in western Connecticut. The Lover's Leap is one of six surviving **lenticular truss** bridges in Connecticut. More than 1,000 lenticular trusses were built nationwide during the late 1880s and 1890s. The Berlin Iron Bridge Company (in Connecticut), which employed Douglas, claimed that it had built more than 90 percent of the highway bridges in New England. Many bridges in the United States are called "Lover's Leap," but this one may be the only bridge that officially bears the name.

Additional Information: DeLony, Eric. *Landmark American Bridges*. Boston: Bullfinch Press, 1993.

Lower Trenton Bridge (Trenton, New Jersey, to Morrisville, Pennsylvania; 1930)

Of the 19 bridges operated by the Delaware River Joint Toll Bridge Commission, the Lower Trenton is the most well known. Essential to Trenton's manufacturing economy, the bridge bears an electric sign that unabashedly declares "Trenton Makes. The World Takes." This bridge across the Delaware has been rebuilt several times. The original **abutments**, founded in stone, date from 1803 and served to support a wooden arch covered bridge built in 1805-06. The bridge was remodeled and strengthened in 1848. An iron bridge was constructed in 1876, but removed in 1929-30 and replaced by the present five-span, steel Warren truss. The current bridge used 3,150 tons (2,835 metric tons) of riveted steel. The five span lengths total 980 feet (299 meters).

Additional Information: Bishop, Gordon. *Gems of New Jersey*. Englewood Cliffs, NJ: Prentice-Hall, 1985; information also supplied by the Delaware River Joint Toll Bridge Commission.

Lu Gou Bridge, also known as the Marco Polo Bridge (Beijing, China; 1189)

Marco Polo, the thirteenth-century Italian traveler, saw and wrote in amazement about this remarkably long bridge across the Yongding River at Beijing. The bridge has 11 semicircular arches with clear spans of 12.3 to 13.4 meters (40.4 to 44 feet) and a total distance of 266 meters (872 feet). It has been refurbished and restored, and still stands; in 1937, it was the site of a key incident in Japan's aggression against China.

Additional Information: Lay, M.G. *Ways of the World*. New Brunswick, NJ: Rutgers University Press, 1992.

Luiz I Bridge (northern Portugal; 1885)

T. Seyrig designed this double deck **arch bridge** across the Douro River in northern Portugal. The 562-foot (171-meter) arch carries one deck on top of the arch and a second deck running from the feet of the arch. Both decks carry vehicular traffic. A student of the great Alexandre Gustave Eiffel, the designer of the EIFFEL TOWER in Paris, Seyrig created a design for this bridge that looks as if it might have been done by Eiffel himself. The Luiz I Bridge bears a remarkable resemblance to Eiffel's single-deck design for the GARABIT BRIDGE in southeastern France.

Additional Information: Hawkes, Nigel. *Structures: The Way Things Are Built*. New York: Macmillan, 1990.

M

Mackinac Straits Bridge (St. Ignace to Mackinaw City, Michigan; 1954-1957)

David B. Steinman built this 3,800-foot (1,159-meter) **suspension bridge** connecting the upper and lower peninsulas of Michigan. The bridge spans the Straits of Mackinac, which connect Lake Michigan to the west with Lake Huron to the east. A bridge was first proposed for this site in 1884, but the current structure was not begun until 70 years later. The structural steel used in the bridge weighs 55,000 tons (49,500 metric tons), and the cables weigh 11,000 tons (9,900 metric tons). With side spans of 1,800 feet (549 meters) each, the Mackinac Bridge is one of the longest suspension bridges in the world, with a total length of 8,614 feet (2,627 meters) from anchorage to anchorage and a main span of 3,800 feet (1,159 meters). Steinman was concerned that the bridge might be too stiff, so he used steel grid rather than solid decking for the roadway. The bridge is traditionally open to pedestrian traffic each year on Labor Day.

Additional Information: Brown, David J. *Bridges*. New York: Macmillan, 1993; Condit, Carl W. *American Building Art: The Twentieth Century*. New York: Oxford University Press, 1961; DeLony, Eric. *Landmark American Bridges*. Boston: Bullfinch Press, 1993.

Madison Swing Bridge. *See* Fort Madison Railroad Bridge.

Madruzza Dam. *See* Ponte Alto Dams.

Maentwrog Scheme (North Wales, 1928)

In 1925, the North Wales Power Company began this arrangement of four dams which together act to dam up Lake Trawsfynd. The largest of these dams is the first large arch dam ever built in Britain. A concrete structure, it is 96 feet (29 meters) high at its peak and 37 feet (11.3 meters) thick at its base.

Additional Information: Smith, Norman. *A History of Dams*. Secaucus, NJ: Citadel Press, 1972.

Main Bridge (Hassfurt, Germany; 1867)

Heinrich Gerber built this bridge across the River Main at Hassfurt in west central Germany. The first modern **cantilever bridge**, the Main Bridge has a central span of 425 feet (130 meters).

Additional Information: Steinman, David B. *Famous Bridges of the World*. London: Dover Publications, 1953, 1961.

Malpas Tunnel (southern France; 1681)

Part of the CANAL DU MIDI, the 528-foot (161-meter) Malpas Tunnel was the first canal tunnel to be built, the first tunnel to be constructed for any type of transportation, and the first tunnel ever to be excavated with gunpowder. The canal itself was the first great canal in Europe, linking the Atlantic with the Mediterranean.

Additional Information: Hawkes, Nigel. *Structures: The Way Things Are Built*. New York: Macmillan, 1990.

Malpasset Dam (Fréjus, France; 1954)

Built between 1952 and 1954, the Malpasset Dam on the Argens River in southeasten France collapsed in December 1959, flooding the town of Fréjus and killing more than 300 people. The presumed cause of failure was excessive pressure of the arch on its rock foundations.

Additional Information: Smith, Norman. *A History of Dams*. Secaucus, NJ: Citadel Press, 1972.

Manayunk Bridge (now Washington, D.C.; 1845)
Built by Richard B. Osborne, the Manayunk Bridge, constructed in 1845, was the first all-metal **railroad bridge** in the United States. Only a single surviving truss from this 34-foot-long (10.4-meter) bridge survives at the Smithsonian Museum in Washington, D.C. as part of the "John Bull" locomotive exhibit.

Additional Information: DeLony, Eric. *Landmark American Bridges*. Boston: Bullfinch Press, 1993.

Manchester Ship Canal (Manchester, England; 1887-1894)
Connecting the Atlantic Ocean with the city of Manchester in northwestern England, this 35-mile-long (56-kilometer) canal was tremendously successful in reviving Manchester's economy. The Manchester Ship Canal, which made use of the River Irwell and crossed in part the earlier BRIDGEWATER CANAL, was the largest canal built in Britain in the nineteenth century. Up to 17,000 workers were employed on the project at the height of construction. Manchester, which had been a thriving cotton mill town earlier in the nineteenth century, enjoyed one of the most successful economic rebirths of the later nineteenth century. So impressive was the city's revival, many newer cities in the United States named themselves Manchester in the hope of obtaining some of the good luck associated with the name.

Additional Information: Hadfield, Charles. *World Canals*. Newton Abbot, England: David & Charles, 1986; McNeil, Ian, ed. *An Encyclopedia of the History of Technology*. London: Routledge, 1990; Russell, Solveig Paulson. *The Big Ditch Waterways*. New York: Parents' Magazine Press, 1977.

Mangfall Bridge (Mangfall Valley, Germany; 1960)
The Mangfall Bridge carries the autobahn (see entry for AUTOBAHN ROAD SYSTEM) between Munich, Germany, and Salzburg, Austria, across the Mangfall Valley in southeastern Germany. The bridge is an early example of the economical use of prefabricated, **prestressed concrete**, rather than steel, in a lattice **girder** bridge. Mangfall has a 106-meter (348-foot) central span, and two 89-meter (292-meter) approach spans, all supported on **reinforced concrete** columns.

Additional Information: Overman, Michael. *Roads, Bridges and Tunnels*. Garden City, NY: Doubleday & Company, 1968.

Mangla Dam (Pakistan; 1968)
This earth dam in the west of Pakistan is one of the world's largest; it stands 370 feet (113 meters) high, with a length of 11,050 feet (3,370 meters). The dam has a capacity of 850,000,000 cubic yards, and a reservoir capacity of 5,880,000 **acre-feet**. The dam created Lake Mangla, one of the largest man-made lakes in the world.

Additional Information: Smith, Norman. *A History of Dams*. Secaucus, NJ: Citadel Press, 1972.

Manhattan Bridge (Manhattan to Brooklyn, New York; 1909)
Leffert L. Buck and Gustav Lindenthal designed this bridge across the East River that connects the Lower East Side of Manhattan with Brooklyn on Long Island. The original design called for an **eyebar** chain bridge, but the final structure was built with wire cable. With a main river span of 1,480 feet (451 meters), the Manhattan Bridge is shorter than the BROOKLYN BRIDGE, which crosses the East River less than a mile to the west, and the WILLIAMSBURG BRIDGE, which crosses the East River about a mile to the northeast. However, the construction of the Manhattan Bridge made use of important technological advances, and the erection of the steel towers and the air-spinning of the cables was much more mechanized than either process had been during the construction of the two earlier East River spans. Four wire cables, each over 18 inches (47 centimeters) in diameter, carry a 40-foot (12.2-meter) stiffening truss, one of the deepest ever used in a **suspension bridge**. Originally, the bridge's two decks carried rail lines as well as walkways and vehicle lanes. Structural calculations on the bridge were done by Leon Moisseiff, who used the "deflection" theory developed by the Austrian engineer Joseph Melan. Melan proved mathematically that deep trusses were unnecessary for stability; put another way, as the **dead weight** of a span increases, the need for stiffness in the bridge's deck decreases. The Manhattan Bridge was the first suspension bridge to be built using modern suspension bridge techniques.

Additional Information: Brown, David J. *Bridges*. New York: Macmillan, 1993; Condit, Carl W. *American Building Art: The Twentieth Century*. New York: Oxford University Press, 1961; DeLony, Eric. *Landmark American Bridges*. Boston: Bullfinch Press, 1993.

Manicougan Dam No. 5 (Quebec, Canada; 1967)
This multi-arch dam is 704 feet (215 meters) high and 4,200 feet (1,281 meters) long. The dam has a capacity of 2,600,000 cubic yards (1,989,000 cubic meters), and a reservoir capacity of 115,000,000 **acre-feet**.

Additional Information: Smith, Norman. *A History of Dams*. Secaucus, NJ: Citadel Press, 1972.

Marble Palace (St. Petersburg, Russia; 1770s and 1780s)
An interesting building on Nevsky Prospect in St. Petersburg in western Russia, the Marble Palace was built by Catherine the Great in the eighteenth century for her lover, Count Grigori Orlov. Its name comes from the 32 different kinds of marble used in its construction. Orlov never lived there—he lost the site twice in card games and Catherine bought it back for him each

time. When he died in 1783, the building was still unfinished. During the Soviet era, the palace was a branch of the Central Lenin Museum; it is now an independent museum.

Additional Information: http://www/spb.su/fresh/museums/marble.html.

Marco Polo Bridge. *See* Lu Gou Bridge.

Marib Dam, also Yemen Dam (Saba, Yemen; sixth century B.C.)

One of the great ancient irrigation dams of the world, the now unused Marib Dam brought mountain water to fields in the southwestern corner of the dry Arabian Peninsula, which is now part of the Republic of Yemen. Made of boulders, the dam is constructed without mortar. Marib was approximately 1,980 feet (604 meters) long, and probably had masonry **sluices** and spillways to regulate water flow into its irrigation channels. It was destroyed, probably by earthquake, in the seventh century A.D.

Source: Wetterau, Bruce. *The New York Public Library Book of Chronologies.* New York: Stonesong Press, 1990; http://www.ceng.metu.edu.tr/~e78199/Walid/Marib.html..

Marine Parkway Bridge (Brooklyn to Rockaway, New York; 1937)

A **vertical-lift bridge** with a span of 540 feet (165 meters), the attractive Marine Parkway Bridge crosses the Rockaway Inlet on Long Island, connecting Brooklyn with Rockaway Point about 5 miles (8 kilometers) southwest of John F. Kennedy International Airport.

Additional Information: Steinman, David B. *Famous Bridges of the World.* London: Dover Publications, 1953, 1961.

Marsden Reservoir. *See* Huddersfield Canal.

Marseilles Aqueduct (Marseilles, France; 1839-1847)

The Marseilles Aqueduct provides water to the city of Marseilles on the Mediterranean coast of France. The **aqueduct** draws its water from the Durance River, a tributary of the Rhône River, at a point about 28 miles (45 kilometers) north of Marseilles. Because the aqueduct takes a rather circuitous course, it is 51 miles (82 kilometers) long, crosses various bridges, and goes through more than 40 tunnels. It crosses the Roquefavour Aqueduct Bridge, built by De Mont Richer, about 5 miles (8 kilometers) west of Aix; the Roquefavour is a three tiered bridge similar to the Roman PONT DU GARD AQUEDUCT at Nimes, 60 miles (18.3 kilometers) to the west. The Roquefavour Bridge is 1,300 feet (397 meters) long, made of cut stone, and rises 300 feet (91.5 meters) above the Arc River. The Marseilles canal aqueduct is typically 30 feet (9 meters) wide on top and 7 feet (2.1 meters) deep. The aqueduct also supplies water for irrigation to the area around Arles to the northwest, a region that has also been served since the sixteenth century by the CRAPPONE CANAL.

Additional Information: Kirby, Richard Shelton et al. *Engineering in History.* New York: Dover Publications, 1990.

Martorelli Bridge (Catalonia province, Spain; 219 B.C.)

This 37-meter-long (121-foot) pointed-arch span is the largest of all extant Roman **arch bridge**s. The Martorelli crosses the Llobregat River in Catalonia northwest of Barcelona in northeastern Spain. Like many other bridges, the Martorelli is referred to by many people in the area as the "Devil's Bridge," a common name applied to spectacular structures that seem to observers to have been possible only with the devil's help.

Source: Kirby, Richard Shelton et al. *Engineering in History.* New York: Dover Publications, 1990; Lay, M.G. *Ways of the World.* New Brunswick, NJ: Rutgers University Press, 1992.

Mausoleum of Halicarnassus (Bodrum [formerly Halicarnassus], Turkey; 352 B.C.)

Mausolus, the satrap (governor) of the Persian province of Caria in southwestern Asia Minor (modern Turkey) broke with the Persian king Artaxerxes III and governed the independent province as king. In 352 B.C., a year after Mausolus's death, his wife, Artemisia, erected a magnificent tomb or sepulcher in his honor at Halicarnassus, the Carian capital, which had been originally founded by the Greeks. The modern Turkish city of Bodrum now stands on the site of ancient Halicarnassus. One of the SEVEN WONDERS OF THE ANCIENT WORLD, the tomb was a rectangular structure of white marble, approximately 120 by 100 feet (40 by 30.5 meters), and about 140 feet (43 meters) high. The massive base contained the burial chamber and the sarcophagus, which were made of white alabaster decorated with gold. The base was surmounted by a stepped pyramid with a truncated apex on which rested a large podium surrounded by Ionic columns and magnificent statuary. The top of the mausoleum was decorated with a 20-foot-high (6.1-meter) statue of a chariot pulled by four horses. The decorative statuary included representations of people, lions, horses, and other animals. Four well-known sculptors—Praxiteles, Leochares, Scopas, and Timotheus—worked on the project; each was responsible for the art work on one side of the mausoleum. This magnificent structure was disassembled during the crusades of the late Middle Ages. In the early fifteenth century, the region was invaded by the crusading Knights of St. John of Malta, who built a massive crusader castle in the region. In 1496, they fortified the castle with stones from the Mausoleum. The crusader castle still exists in Bodrum, and the stones from the Mausoleum are easily recognizable. In 1846, some of the statues were taken to the British Museum in London where they can still be seen. The fame and beauty of Mausolus's tomb gave his name (mausoleum) to any large architecturally or artistically significant above-ground tomb or sepulcher.

Additional Information: Ashmawy, Alaa K., ed., "The Seven Wonders of the Ancient World," at http://pharos.bu.edu/Egypt/Wonders/mausoleum.html

Mauvoisin Dam (Switzerland; 1958)

This arch dam is one of the world's highest dams. It stands 780 feet (238 meters) high and is 1,706 feet (520 meters) long. The dam's volume is 2,655,000 cubic yards (2,031,000 cubic meters), with a reservoir capacity of 146,000,000 **acre-feet**.

Additional Information: Farb, Peter. *The Story of Dams.* Irvington-on-Hudson, NY: Harvey House, 1961; Smith, Norman. *A History of Dams.* Secaucus, NJ: Citadel Press, 1972.

McCall's Ferry Timber Arch Bridge (McCall's Ferry, Pennsylvania; 1815)

Theodore Burr built this **arch bridge** across the main channel of the Susquehanna River at McCall's Ferry, southwest of Lancaster in southeastern Pennsylvania. With an incredible arch span of 360 feet (110 meters), the McCall's Ferry Bridge was, at the time of its completion, the longest single-span timber bridge ever built. Because the river is 100 feet (30.5 meters) deep and regularly filled with ice during the spring thaw, the bridge was built in sections on shore and then slid along the ice to the site for connection. Although Burr felt that he had made a bridge that "God Almighty can not move!," an unusually heavy ice pack destroyed the bridge two years later, and it was never rebuilt.

Additional Information: Allen, Richard Sanders. *Covered Bridges of the Middle Atlantic States.* Brattleboro, VT: Stephen Greene Press, 1959; Kirby, Richard Shelton et al. *Engineering in History.* New York: Dover Publications, 1990.

McCullough Memorial Bridge, originally Coos Bay Bridge (North Bend, Oregon; 1936)

The McCullough Memorial Bridge carries the Oregon Coast Highway (U.S. Route 101) over Coos Bay near North Bend in southwestern Oregon. Built by Conde B. McCullough, this bridge has a main section composed of a 739-foot (225-meter) **cantilever** truss that allows for passage of ships. The rest of the structure is made of 13 double-ribbed, open-**spandrel**, deck-approach arches. The total length of the bridge is 4,566 feet (1,393 meters). This magnificent bridge completed one of the last major links in the Oregon Coast highway system; McCullough built many of the bridges in that system. In recognition of the engineer's contribution to the bridges of Oregon, the state named the McCullough Memorial Bridge after him; the bridge is

one of the few engineering structures in the world to be named after its maker.

Additional Information: DeLony, Eric. *Landmark American Bridges.* Boston: Bullfinch Press, 1993.

Meadowbrook Parkway (Long Island, New York; 1934)

Although an older highway, built at the same time as the AUTOBAHN ROAD SYSTEM was being started in Germany, the Meadowbrook Parkway is an amazing road-building achievement. Built under the direction of Robert Moses, who was responsible for many bridges and roads in and around New York City, this parkway on Long Island was specifically designed to keep the area green and attractive. Groups of trees and planted flowers adorn the wide median. The parkway has weathered-wood lamp standards and guardrails, wide and gently sloping shoulders, and stone facings on its overpasses. It crosses bridges with **Art Deco** guard stations. Moses even went so far as to require bridges crossing the parkway to be low enough to make bus travel on the parkway impossible, thus reserving the roads for the automobile driver, which in the 1930s meant the middle class. The parkway is still in use today.

Additional Information: Patton, Phil. *Open Road: A Celebration of the American Highway.* New York: Simon and Schuster, 1986.

Meer Allum Dam (Hyderabad, India; c.1790-1810)

The Meer Allum Dam was built to supply water to Hyderabad in south central India. It is the earliest known example of a true buttress dam of the multiple-arch type. It has a maximum height of 40 feet (12.2 meters) and a curved length of 2,500 feet (762.5 meters). The wall is divided into 21 semicircular arches with spans ranging from 70 feet (21 meters) in length at the ends to 147 feet (45 meters) in length near the middle. Each arch is 8.5 feet (2.6 meters) thick, except for a tapered section near the top. The arches are supported by buttresses 42 feet (13 meters) long and 24 feet (7.3 meters) thick. The accompanying reservoir has a capacity of nearly 8,000 **acre-feet** of water. Excess water goes through a spillway at one end and over the crest. Despite the water flowing over the crest, which usually is deadly to a dam, this structure has sustained no damage to either its masonry or its foundation.

Additional Information: Smith, Norman. *A History of Dams.* Secaucus, NJ: Citadel Press, 1972.

Memphis Bridge (Memphis, Tennessee; 1892)

The Memphis Bridge carries the Fort Scott & Memphis Railroad over the Mississippi River at Memphis, Tennessee. Built by George S. Morison and Alfred Noble to span the Mississippi at one of its widest locations, the Memphis Bridge was a major engineering accomplishment. The bridge is an asymmetrical span **cantilever** truss with two spans of 621 feet (189 meters), one span of 790 feet (241 meters), and one span of 226 feet (69 meters). When completed, it was one of the largest bridges of its type in the world. Because of the bridge's size, nine different steel companies had to supply the 8,160 tons (7,344 metric tons) of steel required, an average of 3.5 tons (3.2 metric tons) per linear foot. The trusses are 30 feet (9 meters) apart, 78 feet (24 meters) deep, and 80 feet (24.4 meters) above the high water line.

Additional Information: DeLony, Eric. *Landmark American Bridges*. Boston: Bullfinch Press, 1993.

Menai Straits Bridge (Bangor, Wales; 1820-1826)

Thomas Telford built this two-lane **suspension bridge** across the Menai Straits off northwest Wales to connect the Welsh mainland with the island of Anglesey. In 1850, Robert Stephenson also crossed the Menai Straits with the BRITTANIA RAILROAD BRIDGE. Telford made use of the revolutionary idea of an iron bridge with stone towers, a concept that had only been executed twice before on a smaller scale. With an admirable, though possibly extreme, caution, Telford tested each individual link in the **wrought-iron** chains. The first big suspension bridge ever built, the Menai Straits Bridge has a total length of 1,710 feet (522 meters). The towers are 153 feet (47 meters) high and 579 feet (177 meters) apart; the length of the main span was unheard-of for the early nineteenth century. The bridge has two sets of chains, one on each side of the towers. Although damaged by storms in its first decades, and remodeled and repaired many times, the bridge is still in use. Thomas Telford, acknowledged as the greatest engineer in Britain, is buried in WESTMINSTER ABBEY.

Additional Information: Kirby, Richard Shelton et al. *Engineering in History*. New York: Dover Publications, 1990; Overman, Michael. *Roads, Bridges, and Tunnels*. Garden City, NY: Doubleday & Company, 1968; Sandstrom, Gosta E. *Man the Builder*. New York: McGraw-Hill, 1970; Steinman, David B. *Famous Bridges of the World*. London: Dover Publications, 1953, 1961.

Merrimack Bridge. *See* Essex-Merrimack Bridge.

Mersey Tunnel (Liverpool, England; 1934)

The Mersey Tunnel is one of the largest bore underwater road-tunnels in the world. The tunnel runs for 2 miles (3.2 kilometers) under the River Mersey in northwestern England, connecting Liverpool with Birkenhead. With an inside diameter of almost 50 feet (15 meters), the tunnel carries four traffic lanes, with additional space (never used) for two double-decker tramways below the traffic lanes. The tunnel descends to a maximum depth of 157 feet (48 meters) below water. The main tunnel is 36 feet (11 meters) wide; the side tunnels are 19 feet (5.8 meters) wide. Because pressure

on the bottom of the tunnel would be too high to allow tunnelling through the mud, the decision was made when building the tunnel to cut through sandstone below the river. Waterflow from the river above the tunnel was the major construction problem. Two main **headings**, an upper and a lower, were built, and a third drainage tunnel was constructed underneath the two. The drainage tunnel inclined toward the Liverpool heading, where water was removed at a rate of approximately 2,400 gallons a minute. Waterflow at the time the tunnel was begun was higher, and cement was forced into the walls of the tunnel to fill openings (fissures) in the rock ahead and slow the water. During digging, 1.2 million tons (1.08 million metric tons) of sandstone were removed from the tunnel, and moved at night by truck. The iron lining of the tunnel weighs over 82,000 tons (73,800 metric tons) and is made of over 1 million segments. The tunnel is ventilated through six stations in a manner similar to that used in the HOLLAND TUNNEL. Financed by tolls, the tunnel cost $18 million to build.

Additional Information: Beaver, Patrick. *A History of Tunnels*. Secaucus, NJ: Citadel Press, 1973; Overman, Michael. *Roads, Bridges and Tunnels*. Garden City, NY: Doubleday & Company, 1968.

Metropolitan Railway of Paris. *See* Paris Metro.

Miami Conservancy District (southwestern Ohio; 1922)

The creation of this special governmental unit brought water to southwestern Ohio, and also controlled the dangerous effects of flooding in the Ohio River Valley. Especially interesting was the construction of five earthen dams within the district—the Englewood, the Huffman, the Taylorsville, the Germantown, and the Lockington. All five have variable slope sides. The Germantown Dam is on Twin Creek, 2 miles (3.2 kilometers) northwest of Germantown. The Lockington Dam is on Loramie Creek, 2 miles north of where the creek meets the Miami River. The Englewood Dam is on the Stillwater River, 10 miles (16 kilometers) above Dayton. The Huffman Dam is located on the Mad River, 8 miles (13 kilometers) east of Dayton. The Taylorsville Dam is 10 miles above Dayton on the Great Miami River; it is the highest of the five dams and has the largest concrete structure of the five. Its retaining walls are 90 feet (27 meters) high, its spillway is 132 feet (40 meters) long, and its outlet has four conduits, 19 feet (5.8 meters) high and 15 feet (4.6 meters) wide, all in one place.

Additional Information: Schodek, Daniel L. *Landmarks in American Civil Engineering*. Cambridge, MA: MIT Press, 1987.

Mianus River Bridge (Greenwich, Connecticut; 1958)

This bridge in eastern Connecticut collapsed in June 1983, taking several cars into the Mianus River, killing three people, and injuring three more. The Mianus River Bridge was a **cantilever bridge** with no redundancy, that is, no extra support was provided in case of the failure of the supporting members. This type of bridge had been recognized as unsafe since 1968. In the case of the Mianus Bridge, the immediate cause of failure was rusting and corrosion, coupled with metal fatigue. The roadway drains had been paved over 10 years before the bridge collapsed, allowing water to accumulate and corrode pins and connections. Nearby residents reported hearing noises from the bridge (it seemed to be "singing") for several days before it collapsed; for several years before the collapse, people had even picked up and turned in random pieces of concrete and metal to authorities.

Additional Information: Matthys, Levy and Salvadori, Mario. *Why Buildings Fall Down.* New York: W.W. Norton & Company, 1992.

Mica Dam (Revelstoke, British Columbia, Canada; 1973)

This 800-foot-high (244-meter) earth dam is one of the world's highest dams. It is located about 135 kilometers (84 miles) north of Revelstoke in British Columbia. It has a length of 2,550 feet (778 meters) and a volume of 4 million cubic yards (3.060 million cubic meters). This dam across the Columbia River is the largest structure to result from the 1964 Columbia River Treaty between Canada and the United States. The dam was designed to provide flood control in British Columbia and power generation in the U.S. and Canada. The dam and its power facilities are operated by the British Columbia Hydro and Power Authority. The dam's reservoir, Kinbasket Lake, holds 20 million **acrefeet** of water.

Additional Information: Smith, Norman. *A History of Dams.* Secaucus, NJ: Citadel Press, 1972; http://www.inter.chg.use.ca/bchydro...ronment/recreation/site_mica1.html.

Middlesex Canal (Charlestown to Lowell, Massachusetts; 1794-1803)

The Middlesex Canal stretched 27 miles (43 kilometers) across eastern Massachusetts, running northwest from the Boston area to the Merrimack River near Lowell (which was set off from the town of Chelmsford, the original destination of the canal, in 1826). The Middlesex is one of the oldest constructed waterways in the United States; its construction and success demonstrated the practicality of canal transportation and inspired other canal projects elsewhere in the country. The promoters of the canal originally intended it to run into New Hampshire and connect Massachusetts with the trade of Canada. This connection was never made, but the canal did bring the firewood, timber, and other produce of New Hampshire to the markets of Boston.

Gates on a lock of the Middlesex Canal at Lowell, Massachusetts. Photo courtesy of Barbara Berlow.

Water first flowed through the canal on 31 December 1803, and the waterway soon became the cheapest and most popular method of transportation in Massachusetts. The canal boats, each drawn by one horse, were between 40 and 75 feet (12.2 and 23 meters) long, and 9 to 9.5 feet (2.7 to 2.9 meters) wide, with a depth of about 4 feet (1.2 meters) and a maximum capacity of about 20 to 25 tons (18 to 22.5 metric tons). Speed on the canal was restricted to 3 miles (4.8 kilometers) per hour. The canal included seven wooden **aqueducts** and 30 locks made of solid masonry; the locks were 90 feet (27 meters) long and 12 feet (3.7 meters) wide. Near the northern end, three locks accounted for a 28-foot (8.5-meter) drop to the Merrimack River. In an unusual and successful engineering design, the canal's towpaths were set on pontoons; when boats needed to exit the canal along a feeder channel, the pontoons were simply pushed aside. In 1808, Secretary of the Treasury Albert Gallatin gave a report on the roads and canals of the U.S. that mentioned a raft of timber one mile long and weighing 800 tons (720 metric tons) that had been pulled between locks on the Middlesex Canal. Despite such heavy traffic, the canal did not pay dividends until 1819, for the costs of maintaining the waterway always exceeded profits. The development of the cotton industry in Lowell in the 1820s brought the canal a brief period of prosperity, which was over by the late 1840s. Competition from railroads drove the canal out of business in 1853. In 1860, the canal's assets were sold and distributed to the shareholders, who had sued the operating company.

Additional Information: Kirby, Richard Shelton et al. *Engineering in History*. New York: Dover Publications, 1990; Schodek, Daniel L. *Landmarks in American Civil Engineering*. Cambridge, MA: MIT Press, 1987; Spangenburg, Ray and Moser, Diane K. *The Story of America's Canals*. New York: Facts on File, 1992.

Mikhailovsky Castle. *See* Engineer's Castle.

Millennium Tower (Tokyo, Japan; expected completion in first decade of twenty-first century)

When completed, this tower being built by the Obayashi Corporation of Japan will be the tallest building in the world at a height of 800 meters (2,624 feet). According to Shimuzu Keizo, general manager for Obayashi, the building will make use of a magnetic-levitation system for its elevators; faster than conventional elevators, such a system is still slow enough to keep passengers from getting sick. A trip to the 600-meter (1,968-foot) level would take a bare two minutes.

Additional Information: "Edifice Complex," August 1995, *Asia, Inc.*, in Asia, Inc. Online and Keizo, Shimizu. "Sky-High Ambition," August 1995, *Asia, Inc.*, in Asia, Inc. Online (http://www.asia-inc.com/archive/0895sky.html).

Million Dollar Bridge (Cordova, Alaska; 1911)

The Million Dollar Bridge carried the Copper River & Northwest Railway across the Copper River near Cordova in southeastern Alaska. Engineer A.C. O'Neel built the bridge as a string of arches arranged over piers crossing the river; the bridge was partially destroyed when one of the end spans dropped off its pier during the 1964 Alaska earthquake. The span was never repaired.

Additional Information: DeLony, Eric. *Landmark American Bridges*. Boston: Bullfinch Press, 1993.

Mitchell's Point Tunnel (Mitchell's Point, Oregon; 1915)

John A. Elliott built the Mitchell's Point Tunnel, the first important highway tunnel built in the United States, on the Columbia River Highway in northern Oregon in 1915. A 390-foot-long (119-meter) tunnel with a 10-degree curve, it saved about one mile of travel on the highway. Built with a width of 18 feet (5.5 meters) and a height of 10 feet (3 meters) above the roadway, the tunnel included five windows, each 16 feet (4.9

meters) wide, spaced symmetrically near the curved top of the tunnel's midsection.

Additional Information: Rose, Albert C. *Historic American Roads*. New York: Crown Publishers, 1976.

Moffat Tunnel (near Denver, Colorado; 1927)

This 6.1-mile (9.8-kilometer) railroad tunnel under the Rocky Mountains was the first tunnel to cross the Continental Divide. It was named after David H. Moffat, the railroad engineer who first conceived of a tunnel through the Rocky Mountains. Moffat died in 1911, but the city of Denver and the Denver and Salt Lake Railroad did not reach agreement to build the tunnel until 1923. The railway tunnel was planned to be 16 feet (5 meters) wide and 24 feet (7.3 meters) high. About 60 miles (97 kilometers) west of Denver, the Moffat Tunnel, which is still in use, was the longest railroad tunnel in the western hemisphere when it was completed. The tunnel gave Denver and western Colorado a better rail connection to the Midwest.

By passing through the mountains, the tunnel cut transportation costs—only one locomotive, instead of four or five, was now needed to make the trip. Both ends of the tunnel are below the timber line to keep the tracks clear of large snowdrifts in winter. In building the tunnel, a second water tunnel was constructed 75 feet (26 meters) south of the main tunnel to provide access to it; this method of accessing the main tunnel and channeling water had been used in Europe, but was new to the United States. Eventually, the water-service tunnel was lined with concrete and used as an **aqueduct** to carry water from the Colorado River to Denver and its surrounding area. The water conduit was designed to be a circular shaft with a 12-foot (3.7-meter) diameter; it ran 7.5 feet (2.3 meters) above the railway tunnel's grade. Beginning in February 1925, when the tunnel experienced an influx of water from a lake, engineering difficulties arose with great regularity. Other floods brought water into the tunnel at a rate of 3,100 gallons per minute. Power lines to the tunnels were snapped by a blizzard in 1926, cutting off power to the pumps that were controlling the flooding. Large areas of soft ground were found to be under incredible geological pressure and required special support work to hold up tunnel walls. The two sides of the tunnel finally connected on 12 February 1927, and the last barrier between the two sections was destroyed by 24 blasts of dynamite set off by President Calvin Coolidge, who pressed a button at the White House. Over 750 million cubic yards (over 573 million cubic meters) of rock were removed from the mountain, 2.5 million pounds of dynamite was used in 700 miles (1,126 kilometers) of holes, and 11 million feet of board were used for timbering. The final cost of the project was $18 million.

Special care was taken for the safety of the workers in the building of the tunnel. Camps were set up at both drilling entrances; power plants, shops, stores, recreation facilities, sleeping huts with electricity, and showers and laundry facilities were built there for worker use. Married men had cottages, schools, and churches for their families. Doctors and nurses staffed fully equipped hospitals, and men were instructed to report any sign of illness. Safety restrictions were extensive and enforced, and included control of electric lights in the tunnels before and after a blast, and a requirement that workers wear oilskins (raincoats) while working to prevent pneumonia. Food and hot coffee were delivered to the tunnels so that workers could eat at the site; drinking was prohibited. Despite the impressive safety precautions, 11 lives were lost during the project.

Additional Information: Beaver, Patrick. A History of Tunnels. Secaucus, NJ: Citadel Press, 1973; Schodek, Daniel L. *Landmarks in American Civil Engineering*. Cambridge, MA: MIT Press, 1987; Smith, Christopher. "You Can Buy Historic Rail Tunnel—Hole Sale" in online edition of *Salt Lake Tribune*, 1 January 1997 at http://ftp.sltrib.com/97/jan/01/tci/00341717.htm.

Mont Cenis Tunnel, also known as the Fréjus Tunnel (Modane, France, to Bardonecchia, Italy; 1857-1871)

Germaine Sommellier, an Italian engineer, headed an Italian project to drive a railroad tunnel through the Alps between Savoy in France and Piedmont in Italy. Although long desired by railroad companies, this tunnel had long been thought impossible to build because of the depth required to pass under the Alps. King Charles Albert of Sardinia, the small kingdom that controlled both ends of the tunnel in the mid-nineteenth century, assembled teams of planners to figure out how best to build the tunnel. Work on the project began in 1857 under Charles Albert's son, Victor Emmanuel II, who engaged Sommellier as engineer. French and Italian engineers used hand drills and black powder to bore a tunnel nearly 12 kilometers (7.4 miles) through Alpine granite under Mount Cenis. Progress was a painfully slow 22 centimeters (8.6 inches) a day at each end. In 1861, a compressed air drill, invented by Sommeiller, was introduced to the project. Ten of the air drills were set on an 8-meter by 2-meter by 2-meter (26- by 6.6- by 6.6-foot) jumbo. Sommeiller had expected the air pumped out by the drills to ventilate the tunnel, but it was not sufficient, and fans had to be installed to provide ventilation. Dynamite soon replaced gunpowder and the tunneling speed increased to 2 meters (6.6 feet) per day. The Mont Cenis Tunnel project is also distinguished for the care and attention that Sommellier paid to his workers. Careful concern for their health and safety resulted in only 54 injuries, of which 28, a low for the time, were fatalities. In comparison, the American HOOSAC TUNNEL, built at the same time, and with similar methods, cost 75 lives. The finished 12-kilometer (7.4-mile) brick-and-stone-lined shaft, accommo-

dating a double line railroad, was opened in 1871, with many dignitaries signed on for the first rail trip. Sommellier had died several months earlier, supposedly from overwork. The Mont Cenis Tunnel is one of the largest railway tunnels in the world.

Additional Information: Epstein, Sam and Epstein, Beryl. *Tunnels*. Boston: Little Brown & Company, 1985; Kirby, Richard Shelton et al. *Engineering in History*. New York: Dover Publications, 1990; Overman, Michael. *Roads, Bridges and Tunnel*. Garden City, NY: Doubleday & Company, 1968.

Montgomery Bell's Tunnel. *See* Pattison Tunnel.

Mormon Flat Dam. *See* Salt River Project.

Morris Canal (Lake Hopatcong, New Jersey, to New York Harbor; 1831)

James Renwick and David Bates Douglass designed the Morris Canal to connect the Delaware River with New York Harbor, allowing Pennsylvania coal to pass up the Delaware and across northern New Jersey avoiding an open sea voyage to New York around Cape May at New Jersey's southern tip. The Morris Canal made use of 23 locks and 23 inclined planes along its 100-mile (161-kilometer) route to conquer a height of 915 feet (279 meters) above sea level on the way to New York Harbor. Virtually unused by the start of the twentieth century because it was too shallow and too narrow, the Morris Canal was acquired by the state of New Jersey and dismantled.

Additional Information: Kirby, Richard Shelton et al. *Engineering in History*. New York: Dover Publications, 1990; Spangenburg, Ray and Moser, Diane K. *The Story of America's Canals*. New York: Facts on File, 1992.

Morrow, Lacey V. Floating Bridge. *See* Lacey V. Morrow Floating Bridge.

Moscow Tower. See Ostankino Moscow Tower.

Mostar Bridge (Mostar, Bosnia-Hercegovina; 1556)

Until 1993, the Mostar Bridge crossed the Neretva River in Mostar, a city in eastern Bosnia-Hercegovina; Bosnia broke away from the old Republic of Yugoslavia in March 1992, setting off a fierce civil war between various ethnic and religious groups. Mimar Hayruddin built the bridge in the sixteenth century on the orders of Suleiman the Magnificent, the Ottoman sultan. The Ottoman Empire, centered in Istanbul and Turkey, controlled much of the Balkan Peninsula of Europe in the sixteenth century and brought Islam into the region. The Mostar Bridge was an early **muleback bridge**, spanning 29 meters (95 feet). Until November 1993, this **arch bridge** served as a symbol of peace between the Christian Croatians, who lived on one side of the bridge, and the

Bosnian Muslims, who lived on the other. Croatian nationalists destroyed the bridge in 1993 to prevent Muslim fighters from moving supplies across it. At the time, the Croatians hoped to make Mostar the capital of a new mini-state carved out of Muslim Bosnia. In March 1994, an anonymous Croatian contractor offered to rebuild the bridge. Although the Dayton Agreement of 1995 has brought a fragile peace to Bosnia and the other former republics of Yugoslavia, the bridge has not yet been reconstructed.

Additional Information: "Croat Offers to Rebuild the Bridge at Mostar," *New York Times*, 13 March 1994, I:1; Lay, M.G. *Ways of the World*. New Brunswick, NJ: Rutgers University Press, 1992.

Mount Rushmore National Memorial (Shrine of Democracy) (near Rapid City, South Dakota; 1941)

On 3 March 1925, Congress authorized a colossal monument to democracy to be carved on a mountain in the Black Hills in southwestern South Dakota, about

The Mount Rushmore monument in the Black Hills near Rapid City, South Dakota.

25 miles (40 kilometers) southwest of Rapid City. To-day, Keystone is the closest town to the memorial. Congress selected the sculptor Gutzon Borglum, who was experienced in carving granite, to undertake the work. Borglum began in 1927 to carve giant busts of George Washington, Thomas Jefferson, Abraham Lincoln, and Theodore Roosevelt on the face of the mountain. The memorial was dedicated by President Calvin Coolidge in 1927. Washington's statue was ready by 1930, Jefferson's head followed in 1936, Lincoln's in 1937, and Roosevelt's in 1939. Workers stopped carving in October 1941, several months after Borglum died. His son, Lincoln Borglum, finished the detailing on the

heads, which were not dedicated until 1991. Gutzon Borglum believed that a monument should be large and in proportion with the importance of the events and the people it commemorated. In accordance with this belief, Borglum placed quotes from the four presidents nearby. The faces, which are visited by millions of tourists every year, can be seen from as far as 60 miles (97 kilometers) away.

Additional Information: South Dakota Bureau of Tourism, "Mount Rushmore: The Four Most Famous Guys in Rock" at http://www.state.sd.us/tourism/rushmore/rushmore.html; for an extensive discussion of the construction of Mount Rushmore, see Ditzel, Paul C. *How They Build Our National Monuments*. New York: Bobbs-Merrill, 1976.

Mount Vernon Memorial Highway (Washington, DC, to Mt. Vernon, Virginia; 1932)

Landscape architect Gilmore Clark oversaw the landscaping of this project, which provided jobs for many workers left unemployed by the Great Depression. The Mount Vernon Memorial Highway was the first parkway to be constructed and maintained by the federal government. The road commemorates the 1932 bicentenary of the birth of George Washington. Consultants from the Westchester County (New York) Park Commission, who had experience with the construction and maintenance of the BRONX RIVER PARKWAY, consulted on the project. The highway is listed in the National Register of Historic Places.

Additional Information: http://www.cr.nps.gov/phad/ncrobib.html

Mratinje Dam (Foca, Montenegro, Yugoslavia; 1976)

This is one of the world's highest dams with a height above its lowest formation of 220 meters (722 feet) and a crest length of 268 meters (879 feet). The Mratinje is located on the Piva River near the city of Foca. It is an arch dam capable of holding a volume of 742 cubic meters.

Source: Scott K. Nelson, The National Performance of Dams Program (NPDP), Stanford, California.

Myddleton Canal (Hertfordshire to London, England; 1613)

Sir Hugh Myddleton built the Myddleton Canal, one of the earliest canals in England, in 1613. The canal was designed to bring water south from Hertfordshire to London. Forty miles (64 kilometers) long, the canal is 10 feet (3 meters) wide and 4 feet (1.2 meters) deep. Although partially blocked by silt and other obstructions, it is still capable of delivering some water to London. Regrettably, the water is not drinkable; **aqueduct**s at Enfield and Islington are lined with 50 tons (45 metric tons) of lead that contaminate the water. During the German Blitz on London in 1940, parts of the canal were excavated to deliver water to the city for fire-fighting.

Additional Information: Sandstrom, Gosta E. *Tunnels*. New York: Holt, Rinehart and Winston, 1963.

N

Nagarjuna Sugar Dam. *See* Nagarjunasagar Dam.

Nagarjunasagar Dam, also known as the Nagarjuna Sagar Dam (Hyderabad, India; mid-1970s). This major gravity and earth dam is 406 feet (124 meters) high and 15,326 feet (4,674 meters) long. The dam's volume is 73,572 cubic yards (56,283 cubic meters), and the reservoir has a volume of 9,177 **acre-feet**. The dam crosses the Krishna River about 150 kilometers (241 miles) south of Hyderabad in south central India. The dam creates an 85-square-kilometer lake known as Himayat Sagar.

Additional Information: Smith, Norman. *A History of Dams*. Secaucus, NJ: Citadel Press, 1972.

Nantes-Brest Canal (Nantes to Brest, France; 1838)

No longer used, this canal ran from Brest on the tip of the peninsula of Brittany in western France to Nantes on the Loire River southeast of Brittany. The canal had three reservoirs supplying it. The three reservoir dams are all very different. The Vioreau Dam has a simple profile. It is a rectangular dam, 36 feet (11 meters) high, 24.5 feet (7.5 meters) thick, with a small sloping step on the crest. Bosmelac Dam is 50 feet (15 meters) high, 14 feet (4.3 meters) thick at the crest, and 28 feet (8.5 meters) thick at the base. It has a vertical water face and a sloping air face. Glomel Dam is 40 feet (12.2 meters) high, with a vertical air face and a sloping water face. It is 2 feet (0.6 meters) thick at the parapet wall, sloping to 24.5 feet (7.5 meters) thick at the base.

Additional Information: Hadfield, Charles. *World Canals*. Newton Abbot, England: David & Charles, 1986; Smith,

Norman. *A History of Dams*. Secaucus, NJ: Citadel Press, 1972.

Natchez Trace (Nashville, Tennessee, to Natchez, Mississippi; begun 1803)

The Natchez Trace, one of the first roads to be maintained by the federal government, was a route of great commercial and military importance to the southwestern frontier of the United States from the 1780s to the 1830s. The road grew from a series of Choctaw and Chickasaw hunting and war trails that came under increasing use in the eighteenth century by whatever European power controlled the region—the French, English, and Spanish all held or had access to the area at one time or another before it passed to the United States in the 1780s. The route was made an American post road in 1800, and President Thomas Jefferson ordered the army to improve the route in 1803, when the pending purchase of Louisiana from France made good communications with the region vital. Although the army's "improvement" of the roadway consisted only of a few companies of soldiers slashing underbrush and pulling stumps, the Trace served the American military well during the War of 1812. Andrew Jackson marched down the route on his way to defend New Orleans from British attack, and used it again during his various Indian campaigns in the region. At first, American settlers and traders traveled only northward on the route. Having floated their goods down the Mississippi by flatboat to New Orleans, they walked back home via the Natchez Trace to Nashville. Such travelers would often use the "ride and tie" system, whereby two men would share a mule or horse. One would ride ahead, tie up the animal by the side of the road, and continue on foot; his companion would reach the animal, mount it, and catch up to begin the process again. Traveling the Trace was a dangerous venture. In its early days, the route was vulnerable to Indian attack and infested with bandits; the biweekly mail run to Natchez often failed to arrive, and the missing rider was routinely noted in postal records as "presumed lost." The few inns, or "stands," along the Trace were little more than rough cabins offering an opportunity to gamble and drink cheap whiskey. As settlement increased after 1815, the route became more secure and more heavily traveled in both directions. With the coming of the steamboat in the 1820s, traffic over the Trace began to decline, and the road fell into disuse by mid century. Some 400 miles (644 kilometers) of the Natchez trace, over 13,000 acres (5,265 **hectares**), is now the Natchez Trace Parkway, running between Nashville and Natchez. Operated by the National Park Service, the beautifully landscaped two-lane road is dotted with archeological sites and historical landmarks.

Additional Information: Patton, Phil. *Open Road: A Celebration of the American Highway*. New York: Simon and Schuster, 1986; Exploring America's Scenic Highways, prepared by the Special Publications Division, National Geographic Society, Washington, DC, 1985.

Natchez Trace Parkway Bridge (Franklin, Tennessee; 1994)

Designed and built by the Eastern Federal Lands Highway Division of the Federal Highway Administration, the Natchez Trace Parkway Bridge crosses Tennessee Route 96 near Franklin southwest of Nashville, Tennessee. An important bridge on the Natchez Trace Parkway, which closely follows the old NATCHEZ TRACE, the bridge is the first precast, concrete segmental-**arch bridge** in the United States. With a total length of 1,572 feet (480 meters), the bridge's two arches (in eight spans) reach a maximum height of 145 feet (44 meters) above Route 96. Most important of this bridge's several innovations is that the decking is supported only by the top-points of the arches. Conventional arches of this size usually use **spandrels** to distribute the weight of the deck at several points before and after the center of the arch; in this structure, the weight is carried by the tops of the arches alone. In 1995, the bridge was awarded the Outstanding Civil Engineering Achievement Award of Merit by the American Society of Civil Engineers.

Additional Information: Goldstein, Harry. "Triumphant Arches." *Civil Engineering* 65:7, (July 1995), pp. 48-49.

National Pike. See Cumberland Road.

National Road. See Cumberland Road.

National Statuary Hall (Capitol, Washington, D.C.; 1864)

In 1864, Congress invited all states in the Union to provide statues in marble or bronze of up to two of their notable citizens for display in the UNITED STATES CAPITOL BUILDING. By 1933, Congress authorized the architect of the Capitol to relocate some of the statues. Although some statues are on display elsewhere in Washington, any new statues are displayed in the Rotunda of the Capitol for a minimum of six months before being moved permanently to another location in Washington.

Additional Information: http://acs5.bu.edu:8001/~pviles/Hall2.html. For an excellent "virtual tour" of the capitol, including Statuary Hall, see http://xroads.virginia.edu/~CAP/CAP_home.html

Naviglio Grande Canal (Ticino to Milan, Italy; 1179-1258)

Work on the Naviglio Grande Canal, one of the oldest irrigation canals in Europe, began in the late twelfth century. The canal carries water 30 miles (48 kilometers) across the north Italian plain from a dam on the River Ticino in northwestern Italy to the city of Milan. Medieval engineers diverted the alpine waters running to waste from Lake Maggiore to the sea to ensure the growth and prosperity of Milan, which, by the sixteenth century, was one of the most important and powerful cities in northern Italy. In 1269, when Milan Cathedral was being built, the canal was widened and deepened to allow the shipment of construction supplies to the city. In 1438, the canal carried the pink and white marble used for the interior of the cathedral from the Alps foothills 60 miles (97 kilometers) away. The canal also provided water for the city's moat as well as power to run a number of water-mills. The dam on the Ticino has been rebuilt so many times that it is impossible to know exactly what the original looked like, although its dimensions seem typical of medieval dams. At present, it is 920 feet (281 meters) long, 6 feet (1.8 meters) high, and varies in thickness from 6 to 30 feet (1.8 to 9 meters).

Additional Information: Smith, Norman. *A History of Dams*. Secaucus, NJ: Citadel Press, 1972.

Neuilly Bridge (Paris, France; 1770-1772)

Designed and built by Jean Perronet in the eighteenth century, the Neuilly Bridge across the Seine River in Paris was demolished in 1932 to allow better navigation on the river. Perhaps Perronet's most beautiful bridge, the Neuilly used five 37-meter (121-foot) elliptical arches, and was aligned directly with the CHAMPS ELYSÉES.

Additional Information: Lay, M.G. *Ways of the World*. New Brunswick, NJ: Rutgers University Press, 1992.

Neuschwanstein Castle (near Füssen, Germany; 1869-1886)

Ludwig II, the "Mad King" of Bavaria, built this multi-towered, fairytale-like castle in the mountains of southern Germany southwest of Munich. Neuschwanstein is full of representations of Ludwig's favorite bird, the swan; swan pictures and sculptures, and even swan-

The fairytale-like Neuschwanstein Castle near Munich, Germany, was built in the late nineteenth century by Ludwig II of Bavaria.

shaped water faucets can be found in the castle. Ludwig's many eccentricities earned him a reputation for being "mad"; for instance, he ate dinner with his horse, the animal being given a place at the other end of the royal dinner table. The castle was built to match the scenery of three of Richard Wagner's operas: *Lohengrin*, *Parsifal*, and *Tannhäuser*. Ludwig was Wagner's patron and friend, and the composer exercised a powerful influence over the king for many years. The castle was the basis for Walt Disney's castles at Disneyland, including Sleeping Beauty's castle. Although Ludwig was in residence at the castle while it was being constructed, he died by his own hand in 1886, the same year the castle was finally completed.

Additional Information: "Favorite Castles in Europe" at http://www.nectec.or.th/rec-travel/europe/castles; http://www.cde.com/~alpsboy/alpsboy.htm; Oggins, Robin S. *Castles and Fortresses*. New York: Michael Friedman Publishing Group, 1995.

Nevsky Prospect (St. Petersburg, Russia; begun 1710)

Begun in the eighteenth century, the Nevsky Prospect, the central avenue of St. Petersburg, is one of the city's most beautiful streets. Most of the architectural wonders of the city are found along the Prospect, or nearby. Peter the Great ordered the building of St. Petersburg in western Russia in 1703; he wanted his new capital to be a modern Western city. Because the name sounded too German, the city was renamed Petrograd at the start of World War I in 1914. In 1924, the city was renamed Leningrad to honor the Soviet leader Vladimir Lenin. Unlike many other Russian cities, Leningrad was not greatly rebuilt by the Soviets, who moved the capital of the Soviet Union to Moscow after 1917. The city suffered a long and difficult siege by German forces during World War II. Soviet dictator Joseph Stalin, Lenin's successor, installed large, boring buildings in many of the main cities of the Soviet Union; ironically, because Stalin hated Leningrad, the city remained largely undisturbed, its czarist architectural past intact. The name of St. Petersburg was restored to the city after the collapse of the Soviet Union in 1991.

Additional Information: http://www.spb.su/fresh/sights/nevsky.html.

New Cascade Railroad Tunnel (Washington State; 1929)

This 8-mile (12.9-kilometer) tunnel through the Cascade Mountains is one of the longest rail tunnels in the United States. The tunnel is operated by the Burlington Northern Railroad. Workers faced many difficult problems in drilling this tunnel, such as the sudden eruption of underground springs of water. The tunnel was finally completed by using the most modern techniques available at the time. A new drilling rig that could drill 28-foot-deep (8.5-meter) holes in an hour and a half was an important advantage. Air-powered machines loaded up blasted rock and dumped it into small railroad cars to be carried away. Smoke from blasting was cleared in less than half an hour by efficient ventilating fans. At its peak, digging speed on the tunnel reached 40 feet (12.2 meters) in a 24-hour working day.

Additional Information: Sandstrom, Gosta E. *Tunnels*. New York: Holt, Reinholt and Winston, 1963.

New Croton Aqueduct (New York City; 1885-1893)

As New York City's population increased in the late nineteenth century, so did the city's demand for water and its need for a new **aqueduct** to meet that demand.

In the 1880s, work began on augmenting and replacing the 45-mile-long (72-kilometer) CROTON AQUEDUCT, which had carried water to New York City from the upstate Croton River since the 1840s. Opened in 1893, the New Croton Aqueduct connected the Central Park Reservoir with Croton Lake. By the 1890s, the city's population was nearly 2 million and its daily per capita consumption of water was about 100 gallons (380 liters). Croton Aqueduct could deliver about 90 million gallons (342 million liters) daily; the New Croton could deliver 340 million gallons (over 1.2 billion liters) daily. The New Croton Aqueduct followed a less meandering route than the old aqueduct; the new aqueduct was not laid close to the surface as was the previous structure, but was bored straight through rock as a tunnel at the hydraulic grade. The 30 tunnel shafts averaged 127 feet (39 meters) in depth, the deepest being nearly 400 feet (122 meters) below the ground surface. The crossing of the Harlem River Valley required a pressure tunnel, or inverted siphon, that was 300 feet (91.5 meters) deep and nearly 7 miles (11.3 kilometers) long; it carried water at a depth of 420 feet (128 meters) below hydraulic grade. A number of innovations were used during construction: incandescent lamps lighted the tunnel for workers, and the air discharge from compressed-air drills provided ventilation. The average weekly progress during excavation was 25 to 40 feet (7.6 to 12.2 meters).

Additional Information: Kirby, Richard Shelton et al. *Engineering in History*. New York: Dover Publications, 1990.

New Croton Dam, also known as the Cornell Dam (New York City; 1906)

Alphonse Fteley designed and helped to build the New Croton Dam several miles down the Croton River from the original CROTON DAM, which had been built in the 1840s to supply water to New York City. The new structure was intended to replace the old dam, which it eventually submerged by raising the level of the reservoir

The New Croton Dam is part of New York City's water system. Photo courtesy of Mel Wittenstein.

(Croton Lake) 36 feet (11 meters). A masonry gravity dam, the New Croton is 297 feet (91 meters) high and over 2,000 feet (610 meters) long, including its spillway. When built, it was the tallest dam in the world. Although now dwarfed by more modern dams, it remains an impressive structure. The New Croton continues to play an important part in New York City's water supply system.

Additional Information: Jackson, Donald C. *Great American Bridges and Dams*. Washington, DC: The Preservation Press, 1988; Kirby, Richard Shelton et al. *Engineering in History*. New York: Dover Publications, 1990.

New Jersey Turnpike (New Jersey; 1951)

Running north-south through New Jersey, the New Jersey Turnpike is one of the best known roads in the United States. Even Paul Simon and Art Garfunkel sang about "searching for love on the New Jersey Turnpike." Built in the early 1950s, this 142-mile (229-kilometer) stretch of quasi-public toll road was one of the largest public works projects of its time, and one of the most profitable in history. The road runs from the Delaware River near New Castle in the southwest, through the rural areas of southern New Jersey and the industrial centers in the north around Newark, to the New Jersey entrance to the GEORGE WASHINGTON BRIDGE over the Hudson, a major access point into New York City. A major truck route, the turnpike links up all portions of the state. The turnpike has its own corps of police (a dedicated barracks of the New Jersey State Police), wider lanes and shoulders than almost any other road in the United States, rest-stop service areas named after both known and obscure Americans (e.g., Vince Lombardi and Molly Pitcher), and exits to places in New Jersey that owe their existence solely to the need to provide various services for turnpike users.

Additional Information: Gillespie, Angus K. and Rockland, Michael Aaron. *Looking for America on the New Jersey Turnpike*. New Brunswick, NJ: Rutgers University Press, 1989.

New London Bridge (originally London, England; now Lake Havasu City, Arizona; 1831)

Designed and built by John Rennie and refurbished by his son Sir John Rennie, this bridge across the Thames in London replaced Peter Colechurch's OLD LONDON BRIDGE, built in the thirteenth century. The first pile of this New London Bridge was driven on 15 March 1824, and the last stone was placed on 1 August 1831. With a total length of 1,005 feet (307 meters) and a weight of 130,000 tons (117,000 metric tons), Rennie's London Bridge was made of five semi-elliptical stone arches curving out over the water. In 1968, when its foundations proved inadequate, the bridge was sold to American developers, who took it apart piece by piece and reconstructed it in Lake Havasu on the Colorado River in western Arizona. The sale and transport of the bridge caused a mild controversy in England, where some people objected to allowing a London landmark to become a tourist attraction in the American desert. The bridge today attracts thousands of visitors to Lake Havasu City, Arizona, every year.

Additional Information: Brown, David J. *Bridges*. New York: Macmillan, 1993; Lay, M.G. *Ways of the World*. New Brunswick, NJ: Rutgers University Press, 1992; Steinman, David B. *Famous Bridges of the World*. London: Dover Publications, 1953, 1961.

New River Gorge Bridge (near Fayetteville, West Virginia; 1978)

The New River Gorge Bridge is one of the world's longest and highest steel **arch bridges**. When completed, it was the longest such bridge in the world at almost 1,700 feet (518 meters), surpassing the BAYONNE BRIDGE, the previous record-holder. The New River Gorge Bridge rises a breathtaking 876 feet (267 meters) above the New River Gorge. A **suspension bridge** had been considered for the site, but its height would have been too dangerous for aircraft flying nearby. The bridge was built of stainless steel to eliminate the need for painting on the dangerous height. The arch was built in sections from either side, in the traditional method of arch construction. With a confidence born of modern calculation abilities, the designers calculated the exact measurements necessary to have the two halves meet correctly to form the completed arch. Previous arch bridges had been jacked apart so that a center portion could be inserted and the two sides were then lowered against the center portion.

Additional Information: Brown, David J. *Bridges*. New York: Macmillan, 1993.

New York City Water System. *See* City Tunnel No. 3.

New York State Barge Canal (Buffalo, New York, to Albany, New York; restored 1910)

The New York State Barge Canal is the restored and still functioning remnant of the nineteenth-century ERIE CANAL. The Barge Canal makes use of a slightly redirected and improved eastern section; the remaining western third is essentially the same as the original Erie route. At its western end, the canal enters the Niagara River at Tonawanda, New York. The complete contemporary system has 35 locks and raises traffic an impressive 564 feet (172 meters) between Albany along the Hudson (sea level) and Buffalo along the Niagara.

Additional Information: Freedgood, Seymour. *The Gateway States*. New York: Time, Inc., 1967, 1970; http://www.history.rochester.edu:80/canal/index.htm.

New York State Thruway (New York; 1960s)

The New York State Thruway, a limited access toll road, runs north from New York City along the west bank of the Hudson to Albany, where it bends west across the

The Mohawk River Bridge on the New York State Thruway crosses the Mohawk River in Herkimer County. Photo courtesy of the Bethlehem Steel Corporation.

length of the state to pass through Utica and Syracuse before ending in Buffalo in western New York. The thruway provides a quick, direct route from New York City to Canada, which lies just across the Niagara River from Buffalo. At Albany, the New York State Thruway connects with a freeway that runs north along the west side of Lake Champlain to the Canadian border.

Newark Dyke Bridge (near Newark, England; 1853)

Joseph Cubitt built and Charles Wild designed the Newark Dyke Bridge over the Trent near Newark in northeastern England. A combination of **cast-iron** and **wrought-iron**, the Newark Dyke Bridge is the first example of a Warren truss in England. The bridge had a clear span between **abutments** of 140.5 feet (43 meters); each side of the triangles in the truss was 18.5 feet (5.6 meters) long.

Additional Information: Hopkins, H.J. *A Span of Bridges: An Illustrated History*. New York: Praeger, 1970.

Newburyport Suspension Bridge. *See* Essex-Merrimack Bridge.

Newport Transporter Bridge (Gwent, South Wales; 1906)

This bridge over the River Usk in South Wales has two lattice steel towers that support a 646-foot (197-meter) span that is part suspended and part **cable-stayed**. A moving frame rides on the span. The frame suspends a gondola that carries passengers and vehicles. The Transporter was built by Alfred Thorne who was guided by a patent by the frenchman Ferdinand Arnodin.

Additional Information: Brown, David J. *Bridges*. New York: Macmillan, 1993.

Niagara Falls Railroad Suspension Bridge

(crosses the Niagara River Gorge between western New York and Canada; 1851-1855)

Charles Ellet was the original contract holder for this important **suspension bridge**; he had gone as far as the construction of a temporary service structure when he resigned from the project in 1848 over an argument with the construction company. A consummate showman as well as an engineer, Ellet had offered a prize to the first boy who flew a kite across the Niagara Gorge; he then used the kite string to hoist a wire and a cable across the gorge to build the temporary connection. After Ellet's resignation, the contract was assigned to John Roebling, who finished the bridge according to plan. Although not a record in length, the bridge was a magnificent structure. Its four cables were made of 3,640 **wrought-iron** strands, air-spun in the same manner that Roebling would later use on the BROOKLYN BRIDGE. Up until this time, suspension bridges for railroads were considered impossible to build. The bridge stood for 42 years until it was replaced with a bridge that could carry even heavier railroad weights.

Additional Information: Brown, David J. *Bridges*. New York: Macmillan, 1993.

Nina Tower (Hong Kong, China; expected completion in 1998)

Foundation work for this 468-meter-high (1,535-foot) tower has already begun, pending approval by the Hong Kong government. If approved, it will be one of the tallest buildings in the world, and the tallest in Hong Kong, surpassing the 374-meter-high (1,227-foot) CENTRAL PLAZA BUILDING in Hong Kong, which was completed in 1992.

Additional Information: "Edifice Complex," August 1995, *Asia, Inc.*, in Asia, Inc. Online (http://www.asia-inc.com/archive/0895edifice.html).

Nineveh Canal (Nineveh, Assyria [now northern Iraq]; seventh century B.C.)

For centuries, all that was known of this grand, ancient water-delivery system was the boast left inscribed on a rock by its builder, King Sennacherib of Assyria: "By Assur, my Great God, I swear that with these few men I dug the canal, and in a year and three months I finished it." In 1932, another inscribed rock was found in Iraq; the translation of its inscription, along with archaeological excavations in the area, led to the discovery of the canal and a portion of a related **aqueduct**. The canal was more than 49 miles (79 kilometers) long,

and 65 feet (20 meters) wide. The bottom of the canal was lined with approximately 2 million limestone blocks measuring 25 by 17 by 17 inches (63.5 by 43 by 43 centimeters). Underneath the blocks, for waterproofing, was a one-inch layer of bitumen and a 16-inch-thick (41-centimeter) layer of concrete.

Additional Information: Sandstrom, Gosta E. *Tunnels*. New York: Holt, Rinehart and Winston, 1963.

Nonesuch House. *See* Old London Bridge.

Normandy Bridge (Honfleur to Le Havre, France; 1995)

The Normandy Bridge crosses the mouth of the Seine River in northwest France, connecting the Norman ports of Honfleur and Le Havre. One of the world's longest **cable-stayed bridge**s, the 2,141-meter-long (7,022-foot) Normandy Bridge carries Road A29. Built at a cost of $470 million, the bridge was predicted at its opening to carry 8,000 cars per day. Each cable of the bridge is made of up to 51 individual strands, each 15 millimeters (0.6 inches) in diameter; the strands are woven together and extensively treated for corrosion resistance. The 81 cables are connected to 184 cable stays. Much of the final connection work was performed by workers from near Grenoble in mountainous southeastern France; these workers specialized in dangerous construction tasks, having gained their experience as mountaineers, divers, or speleologists (scientists who explore and study caves). In all, approximately 2,000 kilometers (1,240 miles) of cable were strung. The two towers of the bridge are each 215 meters (705 feet) high. A mass damper was placed across the deck, and other dampers were placed at the base of each cable, along with still other precautions, to protect against swaying and vibration caused by high winds.

Additional Information: "Travel Advisory: Normandy Bridge Cuts Travel Time to North," *New York Times*, 26 March 1995, V:1; "News: Record-Setting French Bridge Set to Open." *Civil Engineering* 65:2 (February 1995), p. 18.

Norris Dam (Norris, Tennessee; 1933-1936)

Roland Wank designed this first dam constructed by the TENNESSEE VALLEY AUTHORITY. The dam crosses the Clinch River near Norris in northeastern Tennessee. This 1,872-foot-long (571-meter) dam provides power, flood control, and recreation to the area. A **concrete gravity dam** with a height of 256 feet (78 meters), the dam's spillway is in the middle rather than on the side as is more conventional. The dam is named after former Nebraska Senator George W. Norris, who was a major supporter of the concept of large public-works programs during the Depression of the 1930s. The dam's two generators can produce 100,800 kilowatts of electricity. Despite its benefits, the dam's creation required the controversial dislocation of many residents from the upper Clinch River Valley, and permanently flooded many of those residents' established farms.

Additional Information: Farb, Peter. *The Story of Dams.* Irvington-on-Hudson, NY: Harvey House, 1961; Jackson, Donald C. *Great American Bridges and Dams.* Washington, DC: The Preservation Press, 1988.

North Woodstock Covered Bridge-Clark's Trading Post #64 (North Woodstock, New Hampshire; 1904)

This **railroad bridge** crosses the Pemigewasset River, just east of U.S. Route 3, at Clark's Trading Post in North Woodstock in north central New Hampshire. This bridge is the last remaining **Howe truss** railroad bridge in the world, and possibly the last remaining railroad covered bridge still in use. The bridge was originally built in Barre, Vermont, for use on a small line that ran between Barre and Montpelier, Vermont. Abandoned in 1960, the bridge was disassembled by Ed and Murray Clark and brought to New Hampshire, where it was reconstructed on dry land. Railroad tracks were placed in the riverbed, and two railroad cars were placed on the tracks. Cribwork was built up from the two railroad cars, and, in 1965, after a foundation had been placed (granite **abutments** taken from an old railroad bridge in Connecticut), the bridge was pulled into place with a trailer.

Additional Information: http://vintagedb.com/guides/covered3.html.

Northampton Canal. *See* Farmington Canal.

Northumberland Strait Crossing. *See* Confederation Bridge.

Notre Dame de Paris (Paris, France; 1230)

Notre Dame de Paris ("Our Lady of Paris") is the cathedral church of Paris and a brilliant example of early Gothic architecture in France. The cathedral stands upon the Île de la Cité, a small island in the Seine River upon which a Roman temple once stood. Christian churches had stood on the site since A.D. 375, but these were in ruins by the twelfth century, when they were demolished by Maurice de Sully to make room for a new cathedral large enough to meet the needs of the growing city. Pope Alexander III laid the cornerstone for the cathedral in 1163, and by 1183 work was sufficiently advaced for the high altar to be consecrated. Except for the spires intended for the cathedral's twin towers, the building was completed about 1230. The spires were never added. In the 1790s, during the French Revolution, anti-clerical rioters in the city converted the ca-

thedral into a "Temple of Reason," and destroyed the sculptures on the west facade. Restoration and renovation of the cathedral was undertaken in 1845. Almost

A view of the thirteenth-century cathedral of Notre Dame ("Our Lady") in Paris, France.

wholly Gothic in style, the cathedral has a wide central nave rising 110 feet (33.5 meters) high; the nave is flanked by double aisles and a transept that projects only slightly from the nave. Famous flying butresses brace the external walls of the choir and the side aisles. Three sculptured portals are deeply recessed in the west front and topped by a row of sculptures in niches extending across the facade. Centered above this facade is a huge traceried wheel window.

Additional Information: Sandstrom, Gosta E. *Man the Builder.* New York: McGraw-Hill, 1970; http://www.focusmm.com.au/~focus/fr_re_02.html.

Nourek Dam. *See* Nurek Dam.

Nurek Dam (Nurek, Tajikistan; 1980)

One of the highest dams in the world, the Nurek Dam is made of earth and rock. Although today Tajikistan, which is located in central Asia north of Afghanistan, is an independent republic, it was a part of the Soviet Union at the time the Nurek Dam was built. The dam is 1,040 feet (317 meters) high and 2,624 feet (800 meters) long. The dam's volume is 6,600,000 cubic yards (5,049,000 cubic meters), and the reservoir's capacity is 8,500,000 **acre-feet**. The dam is 47 miles (76 kilometers) south of Dushanbe, the capital of Tajikistan. The new town of Nurek was built in 1960, on the banks of the River Vakhsh. The dam can output 2.7 million kilowatt hours of electricity.

Additional Information: Louis, Victor and Louis, Jennifer. *Complete Guide to the Soviet Union.* New York: St. Martin's Press, 1991; Smith, Norman. *A History of Dams.* Secaucus, NJ: Citadel Press, 1972.

O

Oahe Dam (Pierre, South Dakota; 1963)

This earth dam is 246 feet (75 meters) high and 9,250 feet (2,821 meters) long. The dam has a capacity of 92,000,000 cubic yards (70,380,000 cubic meters), and its reservoir, the manmade Lake Oahe, holds 23,600,000 **acre-feet**. The dam's power capacity is 420,000 kilowatts. Lake Oahe runs 231 miles (372 kilometers) from Pierre, South Dakota, to Bismarck, North Dakota.

Additional Information: Farb, Peter. *The Story of Dams.* Irvington-on-Hudson, NY: Harvey House, 1961; Smith, Norman. *A History of Dams.* Secaucus, NJ: Citadel Press, 1972.

Oakland-San Francisco Bay Bridge, known officially as the James "Sunny Jim" Rolph Bridge (Oakland to San Francisco, California; 1936)

The Oakland-San Francisco Bay Bridge carries Interstate 80 over San Francisco Bay. The West Bay crossing connects San Francisco with Yerba Buena Island in the middle of the bay; it consists of two suspension spans with a total length of 9,200 feet (2,806 meters). The East Bay crossing consists of a 1,400-foot-long (427-meter) **cantilever**, five **through-trusses**, and 14 deck trusses for a total length of 11,000 feet (3,850 meters), not including the Oakland approach spans. A 1,800-foot (549-meter) connecting tunnel runs through Yerba Buena Island. The idea for the **suspension bridge** had been suggested as early as 1850, but the project did not truly begin until Congress approved a $73 million loan for the bridge's construction. Before the bridge was built, traffic was served by ferry boats that quickly lost business once the bridge opened in 1936, and especially after bridge tolls were dropped. The ferries stopped operating in 1940 when bridge traffic reached more than 30,000 vehicles daily. The bridge's official name—the James "Sunny Jim" Rolph Bridge—comes from James Rolph, the mayor of San Francisco from 1911–1931 and governor of California from 1931–1934.

Additional Information: DeLony, Eric. *Landmark American Bridges.* Boston: Bullfinch Press, 1993; Steinman, David B. *Famous Bridges of the World.* London: Dover Publications, 1953, 1961; Rose, Albert C. *Historic American Roads.* New York: Crown Publishers, 1976.

The Oakland-San Francisco Bay Bridge looking west toward San Francisco. Photo by Joy Padayhag, courtesy of the State of California Department of Transportation.

Oderteich Dam and Reservoir (Oberharz, Germany; 1721)

The Oderteich Dam was constructed between 1714 and 1721 in Oberharz in north central Germany, south of Braunschweig (Brunswick). The largest of what would eventually be 60 reservoirs and dams in the region to support mining operations, the Oderteich Dam is 495 feet (151 meters) long, 72 feet (22 meters) high, 52.5 feet (16 meters) thick at the crest, and 144 feet (44 meters) thick at its base. The dam is made of granite blocks sandwiching granitic sand; the spaces between the blocks are filled with earth and moss. Oderteich may be the oldest **rockfill dam** in Europe.

Additional Information: Smith, Norman. *A History of Dams.* Secaucus, NJ: Citadel Press, 1972.

Ogden-Lucin Cutoff Trestle (Promontory, Utah; 1904)

Designed and built by William Hood, this amazing 20-mile-long (32-kilometer) timber trestle carries the Southern Pacific Railroad straight across the Great Salt Lake in northern Utah. The bridge is built upon 38,256 timber piles. The bridge was raised in 1910 to accommodate the rising level of the lake, and is due to be raised again.

Additional Information: DeLony, Eric. *Landmark American Bridges.* Boston: Bullfinch Press, 1993.

Öland Island Bridge (Öland Island, Sweden; 1972)

This massive bridge connects Öland Island in the Baltic Sea off southeastern Sweden with the city of Kalmar on the mainland. With a length of 19,882 feet (6,064 meters), the Öland Island Bridge is one of Europe's longest spans.

Source: http://www.destination-oland.se/engelska/index-e.htm.

Old Bridge at Mostar. *See* Mostar Bridge.

Old London Bridge (London, England; 1176-1209)

Built by a monk named Peter Colechurch in the late twelfth and early thirteenth centuries, Old London Bridge over the River Thames was the first major masonry bridge. The bridge was architecturally unremarkable, but notable for its longevity—it carried traffic across the Thames for six centuries. The bridge had 19 stone arches, varying in length from 15 to 34 feet (4.6 to 10.4 meters). The total length of the structure was 936 feet (285 meters). The seventh span from the Southwark (southern) side was a **drawbridge** that could be hauled up and down to allow traffic to pass. A portcullis, an iron gate that could be raised or lowered, was built along one end of the bridge to close the city off from beseigers in time of war or internal disorder. The bridge's piers were wide and so thick that they caused fast **cutwaters** on either side, making the navigation of

a boat through the arches difficult; indeed, attempting to pass through the arches was known as "shooting the bridge." Constant currents in the Thames caused continuous erosion, and the piers were regularly in need of repair. The bridge was started in 1176 and finished in 1209, four years after Colechurch's death. He was buried in the chapel on the bridge. Isembert, a French monk who was renowned as a bridge builder, completed the project. Originally, the only buildings on the bridge were the gatehouses at each end, and the chapel in the middle. Eventually, about 100 houses and numerous shops were added along the sides of the bridge. In Elizabethan times, in the late sixteenth century, a London Bridge address was extremely fashionable, and many nobles lived in Nonesuch House, the best known of the bridge buildings. Built in 1579, Nonesuch House was brought from Holland piece by piece and placed on the drawbridge pier—probably the first prefabricated house. The parts of the house were fastened together with wooden pegs. Old London Bridge was replaced in the early nineteenth century by John Rennie's NEW LONDON BRIDGE.

Additional Information: Steinman, David B. *Famous Bridges of the World.* London: Dover Publications, 1953, 1961; for a description of both this bridge and earlier bridges at the site, see Lay, M.G. *Ways of the World.* New Brunswick, NJ: Rutgers University Press, 1992.

Old Mine Road (northeastern New Jersey to Kingston, New York; 1620-1650)

Probably the first American highway, this 105-mile (170-kilometer) road connected copper mines in Warren and Sussex counties in northeastern New Jersey with Kingston, New York, a port on the Hudson River about half way between New York and Albany. In the 1880s, the road was still considered to be part of the best route between Philadelphia and Boston. The road remained in use until 1940, well into the beginning of the automobile age.

Additional Information: Lay, M.G. *Ways of the World.* New Brunswick, NJ: Rutgers University Press, 1992.

Oldfield Covered Bridge (Newtown, Kings County, New Brunswick, Canada; 1910)

Constructed on Oldfield family property, this 98-foot-long (30-meter) **Howe truss** across Smith's Creek was designed to make traveling easier for local farmers who came to a grist mill operated by the Oldfield family. An earlier bridge had washed away in a storm. As a consequence, this bridge was built using cement **abutments**. The Oldfield Bridge was selected to appear in 1992 on a 25-cent piece issued by the Canadian Mint; each province chose its own theme for a coin to commemorate the 125th anniversary of the Canadian Confederation.

Additional Information: http://www.mi.net/users/alstondj/bridges.html.

Olympian Zeus (Olympia, Greece; after 450 B.C.)
The Greek sculptor Phidias, who also worked on the PARTHENON, created this statue of the god Zeus for the temple of Zeus at Olympia in northern Greece. The temple had been designed by the architect Libon, and the statue was intended to amplify the temple's majesty. The magnificence of the statue was recognized immediately, and it was soon considered one of the SEVEN WONDERS OF THE ANCIENT WORLD. In the second century A.D., the Greek writer Pausania described the statue as follows: "In his right hand a figure of Victory made from ivory and gold. In his left hand, his scepter inlaid with all metals, and an eagle perched on the scepter. The sandals of the god are made of gold, as is his robe." In the first century A.D., the Roman emperor Caligula attempted to move the statue to Rome; when the scaffolding around the statue collapsed, the effort was abandoned. The temple itself survived until the fifth century A.D. when it was destroyed by fire. The statue survived the fire, and was moved by wealthy Greeks to a palace in Constantinople, where it was destroyed by a fire in A.D. 462. Phidias had developed a method of building enormous gold and ivory statues by erecting a wooden frame that held sheets of metal and ivory. His workshop in Olympia still exists and is identical in size and orientation to the temple of Zeus. Phidias created the different parts of the statue in his workshop and assembled them in the temple. The statue's base was 3 feet (0.9 meters) high and approximately 20 feet (6.1 meters) long on each side. The statue itself was about 40 feet (12.2 meters) high.

Additional Information: Ashmawy, Alaa K., ed., "The Seven Wonders of the Ancient world," at http://pharos.bu.edu/Egypt/Wonders/zeus.html.

Olympic Tower (Munich, Germany; 1972)
Built in honor of the 1972 Olympics, which were held in Munich, the 290-meter (951-foot) tower stands in the middle of the 850,000-square-meter Olympic Park. Since its construction, it has been used as a television transmission tower.

Additional Information: http://www.centrepoint.com.au/tower/index.html.

Oresund Link (between Denmark and Sweden; expected completion in 2000)
The Oresund Link is a complex trans-water connection between Denmark and Sweden that will comprise a tunnel, a bridge, and a viaduct to connect the two other structures. The link will cross the sound that connects the Danish island of Sjaelland, on which Copenhagen is located, with the southern tip of Sweden. The overall length of the link will be 16.2 kilometers (10 miles). Oresund Bridge is being built by two consortiums—Oresundkonsortiet on the Danish side, and SVEDAB on the Swedish side. The bridge will reach a maximum height of 57 meters (185 feet) and will be 300 meters (984 feet) wide to allow for four lanes of motor traffic and two railroad tracks. The connecting viaduct will be 3.9 kilometers (2.4 miles) long. The 3.7-kilometer (2.3-mile) tunnel will be one of the largest immersed tunnels in the world. The tunnel will be made of 20 concrete sections, each 175 meters (574 feet) long and 40 meters (131 feet) wide. The concrete sections will be constructed off-site, and then taken to an adjacent basin of water that will act like a canal lock to raise the water level to the level of the adjacent sea. The sections will then be towed to the site, carefully sunk into a prepared trench, and connected underwater. The tunnel will carry a two-track railroad, two lanes for vehicular traffic, and a service gallery.

Additional Information: "News: Nordic Neighbors to Sink Link." *Civil Engineering* 65:11 (November 1995), pp. 20-22; http://www.oresund.com/bron/broinfo.htm.

Oroville Dam (Oroville, California; 1967)
At 770 feet (235 meters), this earth and rock dam is one of the highest dams in the world. The dam is 6,850 feet (2,089 meters) long, has a volume of 80,600,000 cubic feet (20,553,000 cubic meters), and a reservoir capacity of 3,500,000 **acre-feet**. The dam is located about 70 miles (113 kilometers) north of Sacramento in northern California.

Additional Information: Smith, Norman. *A History of Dams.* Secaucus, NJ: Citadel Press, 1972.

Ostankino Moscow Tower (Moscow, Russia; 1967)
At 540 meters high (1,771 feet), the Moscow Tower is one of the highest towers in the world. The tower boasts an observation deck at 337 meters (1,105 feet), and three revolving restaurants with spectacular views of the city and its environs.

Additional Information: http://www.centrepoint.com.au/tower/index.html.

Otira Railroad Tunnel (through Arthur's Pass, New Zealand; 1923)
Begun in 1907, this 5.5-mile (8.8-kilometer) tunnel through the mountains of the South Island of New Zealand presented an interesting problem because the site was so difficult to reach; in winter, Arthur's Pass was inaccessible. At each of the two headings, small towns were built for the workers and engineers. Problems with the weather began even before work started, with surveyors frequently stranded by snowfall in the winter, flooding in the spring, and intense heat in the summer. After work began, far more timbering and support was needed than had been foreseen. After a series of worker strikes, the original contractors quit the project. New Zealand's Public Works Department took over construction and progress continued slowly until

the tunnel was holed through in July 1918. It took another five years to complete the project. Despite the construction difficulties, the tunnel is a major achievement, dropping from east to west at a 1 to 33 decline from 2,435 feet to 1,586 feet (743 to 484 meters) above sea level.

Additional Information: Beaver, Patrick. *A History of Tunnels.* Secaucus, NJ: Citadel Press,1973.

Outerbridge Crossing (Staten Island, New York, to Perth Amboy, New Jersey; 1928)

The Outerbridge Crossing crosses the Arthur Kill, a narrow river tributary flowing from New York Harbor to Raritan Bay, and connects Perth Amboy, New Jersey, with Tottenville on New York's Staten Island. Designed by Othmar Ammann, this steel truss **cantilever** is a design cousin to the GOETHALS BRIDGE, which crosses the Arthur Kill about 6 miles (9.7 kilometers) to the north. The bridge was named after Eugenius H. Outerbridge, the first chairman of the Port Authority of New York and New Jersey, which built and operates the bridge. The Outerbridge Crossing has a center span of 750 feet (229 meters), a main span of 62 feet (19 meters) in width, and an overall length of 10,140 feet (3,093 meters).

Additional Information: Bishop, Gordon. *Gems of New Jersey.* Englewood Cliffs, NJ: Prentice-Hall, 1985.

Owyhee Dam (near Adrian, Oregon; 1932)

The Owyhee Dam was the first dam in the United States to exceed 400 feet (122 meters) in height. It crosses the Owyhee River in eastern Oregon. A concrete, gravity arch dam, Owyhee's reservoir, Lake Owyhee, has a maximum capacity of 1,183,300 **acre-feet.** The dam was designed to carry about three-quarters of the reservoir's hydraulic loading by arch action, and the rest by gravity action. Beginning with a width at its base of 265 feet (81 meters), the dam tapers upward to a width of 30 feet (9 meters) at its crest. The dam is operated by the Bureau of Reclamation.

Additional Information: Kirby, Richard Shelton et al. *Engineering in History.* New York: Dover Publications, 1990; http://www.usbr.gov/cdams/dams/owyhee.html.

P

Palace of Fine Arts-Exploratorium (San Francisco, California; 1915)

Designed by Bernard R. Maybeck to showcase modern art, the Palace of Fine Arts in San Francisco was built as a temporary structure for the 1915 Panama-Pacific International Exhibition. Maybeck started with the idea of a Roman ruin and amplified it with his ideas of art and architecture to create a model for his structure. Covering more than 3 acres (1.2 **hectares**), the structure was an immediate hit. A Roman style building, although with more modern romantic elements, the beautiful structure was reflected in a nearby lagoon. Because it was essentially a temporary structure, the building was made of wood covered with "staff," a pliable mixture of plaster and fiber. Various artistic elements of the building were designed by others. William Merchant, an architect in Maybeck's employ, contributed several of the decorative elements. Bruno L. Zimm was responsible for the eight panels in low relief beneath the dome of the impressive rotunda; the panels symbolize Greek culture and its love for artistic expression. Boxes on the colonnade were surrounded by weeping figures, the work of sculptor Ulric Ellerhusen; the figures suggest contemplation to some, and the melancholy of life without art to others. Maybeck had intended the Palace to be surrounded by redwood trees, but they were never planted. A movement to preserve the building took shape even before the end of the Exposition in December 1915, and the building continued in service for several more years. For a while, the Palace housed an art exhibit, and, during the Depression in the 1930s, WPA (Works Progress Administration) artists were hired to replace the murals on the ceiling of the rotunda. The building found other uses as time went by. In 1934, it became the home for 18 tennis courts. During World War II, the U.S. Army used it as a garage for trucks and jeeps. After the war, it was used as a Parks Department warehouse, a telephone book distribution center, and a temporary Fire Department Headquarters. Maybeck himself said that he would like the building torn down, and the rotunda left, surrounded by redwood trees. He suggested an altar on the rotunda with the figure of a praying maiden, so that the rotunda would become its own grave. Maybeck changed his mind before his death, and the late 1950s saw a concerted effort to save the structure. The Palace was demolished in 1964 and rebuilt to original specifications, with sturdier materials. In 1969, under the guidance of physicist Frank Oppenheimer, the Palace began to house a series of science exhibits, and is today one of the finest attractions in San Francisco.

Additional Information: http://www.exploratorium.edu/ Palace_History/Palace_History.html#Maybeck. This is an extensive, detailed, and fascinating history of the Palace and of the Exploratorium.

Pan American Highway System (North and South America; begun 1920s)

At the Fifth International Conference of American States in 1923, a suggestion was made to build a system of highways, over 15,000 miles (24,000 kilometers) long, to link the capitals of all the American nations. This roadway is now complete from central America to Canada, and largely complete in South America. The section of the system between the United States and the PANAMA CANAL is called the Inter-American Highway. The road has not yet been built from the South American side of the canal into Colombia.

Additional Information: *Academic American Encyclopedia,* (electronic version). Danbury, CT: Grolier Electronic Publishing, 1991.

Panama Canal (across the Isthmus of Panama; 1914)

In 1914, George Washington Goethals succeeded in building the long-desired canal across the narrow isthmus of Panama to connect the Atlantic Ocean with the

The east chamber of the Miraflores Lock on the Pacific end of the Panama Canal. Photo courtesy of the Library of Congress, World's Transportation Commission Photograph Collection.

Pacific. Ferdinand de Lesseps, the French engineer who began the SUEZ CANAL in Egypt, started work on a Panama canal in 1881, but unforeseen difficulties, a lack of funds, and, especially, disease among the workers, led to the collapse of the project. In 1901, the U.S. Congress, in an action strongly endorsed by President Theodore Roosevelt, authorized the purchase of the rights to the Panama route from the French company that held them. In January 1903, the United States negotiated a treaty with Colombia, which controlled the isthmus; the treaty, which gave the U.S. a strip of land on the isthmus for an initial cash payment of $10 million and a yearly annuity, was rejected by the Colombian senate. Pro-canal groups in Panama led an insurrection against the Colombian government in November 1903; the U.S. sent warships to the region to prevent Colombian troops from suppressing the uprising, and the independent Republic of Panama was born. Within weeks of the uprising, the U.S. had signed the Hay-Bunau-Varilla Treaty with the new republic; the treaty restated the terms of the earlier Colombian agreement and gave the U.S. control of the canal zone in perpetuity. From 1904 to 1907, Goethels and the commission he headed built construction facilities, conducted surveys, and worked to eradicate malaria and yellow fever in the region, a vital accomplishment, for disease among the workers had doomed the earlier French effort. Construction of the canal itself took seven years; the waterway opened on 15 August 1914, when the U.S.S. *Ancon* became the first American cargo ship to transit the Panama Canal. The canal cost over $336 million to build and required the excavation of 240 million cubic yards (over 183 million cubic meters) of earth. The canal runs south-southeast from Limon Bay at Colon on the Atlantic to the Bay of Panama at Balboa on the Pacific, a distance of over 40 miles (64 kilometers) from shore to shore and of almost 51 miles (67 kilometers) between channel entrances. Oddly, the Pacific terminus is about 27 miles (43 kilometers) east of the Atlantic end of the canal. From Limon Bay, a ship is raised by the three Gatun locks to an elevation of 85 feet (26 meters) above sea level. The ship then crosses Gatun Lake to reach the Gaillard Cut, where the vessel crosses the continental divide. After that, the ship is lowered by Pedro Miguel lock to Miraflores Lake, and then by the two Miraflores locks to sea level. Passage through the canal can take six to eight hours. Two of the locks, Miraflores (on the Pacific side) and Gatun (on the Atlantic side), have spectator stands and bilingual commentators to explain what is happening as ships pass through. The canal lost some of its strategic importance in the 1950s with the development of supertankers and other ships too large to pass through it. Until 1979, the 553-square-mile (1,432-square-kilometer) Canal Zone was administered by the United States. Two U.S-Panamanian treaties negotiated by the Carter administration and narrowly ratified by the U.S. Senate in 1978 turned administration of the zone over to Panama. In 2000, control of the canal itself will, by action of the 1978 treaties, pass to Panama, which will guarantee the continued neutral operation of the waterway.

Additional Information: http://iaehv.iaehv.ni/users/grimaldo/canal.html; Haskins, Frederic J. *The Panama Canal.* Garden City, NY: Doubleday, Page & Company, 1913.

Paris Metro (Chemin de Fer Métropolitain or Metropolitan Railway of Paris) (Paris, France; 1900)

The first line of this underground railroad was opened only two years after work began in 1898. Much of the construction was cut and cover, and several unique problems and solutions developed during the line's construction. Workers occasionally cut into quicksand and had to dig extremely deep foundations for supporting columns. The famed PARIS SEWER SYSTEM was also in the way of construction from time to time, and many sewer lines had to be shifted. In some cases, workers found

A train on the Paris Metro underground rail system.

themselves operating in marshy grounds near the River Seine that were too soft to support even the tunneling work being done, and refrigeration plants had to be built to freeze the ground before it could be worked.

Additional Information: Epstein, Sam and Epstein, Beryl. *Tunnels.* Boston: Little Brown & Company, 1985.

Paris Opera House (Paris, France; 1875)

Designed and built by Charles Garnier, the famed Paris Opera House is the setting for the famous "Phantom of the Opera." The structure is the largest opera house in the world, although seating is limited to only 2,156 people because other parts of the building were made for uses other than performances. Its foyers, lounges, and grand staircase were intended for use on state occasions, and for balls, festivals, and feasts. The famed building sits on a site of about 3 acres (1.2 **hectares**). Since the founding of the Royal Academy of Music in 1669, Paris has had 13 opera houses. The New Opera House was planned by Napoleon III and his city designer, Baron Haussmann, in 1858. Haussmann constructed the Avenue de l'Opéra, with the Opera House at one end, and the LOUVRE at the other. In 1860, the romantic young architect, Charles Garnier, won the competition to build the new structure. Garnier's grand designs were both revolutionary and successful, including his design of the great staircase, a steel framework with a gigantic amount of marble **cantilevered** outward. The magnificent stage, 175 feet (53 meters) wide and 85 feet (26 meters) deep, can be deepened another 65 feet (20 meters) by opening up the dance salon behind it. When used for dance, the room is dominated by a gigantic St. Gobain mirror, 33 feet (10 meters) across and 23 feet (7 meters) high with a ballet barre in front of it. One of the sub-basements (there are at least five) had accommodations for up to 20 horses for use in spectacular performances. An underground stream ran through the property. Garnier drained the stream and then laid a foundation; he next let the stream form a lake underneath the opera house. For a while, the lake was a source of hydraulic power for some of the machinery. Today, the lake is kept as a source of water in case of fire, and is drained every several years so the foundation can be inspected. In 1896, the magnificent chandelier of the auditorium fell, though perhaps not because of the work of any phantom. The opera house is a romantically exotic structure that has become the subject of many legends and tall tales.

Additional Information: George Perry, "The New Opera House," from *The Complete Phantom of the Opera*, Wordsworth Books, excerpted at http://phantom/skywalk.com/operahouse/opera_house.html.

Paris Sewer System (Paris, France; mid-nineteenth century)

The Paris Sewer System was constructed primarily during the time of Napoleon III (reigned 1852-1870). The system has 1,305 miles (2,100 kilometers) of underground tunnel-sewers. The sewers are aimed primarily at a treatment plant at Achères, the largest biological purification station in Europe. The tunnels are also used to transport drinking and industrial water, and as conduits for telephone cables. The sewers are historically known as large enough to live in; during times of revolution, the sewers became the roads for rebels to travel through the city. Tours are available showing an overflow outlet, regulatory reservoirs, and other parts of the engineering. The tours are not available during storms, heavy rainfalls, or when the Seine is flooding.

Source: *Michelin Paris.* Paris: Michelin et Cie, 1992.

Parthenon (Athens, Greece; 447-432 B.C.)

The Parthenon, a temple sacred to the goddess Athena, is one of the masterpieces of ancient Greek architecture and sculpture. The Parthenon, today in partial ruin, stands upon the Acropolis, a rocky hill in Athens upon which the Athenians in the sixth and fifth centuries B.C. built some of their most important and most beautiful public buildings and monuments. The Acropolis had acquired walls some time before the sixth century B.C., and seems to have served earlier as a fortress and place of refuge. The hill was about 260 feet (79 meters) high and had a flat oval top that covered about 8 acres (3.2 **hectares**). The Parthenon, which in Greek means "the virgin's place," replaced an earlier temple to Athena Parthenos that had been destroyed by the Persians, along with many other temples and monuments, in 480 B.C. when they burned the Acropolis. The citizens of Athens, under the leadership of Cimon and Pericles, rebuilt the temples and monuments. In 447 B.C., work began on the Parthenon, with Ictinus and Callicrates selected as architects for the structure, and Phidias as supervisor of sculpture. Phidias's 40-foot-high (12.2-

meter) gold and ivory statue of Athena Parthenos was dedicated in 438 B.C., and the entire temple was completed in 432 B.C. The rectangular temple has eight Doric columns at each end and 17 on each side. The building

The Parthenon, a temple of the goddess Athena, was built in Athens in the fifth century B.C.

stands on a base three-steps high. The temple is 228 feet (70 meters) long, 101 feet (31 meters) wide, and 65 feet (20 meters) high. At the front and rear of the building, the outer colonnade gave way to two porticos. The cella, or inner structure of the temple, contained a two-tiered colonnade that held up the roof timbers and divided the interior space into a high central nave bound on three sides by narrower aisles. The statue of Athena stood at the west end of the nave. The upper part of the cella walls and the friezes above the porticos formed a continuous band of sculpture around the whole building; the frieze represented the Pananthenaic procession held every fourth year in honor of the goddess. Thirty-nine feet (12 meters) above the portico floor, this magnificent frieze was 524 feet (160 meters) long and over 3 feet (almost one meter) wide. The procession includes 350 human figures and 125 horses. Part of the frieze, about 335 feet (102 meters), still exists; the western portion is still in place and the rest of the existing section can be seen in the British Museum in London. In the fifth century A.D., with Greece part of the Christian Byzantine Empire, the Parthenon became a church. In the fifteenth century, with Greece under the rule of the Ottoman Turks, the temple served as a mosque. A good portion of the temple was destroyed in 1687. When a Venetian cannonball hit the gunpowder the Turks had stored in the building, the temple exploded, killing 300 men and doing considerable damage to the structure. In 1801, Lord Elgin, the British Ambassador to the Ottoman Empire, received permission to remove "a few blocks of stone with inscriptions and figures." What he actually took was a good part of the frieze. These por-

tions of the pediment, sculptures, and reliefs became known as "The Elgin Marbles," and are displayed in sequence in the British Museum. Efforts to rebuild the Parthenon, and the Acropolis, began as soon as Greece achieved independence in the early nineteenth century. The current reconstruction project, which involves efforts to protect the building and the remaining statuary and frieze sections from the effects of urban pollution, should be finished by 2005.

Additional Information: Mazie, David. "The Glory That Was Greece; Restorers Battle Time and Elements to Save Ancient Acropolis." *Los Angeles Times*, 7 February 1993, A10; http://www.elibrary.com/cgi-bin/hhweb/hhfetch?32411045X0Y670:QOO2:D013.

Pas du Riot Dam (St Etienne, France; c. 1875)

The Pas du Riot Dam crosses a tributary of the Loire River about 2 miles (3.2 kilometers) upstream from the well-known FURENS DAM, of which it is a replica. Both dams are located in southeastern France near the town of St. Etienne. The Pas du Riot Dam is 113 feet (35 meters) high; it was built to add to the water-supply system of Saint Etienne.

Additional Information: Smith, Norman. *A History of Dams.* Secaucus, NJ: Citadel Press, 1972.

Pathfinder Dam (near Alcova, Wyoming; 1906-1909)

The Pathfinder Dam, along with the BUFFALO BILL DAM near Cody, Wyoming, is a large masonry-arch dam that was built for irrigation reservoirs by the U.S. Reclamation Service. The Pathfinder Dam crosses the North Platte River in east central Wyoming, southwest of Casper near Alcova. For political reasons, the Reclamation Service sought, at the time, to build in and affect as many states as possible; the dam thus provides irrigation water to both Wyoming and Nebraska. The Pathfinder Dam is 218 feet (66 meters) high with a maximum base width of 94 feet (29 meters); it stands in a deep, narrow gorge of granite.

Additional Information: Jackson, Donald C. *Great American Bridges and Dams.* Washington, DC: The Preservation Press, 1988; Smith, Norman. *A History of Dams.* Secaucus, NJ: Citadel Press, 1972.

Pattison Tunnel, also known as Montgomery Bell's Tunnel (Cheatham County, Tennessee; 1818)

Montgomery Bell cut the Pattison Tunnel across a loop in the Harpeth River in Cheatham County in north central Tennessee. The Pattison is the first American tunnel dug through rock. Built with slave labor, the tunnel was built to create a waterfall that would drive a steel forge. The tunnel is about 300 feet (91.5 meters) long, and creates a waterfall of about 18 feet (5.5 meters).

Additional Information: Schodek, Daniel L. *Landmarks in American Civil Engineering*. Cambridge, MA: MIT Press, 1987.

Peace Bridge (Buffalo, New York, to Fort Erie, Ontario, Canada; 1927)

Commemorating 100 years of peace between the United States and Canada, this vehicular bridge made Buffalo the major port of entry into the United States from Canada. The bridge crosses the Niagara River near its mouth as it flows north from Lake Erie toward the Niagara Falls 25 miles (40 kilometers) away.

Additional Information: http://www.grasmick.com/peacepoe.htm.

Peak Forest Canal (England; 1800)

Benjamin Outram built the Peak Forest Canal between 1794 and 1800; the canal includes two reservoirs and two dams, both near Whaley Bridge. Todd Brook Dam, west of Whaley Bridge, has the greatest height of all the late eighteenth-century English canal dams; it is 70 feet (21 meters) high at its center. Its crest is 7 feet (2.1 meters) wide and the dam is approximately 700 feet (214 meters) long. The dam is a huge earth embankment with a 45-degree slope on both faces, and a base that is about 200 feet (61 meters) thick. The dam has a spillway on one side, and two low-level outlets for feeding the canals. One of the outlets has escapes so that extra water can be sent into the spillway channel. Coombs Dam is 2 miles (3.2 kilometers) southeast of Whaley Bridge. An earth bank dam, Coombs is 52 feet (16 meters) high, 1,000 feet (305 meters) long, and slightly curved. The dam has a single low-level outlet for feeding the canal, and two overflows. Both overflows can be adjusted with **sluices**, an unusual feature. The canal was built to carry limestone from the Derbyshire hills. The canal is about 21 miles (34 kilometers) long, running from Marple south to Whaley Bridge, and then north to Dukinfield Junction.

Additional Information: Smith, Norman. *A History of Dams*. Secaucus, NJ: Citadel Press, 1972; http://www.blacksheep-org/canals.

Pei Yun Ho. *See* Grand Canal [China].

Pennsylvania Canal (Columbia to Pittsburgh, Pennsylvania; 1834)

This important industrial canal on the Susquehanna River crossed the Allegheny Mountains, running from Columbia in southeastern Pennsylvania to Pittsburgh in western Pennsylvania. Because the mountains were so steep, the canal did not use locks. Rather, railroad tracks ran directly into the canal on either side of the

mountains. Trains pulled several boats on carriages over the mountain, and deposited the boats back into the water on the other side. The canal itself did not go through the mountains.

Additional Information: Russell, Solveig Paulson. *The Big Ditch Waterways*. New York: Parents' Magazine Press, 1977.

Pennsylvania Turnpike (Harrisburg to Pittsburgh, Pennsylvania; 1940)

The first stretch of the Pennsylvania Turnpike, a toll road running from Harrisburg in southeastern Pennsylvania to near Pittsburgh in western Pennsylvania, opened to traffic on 1 October 1940. The first superhighway built in the United States, construction was handled by the Pennsylvania Turnpike Commission, which sold bonds and acquired a state loan to build the 159-mile (256-kilometer) stretch of road. Given the designation I-76 (Interstate 76), the turnpike was the first link in what eventually became the DWIGHT D. EISENHOWER INTERSTATE AND DEFENSE HIGHWAY SYSTEM. Today the turnpike runs the length of the state, leaving Pennsylvania on the west near the town of Petersburg, Ohio, and crossing the Delaware River into New Jersey near the town of Levittown, Pennsylvania. The turnpike runs through several old railroad tunnels along its route. The twin Allegheny Tunnels in the southwestern part of the state near New Baltimore are each 6,072 feet (1,852 meters) long, and are among the longest vehicular tunnels in North America. The turnpike passes through the Tuscarora Tunnel in south central Pennsylvania to cross the Tuscarora Mountains and, about 10 miles (16 kilometers) to the east, through the Kittatinny Tunnel and the Blue Mountain Tunnels to cross the Kittatinny Mountains. With lengths of 4,435 feet (1,353 meters) each, the twin Blue Mountain Tunnels are also among the longest land vehicular tunnels in North America. No speed limits were imposed when the turnpike was first built, and highway speeds in the fast lane reached 90 miles (145 kilometers) per hour. Today, the maximum posted speed on the turnpike is 55 miles (88 kilometers) per hour.

Additional Information: *Academic American Encyclopedia*, (electronic version). Danbury, CT: Grolier Electronic Publishing, 1991; Cupper, Don. *The Pennsylvania Turnpike*. Lebanon, PA: Applied Arts, 1990; Patton, Phil. *Open Road: A Celebration of the American Highway*. New York: Simon and Schuster, 1986.

Pennypack Bridge. *See* Frankford Avenue Bridge.

Pentagon (Arlington, Virginia; 1943)

Perhaps the world's most recognizable building, the Pentagon, located in northern Virginia across the Potomac from Washington, D.C., houses the U.S. Department of Defense. Construction began in the 1930s during the Depression, when the site was expected to become a hospital. During World War II, the Pentagon was selected to be the consolidation point for 14 different Defense Department offices in Washington. Actually 25 different buildings connected by corridors to form a five-sided structure, the Pentagon houses more than 25,000 employees and covers an area of 34 acres (13.8 **hectares**). The structure has five stories above ground, and an unknown number below (the exact details are classified). Having been rushed into service during the war, the Pentagon is constantly undergoing rebuilding and modernization.

Additional Information: http://www.dgsys.com/-mwardel/pent.html.

Permanent Bridge (Philadelphia, Pennsylvania; 1806)

Built by Timothy Palmer, the Permanent Bridge carried Market Street in Philadelphia over the Schuylkill River. Built to replace an earlier bridge of floating logs, the Permanent Bridge was notable for its extensive use of timber. It had three spans—two 150-foot (46-meter) side spans and a 185-foot (56-meter) center span. Because of the size of the **abutments** and other elements, the total length of the bridge was 1,300 feet (397 meters). The Permanent Bridge may also have been the first covered bridge in the United States. Palmer had not planned on covering the bridge, but when Judge Richard Peters, president of the bridge company, suggested it, Palmer quickly agreed. The western pier was unusual in that it went to bedrock nearly 42 feet (13 meters) below the river's surface. The bridge required significant alterations in the 1850s and was destroyed by fire in 1875.

Additional Information: Kirby, Richard Shelton et al. *Engineering in History.* New York: Dover Publications, 1990; Lay, M.G. *Ways of the World.* New Brunswick, NJ: Rutgers University Press, 1992; Overman, Michael. *Roads, Bridges and Tunnels.* Garden City, NY: Doubleday & Company, 1968; Steinman, David B. and Watson, Sara Ruth. *Bridges.* New York: Dover Publications, 1941, 1957.

Petronas Towers (Kuala Lumpur, Malaysia; 1996)

At 1,483 feet(452 meters), the Petronas Towers in Kuala Lumpur are taller than the SEARS TOWER in Chicago or the twin towers of the WORLD TRADE CENTER in New York City. A double tower linked by a concourse level at the base, the structure was not originally designed to be the tallest in the world. However, as the design was being finalized, the Malaysian government asked for spires on the buildings. The addition of 20 feet (6.1 meters) of spires made the building the tallest in the world. According to Malaysia's prime minister, the building is part of an attempt to "transform Malaysia into a developed nation by the year 2020, and it will definitely put Kuala Lumpur on the world map." The structure is named after the Malaysian oil company, Petronas, which will occupy one of the two towers. The structure has slightly less than 240,000 square meters of usable space for offices and related facilities. The tow-

Petronas Towers in Kuala Lumpur, Malaysia. Photo courtesy of J. Apicella/Cesar Pelli & Associates, Inc.

ers are linked by a 58.4-meter (192-foot) double-decker sky-bridge at levels 41 and 42, 175 meters (574 feet) above street level. The towers also include an 850-seat concert hall designed to the highest professional standards, a Petroleum Discovery Center, a reference library for energy- and petroleum-related topics, and an art gallery. Building the foundation for the massive structure was an engineering challenge. The foundation was built on 208 "barrette piles" of 2.8 by 1.2 meters (9.2 by 3.9 feet), piled to depths ranging from 60 to 115 meters (197 to 377 feet). On top of that, the builder constructed a 4.5-meter-thick (15-foot) raft-foundation. The foundation for Tower One took a continuous pour of concrete that lasted a record-breaking 54 hours; the foundation for Tower Two took 10 hours less.

Additional Information: Petronas press release reprinted in http://www.jaring/my/petronas/latest/press2.html; Robison, Rita. "Malaysia's Twins: High Rise, High Strength." *Civil Engineering* 64:7, (July 1994), pp. 63-65; http://www.jaring/my/petronas/compro/twintwrs/twintwrs.html; Pacelle, Mitchell. "U.S. Architects in Asia: Only Way to Go Is Up." *Wall Street Journal*, 21 March 1996, B:1.

Pharos at Alexandria. *See* Lighthouse at Alexandria.

Philadelphia-Camden Bridge. *See* Benjamin Franklin Bridge.

Pian de Setta Tunnel (Apennine Mountains, Italy; 1928)

Like the GREAT APENNINE TUNNEL, the Pian de Setta Tunnel is part of the trans-Apennine rail line running from Grizzana to Vernio in north central Italy. Over 10,000 feet (3,051 meters) long, the tunnel was drilled through difficult ground. The workers encountered pockets of methane gas which slowed the advance tremendously. Work began in 1921 from the northern heading, and in 1925 from the southern heading. The tunnel was finally holed through in 1928.

Additional Information: Beaver, Patrick. *A History of Tunnels.* Secaucus, NJ: Citadel Press, 1973.

Pine Brook Arch (Central Park, New York City; 1861)

Designed by Calvert Vaux and Jacob Wrey Mould, Pine Brook Arch is one of five **cast-iron** bridges in Central Park. Except for DUNLAPS CREEK BRIDGE in Pennsylvania, these five Central Park structures are the oldest cast-iron bridges in the United States.

Additional Information: DeLony, Eric. *Landmark American Bridges.* Boston: Bullfinch Press, 1993.

Plauen Stone Arch Bridge (Plauen, Germany; 1903)

The Plauen Stone Arch Bridge crosses a tributary of the Elbe in east central Germany. When constructed, the bridge's 295-foot (90-meter) length made it the longest stone **arch bridge** ever built.

Additional Information: Kirby, Richard Shelton et al. *Engineering in History.* New York: Dover Publications, 1990.

Plougastel Bridge, also known as the Albert-Louppe Bridge (Brest, France; 1930)

Eugene Freyssinet's masterpiece, the Plougastel Bridge, crosses the Elorn River as it flows into the harbor of Brest on the western tip of the Breton peninsula. The attractive Plougastel Bridge has three 612-foot (187-meter) arches. When completed in 1930, its three arch spans were the longest ever built in concrete. The bridge carries a single-track railway below and a 26.2-inch (67-centimeter) wide roadway above. Its piers and **abutments** rest on rock. The roadway on top of the arches is connected to the arches with four vertical slabs for each half arch. The bridge is made of extremely large, hollow concrete-box arches that reach a height of 90 feet (27 meters) at the centers. The arches were cast on top of a floating wooden arch that was used as a frame for the arches; after the concrete was poured and dried, the wooden arch was removed and then taken to the location for the next arch.

Additional Information: Brown, David J. *Bridges.* New York: Macmillan, 1993; Kirby, Richard Shelton et al. *Engineering in History.* New York: Dover Publications, 1990.

Poinsett Bridge (Tigerville, South Carolina; 1820)

Poinsett Bridge carries State Route 42 over Little Gap Creek in the up-country of western South Carolina. Built by Joel Poinsett for the South Carolina Board of Public Works, this stone and masonry Gothic arch was originally built on the road linking Charleston and Greenville. The bridge supports are made of wedge-shaped rocks that are so tightly fitted no concrete is needed to hold them together. Poinsett left his position before the bridge was completed to serve in a number of federal positions, including ambassador to Mexico and U.S. Secretary of War under President Martin Van Buren from 1837 to 1841. (On his return from Mexico, Poinsett introduced the United States to a flower native to Mexico; the flower is now known as the Poinsettia.) The bridge was completed under the supervision of Abram Blanding, acting commissioner of the South Carolina Board of Public Works.

Additional Information: DeLony, Eric. *Landmark American Bridges.* Boston: Bullfinch Press, 1993.

Point Pleasant Bridge, also known as the Silver Bridge (Point Pleasant, West Virginia; 1928)

The Point Pleasant Bridge crossed the Ohio River, connecting Point Pleasant, West Virginia, with southern Ohio. It was also known as the Silver Bridge because of the aluminum paint used on the bridge to prevent rusting. The bridge collapsed at 5:00 p.m. on 15 December 1967, when it was loaded with both rush-hour traffic and Christmas shoppers. Three spans, totalling 1,460 feet (445 meters), began to fall, carrying 37 vehicles into the water below. The collapse started at the western span, twisting the bridge in a northerly direction (indicating that a member of the north-side truss had failed). The span crashed and folded over on top of the fallen cars and trucks. The east tower, overloaded with the weight of the center span, fell westward into the center of the river, carrying the center span with it. The west tower then collapsed backward toward Point Pleasant. The disaster took 46 lives. The immediate cause of the failure was the weakening of a corroded **eyebar**; the chains of the bridge were made of 50-foot (15-meter) links with eyebars. Unlike modern bridges made of steel cables that can survive damage to an individual cable, the Point Pleasant Bridge collapsed because of a single failed connection. The Saint Mary's Bridge, a twin structure that also crossed the Ohio, was dismantled in 1969 to prevent a similar disaster.

Additional Information: Matthys, Levy and Salvadori, Mario. *Why Buildings Fall Down.* New York: W.W. Norton & Company, 1992.

Pons Aelius. *See* Ponte Sant'Angelo.

Pons Fabricus, also known as the Ponte Fabrico or Ponte a Quattro Capi (Rome, Italy; 62 B.C.)

The Pons Fabricus, now known as the Ponte a Quattro Capi, connects the banks of the Tiber River in Rome

The Pons Fabricus, a first-century B.C. bridge, crosses the Tiber at Rome.

with the island of Aesculapius. Built by Lucius Fabricus in the first century B.C., the bridge is the oldest in Rome and is still in service. It has two arches, each 90 feet (27 meters) long, an exceptional length by Roman standards. Between the two arches is a smaller, weight-relieving central arch. Other supporting arches once stood at the ends, but are now replaced by **abutments**. The structure's current name derives from a statue on the bridge of Janus, the four-headed god of entrances. Ironically, the Romans spread out payment to Fabricus for the bridge over a period of 40 years, presumably to ensure that the bridge would last.

Additional Information: Brown, David J. *Bridges*. New York: Macmillan, 1993; Lay, M.G. *Ways of the World*. New Brunswick, NJ: Rutgers University Press, 1992; http://www.iol.it/tci.sdp/html/eo706.htm.

Pons Hadriani. *See* Ponte Sant' Angelo.

Pons Sublicus (or Sublicious) (Rome, Italy; seventh century B.C.)

Built by Ancus Martius, the Pons Sublicus, which means "bridge of piles," crosses the Tiber in Rome. This timber-**beam bridge** was made famous by Horatius, who held the bridge against Etruscan invaders in 598 B.C. The Horatius story is retold in the contemporary poem, "Horatio at the Bridge." The Pons Sublicus was replaced with a stone arch in A.D. 350. With the Pons Sublicus, the Romans began the tradition of giving the position of chief bridge builder

to a high priest known as pontifex. One of the current titles of the pope is pontifex maximus or pontiff.

Source: Lay, M.G. *Ways of the World*. New Brunswick, NJ: Rutgers University Press, 1992.

Pont Albert-Louppe. *See* Plougastel Bridge.

Pont Alexandre III (Paris, France; 1900)

This fascinating **arch bridge** across the Seine River in Paris was constructed for the 1900 Paris World's Fair. The simple appeal of the bridge's attractive arch is overwhelmed by the incredible amount of decoration found everywhere on the structure. At least 15 artists contributed the garlands, winged horses, gilded cupids, and other decorative motifs that comprise the bridge's belle epoque style of optimism. Dedicated to the hope of friendship between France and Russia, at a time when both countries were wary of the growing power of imperial Germany, the bridge is named for Russia's Czar Alexander III, who died in 1894. Both the Russian and French coats of arms decorate the bridge's facades.

Additional Information: Dunlop, Fiona. *Fodor's Exploring Paris*. 3rd ed. New York: Fodor's Travel Publications, 1993.

Lamps and statuary lining the Pont Alexandre III in Paris, France.

Pont d'Avignon, also known as Pont Saint Benezet (Avignon, France; twelfth century)

This amazing structure near Avignon in southeastern France was the longest bridge in Europe in medieval times, and the longest masonry **arch bridge** ever built. The bridge has four spans, each more than 100 feet

(30.5 meters) long, still standing out of the original 20 or 21 spans (one authority has suggested that there may have been as many as 28 spans). The total length was

The twelfth-century Pont d'Avignon in Avignon, France.

about 3,000 feet (915 meters) across the Petit-Rhône River, over the island of Barthelasse, and into the middle of the Grand-Rhône where it turned westward in a V shape, presumably to enable the bridge to withstand high spring floods. It was nearly three times as long as the contemporary OLD LONDON BRIDGE, yet had only one or two more spans. Each of the surviving spans are longer than any other remaining Roman arch. The arches were elliptical, each about 98 feet (30 meters) long, rather than semicircular or pointed. This configuration means the piers could be narrower than usual and the arches taller than semicircular arches, lifting the bridge out of the way of flooding. The bridge was narrow, with a maximum width of 16.5 feet (5 meters) tapering to 6.5 feet (2 meters) alongside a chapel on the bridge. According to legend, the bridge was built by Saint Benezet, a shepherd who had a vision in 1178 in which God commanded him to build the bridge. When the bishop of Avignon asked for evidence that God had commanded the construction, the builder demonstrated his divine inspiration by lifting a giant stone and moving it to the place on the riverbank where the bridge work would begin.

Additional Information: *Academic American Encyclopedia* (electronic version) Danbury, CT: Grolier Electronic Publishing, 1991; Brown, David J. *Bridges.* New York: Macmillan, 1993; Lay, M.G. *Ways of the World.* New Brunswick, NJ: Rutgers University Press, 1992.

Pont de Châtellerault (Châtellerault, France; 1898-1899)

Designed and built by François Hennebique, the Pont de Châtellerault crosses the Vienne River, a tributary of the Loire, at Châtellerault in east central France. The bridge is one of the first notable **reinforced concrete**

arches, and the most notable concrete bridge to be built by Hennebique before the turn of the century. The bridge has two segmental arches of 131 feet (40 meters) each on either side of a central span of 164 feet (50 meters). Lower than many other arches, it has a rise to span ratio of only 1 in 10. The arches are so thin at the top that the top thickness of the span is only one one-hundredth of the thickness below it.

Additional Information: Brown, David J. *Bridges.* New York: Macmillan, 1993.

Pont de la Concorde (Paris, France; 1788-1791)

Jean Rodolphe Perronet was 78 years old when he started to build this beautiful structure. This multiple arch bridge is faced with marble to match the ambience of its setting in Paris's Place de la Concorde. The king appointed Perronet engineer general of bridges and roads in 1763, and he achieved fame with the construction of the beautiful NEUILLY BRIDGE in Paris in the early 1770s. So great was his reputation that he retained his post during the French Revolution, and continued building the Pont de la Concorde in the midst of the revolutionary turmoil. Perronet's original design for the bridge was too radical for the French authorities. They objected to the carefully engineered sleekness of the structure, a hallmark of Perronet's work, and demanded that he widen the arches of the bridge.

Additional Information: Lay, M.G. *Ways of the World.* New Brunswick, NJ: Rutgers University Press, 1992.

Pont de Montignies (near St. Christophe, Belgium; date uncertain)

This highly unusual bridge across the River Hantes in southern Belgium is probably Roman, but that is not definite. The masonry bridge has 13 arched spans, with an overall length of about 27 meters (89 feet), and a slender width of only 2.65 meters (8.7 feet) at the piers. On the upstream side, the bridge has recessed double arches; the **spandrel** walls are on arches higher than those of the vault. On the downstream side, the substructure of the bridge acts as a **weir** to divert the river almost 90 degrees.

Additional Information: O'Connor, Colin. *Roman Bridges.* Cambridge, England: Cambridge University Press, 1993.

Pont du Gard Aqueduct (Nîmes, France; 19 B.C.)

In the late first century B.C., the Roman general Agrippa built this **aqueduct** across the Gard River to supply water to Nîmes in the southeastern part of the Roman province of Gaul, now France. The aqueduct still stands, and is perhaps the most famous "bridge" in the world. The Pont du Gard is 885 feet (270 meters) long and 155 feet (47 meters) high; it has three tiers of stone arches, one on top of the other. Swimmers regularly use the aqueduct as a diving board to reach the waters of the river below. A legend says that the devil built the aqueduct in exchange for the soul of the first being to cross the bridge; when a hare was the first to cross, the devil threw himself off the bridge in dismay.

Additional Information: Knowles, Rebecca, ed. *Let's Go: The Budget Guide to France*. New York: St. Martin's Press, 1993.

Pont Flavien (near Miramas, France; c. 12 B.C.)

This highly unusual Roman bridge near Miramas in southeastern France has a clear span of 6.7 meters (22 feet). The bridge's two triumphal arches on either end are especially unique; built on Corinthian columns, the arches reach a height of 6.7 meters (22 feet). The width of each arch is 3.6 meters (12 feet) at the base. The bridge is inscribed to C. Donnius Flavos, a priest, and to the Emperor Augustus.

Additional Information: O'Connor, Colin. *Roman Bridges*. Cambridge, England: Cambridge University Press, 1993.

Pont Saint Benezet. *See* Pont d'Avignon.

Pont Saint Maxence (near Paris, France; late eighteenth century)

The Pont Saint Maxence crosses the Oise River about 60 kilometers (37 miles) north of Paris. Designed and built by Jean Perronet, who built the NEUILLY BRIDGE and the PONT DE LA CONCORDE in Paris, the Pont Saint Maxence has a 12 to 1 span to rise ratio, doubling what had been done before in stone-**arch bridges**. The bridge has been destroyed three times—once by Napoleon in 1804 and twice by the Germans in 1870 and 1914—and has been rebuilt each time.

Additional Information: Lay, M.G. *Ways of the World*. New Brunswick, NJ: Rutgers University Press, 1992.

Pontcysyllte River Aqueduct (Llangollen, Wales; 1805)

The Pontcysyllte River Aqueduct carries the Ellesmere Canal over the Pontcysyllte River near Llangollen in northeast Wales. Built by Thomas Telford, one of the greatest of early nineteenth-century English engineers, the cast-iron **aqueduct** is 1,007 feet (307 meters) long, reaching a maximum height of 127 feet (39 meters) above the river. It has 19 **cast-iron** arches, each 45 feet (14 meters) long, carried by stone piers, constructed solid to a height of 70 feet (21 meters) and then built hollow to reduce weight and allow access for inspection and maintenance. The water trough is partly covered by the canal's tow paths. After 12 years of construction, the canal was already of reduced significance by the time it opened to traffic.

Additional Information: Brown, David J. *Bridges*. New York: Macmillan, 1993; Sandstrom, Gosta E. *Man the Builder*. New York: McGraw-Hill, 1970.

Ponte a Quattro Capi. *See* Pons Fabricus.

Ponte Alto Dams (near Trento, Italy; 1537, 1550, 1613, 1883)

Since the sixteenth century, several dams have been built or rebuilt on this site on the River Fersina about 1.5 miles (2.4 kilometers) from Trento in northern Italy. The dams have been built primarily for flood control; the springtime snow melt in the nearby Alps Mountains turns the Fersina from a mild stream into a raging torrent from which Trento needs protection. At the site of the dams, the river flows through a deep, narrow gorge, ideal for dam building. The first dam at the site was constructed in 1537. A wooden dam, it was destroyed by floods only five years later. The next dam, built in 1550, was made of stone and cement, and held for 32 years. Between 1611 and 1613, a thin arch dam was built on the site, the first arch dam in all of Europe. It was 16 feet (4.9 meters) high and 6.5 feet (2 meters) thick, with a radius of 46.5 feet (14.2 meters). Constructed of cut masonry blocks, it was subject to damage because the joints were not mortared. The height of the dam was raised to 56 feet (17 meters) in 1752, and the crest of the dam was fitted with a wooden overflow lip to prevent damage to the masonry. However, the wood of this lip was itself frequently damaged or destroyed by flooding. The next addition to the dam raised its height to 82 feet (25 meters), with a section that is 14.5 feet (4.4 meters) thick. Additional increases in the dam's height were made in 1847 and in 1850, with the last addition made in 1887 to bring the dam to a height of 124 feet (38 meters). Another dam, the Madruzza Dam, was constructed in 1883, about a mile downstream from the Ponte Alto. This dam is 133 feet (41 meters) high, and impounds water up to 83 feet (25 meters) up the air face of the Ponte Alto, relieving pressure on the Ponte Alto, but allowing only that part of the structure above the 83-foot level to act as a true dam.

Additional information: Smith, Norman. *A History of Dams*. Secaucus, NJ: Citadel Press, 1972.

Ponte de 25 Abril, also known as the Tagus Bridge or the Salazar Bridge (Lisbon, Portugal; 1966)

Named for Portugal's Liberty Day (25 April), the Ponte de 25 Abril across the River Tagus is also know as the Salazar Bridge, after a former premier of Portugal. The bridge was designed in part by David Steinman. The original design for the bridge called for a second deck, in the tradition of the GEORGE WASHINGTON BRIDGE, but the deck was never built. The bridge has a main span of 3,323 feet (1,014 meters) and side spans of 1,586 feet (484 meters) each; it is one of mainland Europe's longest **suspension bridges**. Its towers are an extremely high 625 feet (191 meters) to allow a required 230-foot (70-meter) clearance for navigation underneath the bridge.

Its foundations go down 260 feet (79 meters) below water level to hit bedrock. The bridge was the first use in Europe of a method in which the sinking of **caissons** was operated from compressed-air domes.

Additional Information: Brown, David J. *Bridges*. New York: Macmillan, 1993.

Ponte Fabrico. *See* Pons Fabricus.

Ponte Presedente Costa e Silva Bridge (Rio de Janeiro, Brazil; 1973)

This bridge across Guanabara Bay connects Rio de Janeiro with the city of Niteroi, Brazil. A massive **box girder**, the Ponte Presedente Costa e Silva Bridge has a total length of 6 miles (9.6 kilometers), with two center main spans of 656 feet (200 meters), and a third of 984 feet (300 meters).

Additional Information: Brown, David J. *Bridges*. New York: Macmillan, 1993.

Ponte Sant'Angelo, formerly Pons Aelius and Pons Hadriani (Rome, Italy; A.D. 135)

This second-century Roman bridge across the Tiber River in Rome is still standing. The bridge was built on **cofferdam** foundations set in the river. It has seven arches spanning the river, and connects the city of Rome with the CASTEL SANT'ANGELO and the Vatican. It may be the finest extant example of a Roman **arch bridge**. Originally named Pons Aelius after Emperor Aelius, it was renamed in A.D. 600 when Pope Gregory stood on the bridge and had a vision that a plague then afflicting Rome would end. The approaches were rebuilt between 1892 and 1894. Still used as a footway, the bridge originally had a roadway width of 4.7 meters (15 feet) between two 3.1-meter (10-meter) sidewalks. Sculpted figures may have originally stood on the bridge, but this is not certain. At the present time, the eastern end has

The Ponte Sant'Angelo was built across the Tiber in Rome in the second century A.D.

statues representing St. Peter and St. Paul; these were placed on the bridge by Pope Clement VII in 1527. On the other side are eight statues of angels carrying symbols of the Passion of Christ; these were placed on the bridge between 1667 and 1669. One of the angels was carved by the noted sculptor Bernini.

Additional Information: Brown, David J. *Bridges*. New York: Macmillan, 1993; Lay, M.G. *Ways of the World*. New Brunswick, NJ: Rutgers University Press, 1992; O'Connor, Colin. *Roman Bridges*. Cambridge, England: Cambridge University Press, 1993.

Ponte Santa Trinita. *See* Santa Trinita Bridge.

Ponte Vecchio (Florence, Italy; 1375)

Designed by Taddeo Gaddi, the Ponte Vecchio crosses the Arno River at its narrowest point in Florence in north central Italy. The bridge is a 29-meter-long (95-

The fourteenth-century Ponte Vecchio crosses the Arno in Florence, Italy.

foot) replacement for an earlier Roman bridge. The Ponte Vecchio was the first European **arch bridge** that was built flatter than earlier Roman bridges, which had been semicircular. Shops lined the bridge until the 1500s, when the storekeepers were removed from the bridge in an early and regrettable effort at urban renewal. The stores are now back, most selling gold and leather goods, and contribute to, rather than detract from, the charm of Florence.

Additional Information: Chao, Lorraine S., ed. *Let's Go: The Budget Guide to Italy*. New York: St. Martin's Press, 1993.

Portage Viaduct (Portageville, New York; 1852)

The Portage Viaduct crosses the Genesee River near Portageville in western New York. Designed by Silas Seymour, the Portage Viaduct was one of the most spectacular wooded trestle bridges of the mid-nineteenth century. Each individual **Howe truss** of the **viaduct** spanned 50 feet (15 meters) to carry railroad traffic on an 876-foot-long (267-meter) bridge across a 234-foot-long (71-meter) gorge.

Additional Information: DeLony, Eric. *Landmark American Bridges*. Boston: Bullfinch Press, 1993.

Portland Bridge(s) (Portland Maine; 1916, expected 1998-1999)

A 25-foot-wide (7.6-meter) **drawbridge**, with a toll of 6 cents, was the original bridge at this site across the Fore River. In 1916, this wooden bridge was replaced with the "Million Dollar Bridge," so called because the Maine legislature approved a spending limit of $1 million for the replacement. The newer bridge has a channel width (permissible width for ships navigating through) of 95 feet (29 meters); a 93-foot vessel is reported to have slowly and gingerly made its way through the bridge. The replacement bridge, under construction in 1997, will have a channel width of 198.6 feet (60 meters), and the open draws will reach a height of 200 feet (61 meters) above the water. The new structure will be one of the largest drawbridges in the United States. At an expected cost of $165 million, the new Portland Bridge will be the largest project ever undertaken by the Maine Department of Transportation.

Additional Information: Smith, Malcolm G. "The Multi-Million Dollar Bridge," Coastal Beacon (online version), 14 December 1995 at http://www.mbeacon.com/archive/121495dir/b.html; Naclewicz, Tess. "Final Piece Brings Bridge Together," Portland Press Herald (online version), 24 January 1997 at http://www.portland.com/bridge/012497.htm.

Poughkeepsie-Highland Railroad Bridge. *See* Poughkeepsie Railroad Bridge.

Poughkeepsie Railroad Bridge, also known as the Poughkeepsie-Highland Railroad Bridge (Poughkeepsie, New York; 1888)

Conrail closed the Poughkeepsie Railroad Bridge in 1974 following a fire, but may eventually put it back into light rail or pedestrian service. The bridge crosses the Hudson at Poughkeepsie in southeastern New York, about 70 miles (113 kilometers) north of New York City. With a total length of 6,767 feet (2,064 meters), the bridge once held the length record for a steel structure. Two anchor spans of 525 feet (160 meters) alternate with three **cantilevers** of 548 feet (167 meters) each. The bridge also has two shore spans of 200 feet (61 meters) each, and approaches totalling 3,673 feet (1,120 meters). The bridge was strengthened in 1906 with the

addition of a third truss midway between the original two, and by the addition of new columns in the towers. The deck is 212 feet (65 meters) above water, and trusses vary from 37 to 57 feet deep (11.3 to 17.4 meters). When the bridge opened, it was the most direct link available for railroads between the northeastern United States and the West.

Additional Information: DeLony, Eric. *Landmark American Bridges*. Boston: Bullfinch Press, 1993, http://www.academic.marist.edu

Prospective Road. *See* Avenue Road.

Puente de Alcantara. *See* Alcantara Bridge.

Puente Romano (Mérida, Spain; c. 25 B.C.-A.D. 117)

The Puente Romano crosses the Guadiana River in Mérida in southwestern Spain; at a length (with approaches) of approximately 755 meters (2,476 feet), it is the longest of all existing Roman bridges. A granite masonry bridge, the Puente Romano has three sections. Only 465 meters (1,525 feet) of the bridge's length is between spans; the balance is taken up by the width of the massive piers. The sections have slightly different orientations from each other because each section, when built, changed the flow pattern of the river slightly, and probably toppled the approaches. To protect the bridge piers, each of the piers has **cutwaters** to protect the arches on the northern or upstream side. In addition, several of the piers have floodways which could be opened if the height of the river rose enough to threaten the bridge, as happened in 1942, when a flood submerged part of the bridge. Exact construction dates for the bridge are not known, but evidence points to construction beginning late in the first century B.C. and

A photo (c.1904) of the Poughkeepsie Railroad Bridge across the Hudson at Poughkeepsie, New York. Photo courtesy of the Library of Congress, World's Transportation Commission Photograph Collection.

continuing until about A.D. 117 in the reign of the Spanish emperor Trajan.

Additional Information: O'Connor, Colin. *Roman Bridges.* Cambridge, England: Cambridge University Press, 1993.

Pulaski Skyway (Newark, New Jersey; 1932)

The Pulaski Skyway carries U.S. Routes 1 and 9 over the Passaic and Hackensack rivers to connect Newark, New Jersey, with Jersey City, New Jersey. When built, the skyway was one of the first elevated expressway systems in the country. Named after General Casimir Pulaski, a Polish-born hero of the American Revolution, the skyway, which was modeled after railroad **viaducts**, runs (with approaches) 6.2 miles (10 kilometers) over two separate spans. Construction of the skyway took an incredible 88,461 tons (79,615 metric tons) of steel, and more than 2 million rivets. The excavations for the foundations set a record for depth by going down 147 feet (45 meters) below mean high-water level. Despite its size, the skyway is attractive and won a 1932 award from the American Institute of Steel Construction for "the most beautiful and monumental bridge built in the United States." The skyway bridges over the Passaic River and the Hackensack River are identical; both bridges are swinging anchor spans bordered by cantilevered trusses. The main spans of each bridge are 550 feet (168 meters) long with 75-foot-long (23-meter) **cantilevers** at each end. Construction of the Pulaski was the final link in the New York-Washington, D.C., road connection, the most traveled route in the United States.

Additional Information: Bishop, Gordon. *Gems of New Jersey.* Englewood Cliffs, NJ: Prentice-Hall, 1985; Condit, Carl W. *American Building Art: The Twentieth Century.* New York: Oxford University Press, 1961; DeLony, Eric. *Landmark American Bridges.* Boston: Bullfinch Press, 1993.

Pulaski Street Bridge (Stamford, Connecticut; 1887)

The Berlin Iron Bridge Company built this bridge across the Rippowam River in Stamford in southwestern Connecticut. The company also built the TURN OF RIVER BRIDGE in Stamford. The story of this bridge illustrates the ongoing conflict between lovers of historic sites and the needs of the modern world. The third longest iron bridge in Connecticut, and the longest bridge in Stamford, the Pulaski Street Bridge was a parabola-shaped lenticular **pony truss**, a bridge type invented by the Berlin Iron Bridge Company. Iron bridges were built throughout the country, especially in the Northeast, in the last 30 years of the nineteenth century. In 1987, the city removed the bridge because it was not strong enough nor wide enough to handle modern truck traffic. A steel-and-concrete span that has been characterized as "generic" and having "no great charm or visual interest" replaced the Berlin Company structure. Taken apart with blowtorches, rather than being unbolted, the parts of the Pulaski Street Bridge now lie unused in nearby Kosciuszko Park. The city had intended to reuse the bridge in a marina, but that project was never built, and no other location has shown an interest in the bridge.

Additional Information: Sierpina, Diane Sentementes. "Stamford; A (Sort of Purple) Iron Bridge Gathers Champions." *New York Times,* 4 June 1995, XIIICN:2.

Purple Bridge. *See* West Main Street Bridge.

Q

Quebec Bridge (Quebec, Canada; 1917)

The Quebec Bridge over the St. Lawrence River is one of the world's great bridges. The history of the bridge encompasses two major engineering disasters. First opened in December 1917, this gigantic steel cantilever had collapsed while under construction in 1907 after seven years of work. Its main chords buckled, throwing about 80 men to their deaths in the water below. Planned with an 1,800-foot (549-meter) center span, the bridge was to be the longest in the world. Although the 1907 collapse seemed sudden, various reviews of the bridge records revealed early signs that the structure was not strong enough. Some have suggested that the bridge engineer, Theodore Cooper, was responsible for the collapse because of his inattention, negligence, or arrogance. Hours before the disastrous collapse, Cooper's assistant warned him that the end of the **cantilever** was swaying. Cooper sent a telegraph message ordering all workers off the bridge, but the message arrived too late to stave off the worst loss of lives in bridge-engineering history. The collapse of the bridge destroyed Cooper's career. Phelps Johnson and G.H. Duggan undertook to rebuild the bridge, but had a disaster of their own when the structure's prefabricated central span fell off its jack, causing the death of 11 men. The Quebec Bridge has suffered no other untoward incidents since its 1917 opening, and it remains an admired engineering landmark.

Additional Information: *Academic American Encyclopedia,* (electronic version).

Danbury, CT: Grolier Electronic Publishing, 1991; Steinman, David B. *Famous Bridges of the World*. London, Dover Publications, Inc., 1953, 1961; http://www.civeng.carleton.ca/ECL/files/ecl.270txt.

Queen Elizabeth II Bridge (Dartford, England; 1991)

Built by Helmut Homberg, the Queen Elizabeth II spans the Thames, connecting Dartford with Thurrock in southeastern England. With a central span of 1,476 feet (450 meters), the Queen Elizabeth II is one of Europe's longest **cable-stayed bridges**. The main piers are set on **caissons** that measure a massive 197 feet by 98.5 feet by 75.5 feet (60 by 30 by 23 meters); the piers were precast in the Netherlands and towed across the North Sea. Like most cable-stayed bridges, the Queen Elizabeth is extremely narrow, and was built especially high to allow navigation traffic below it. The massive caissons, required by the engineering demands of allowing

Queen Elizabeth II Bridge crosses the Thames at Dartford in southeastern England. Photo courtesy of Ray Farrar.

wide and high shipping channels underneath the bridge, support similarly massive piers, creating a somewhat unpleasant contrast with the narrow bridge.

Additional Information: Brown, David J. *Bridges.* New York: Macmillan, 1993.

Queen Emma Pontoon Bridge (Willemstad, Curacao; 1888)

Built by Leonard Burlington Smith, the 700-foot-long (213.5-meter) Queen Emma Bridge is the largest floating pedestrian bridge in the world. Curacao is an island off the Venezuelan coast of northern South America. The bridge connects the shopping districts of Punda and Otrobanda. The Queen Emma opens as many as 30 times a day to allow ships to enter St. Anna Bay. Nicknamed "The Lady," the bridge was a commercial enterprise for Smith, who made a profit on the tolls he charged pedestrians. People without shoes, presumably the poor, were allowed to cross the bridge without paying a toll; poor people, however, borrowed shoes to avoid seeming poor, and many wealthy pedestrians removed their shoes to avoid paying the toll.

Additional Information: http://www.interknowledge.com/curacao/ancpn01.htm.

Queensboro Bridge, also known as the 59th Street Bridge (New York City; 1909)

Gustav Lindenthal built this bridge to carry 59th Street over the East River. Because piers could be set in Roosevelt Island (formerly Blackwell's Island) in the middle of the East River, the Queensboro could be a continuous **cantilever bridge**, rather than a long **suspension bridge**. When completed, the Queensboro was the longest cantilever in the United States. Channel spans on either side of Roosevelt Island are 1,182 feet (361 meters) and 984 feet (300 meters). With anchor spans of 630 feet (192 meters) and shore arms of 469 feet (143 meters) and 459 feet (140 meters), the bridge's total length is 3,825.5 feet (1,167 meters). Not the most attractive of bridges, the Queensboro was built with an excess of material to make it as safe as possible. Upon

first seeing the completed bridge, the consulting architect is supposed to have declared: "My God!—It's a blacksmith's shop!"

Additional Information: DeLony, Eric. *Landmark American Bridges.* Boston: Bullfinch Press, 1993.

Qutab Minar (Delhi, India; early thirteenth century)

This 73-meter-high (239-foot) tower of victory was built immediately after the defeat of Delhi's last Hindu kingdom to symbolize Islamic rule of the city. It was built by Qutb-ud-Din Aybak, the first Muslim Indian ruler to choose Delhi as a capital. The tower has five distinct stories, each with its own projecting balcony. The tower tapers from a 15-meter (49-foot) diameter at the base to a bare 2.5 meters (8.2 feet) at the top. The first three stories are made of red sandstone; the top two stories

A photo (c. 1910) of the Queensboro Bridge over the East River in New York City. Photo courtesy of the Library of Congress, World's Transportation Commission Photograph Collection.

are made of marble and sandstone. Extremely steep stairs wind to the top of the tower. Following a deadly stampede in 1979 during a school visit, the inside of the tower has been closed to visitors.

Additional Information: http://www.lonelyplanet.com.au/dest/ind/cal.htm; http:/sun10.vsz.bsme.hu/~S7809sak/qutab.html/.

R

Radio City Music Hall (New York City; 1930s)

The Radio City Music Hall is located at 47th Street and Sixth Avenue in New York City, part of the ROCKEFELLER CENTER complex of buildings. A prime example of 1930s building splendor, the combined music hall and movie theater is still a sight to see. Its magnificent staircase dominates an amazingly large and regal entranceway. At one time, even the bathrooms had murals; the murals now reside in the Museum of Modern Art. The huge auditorium is built to resemble a scalloped shell. The gigantic stage, on which the world famous Rockettes perform, adjoins an amazing orchestra pit with an elevator that can raise the orchestra to stage level for the enjoyment of the audience.

Additional Information: http://www.hotwired.com/rough/usa/mid.atlantic/ny/nyc/city/midtown.html.

Rainbow Bridge (Tokyo, Japan; 1993)

This beautiful **suspension bridge** across Tokyo Bay in Japan is illuminated at night by a pattern of white, green, and coral lights that change with the seasons. Built at a cost of $1.23 billion, the 570-meter-long (1,870-foot) bridge stretches from Tokyo to an expected city development on the east side of Tokyo Bay. Ordinarily, a bridge of this size would have been a **cable-stayed** structure, but various considerations dictated a suspension bridge. Among these constraints was a requirement for a 50-meter (164-foot) clearance over the 500-meter-wide (1,640-foot) shipping channel, and a need for the shorter towers of a suspension bridge because of the closeness of Haneda International Airport 9 kilometers (5.6 miles) away.

Additional Information: Heidengren, Charles R. "Japan Builds 21st Century Monuments." *Civil Engineering* 64:8 (August 1994), pp. 57-59.

Reading-Halls Station Bridge (Muncy, Pennsylvania; c. 1846)

Richard B. Osborne built the Reading-Halls Station Bridge in the 1840s to cross the former Reading Railroad tracks in Muncy in east central Pennsylvania. Although now used for vehicles, the structure is the oldest all-metal railroad bridge still in use in the United States. It is also one of two surviving all-metal railroad bridges in the United States, the other being the HAUPT TRUSS BRIDGE (c. 1854) now disassembled at the Railroaders' Memorial Museum in Altoona, Pennsylvania. The Reading-Halls Bridge has been rebuilt somewhat. The original was a 69-foot (21-meter), 18-panel **Howe truss** with **cast-iron** diagonal-**compression** members, **wrought-iron** vertical **tension** rods, and wrought-iron bars for the top and bottom chords. The bridge's castings contain "Egyptian Revival" motifs.

Additional Information: DeLony, Eric. *Landmark American Bridges*. Boston: Bullfinch Press, 1993.

Red Fort (Lal Qila) (Old Delhi, India; seventeenth century)

This gigantic fort rises 33 meters (108 feet) above the city. Built originally by the Mogul emperor Shah Jahan, who also constructed the TAJ MAHAL and the JAMI MASJID, the fort was walled in 1638 to keep out invaders. The gigantic main gate of the fort, the Lahore Gate, gathers a large throng each year at the time of India's Independence Day celebration in August. A variety of buildings inside the fort include the Drum House, the Hall of Public Audiences, the Hall of Private Audiences, the Pearl Mosque, the Royal Baths, and the Palace of Color.

Additional Information: http://www.lonelyplanet.com.au/dest/ind/del.htm.

133

Reichenau Bridge (Reichenau, Germany; 1755)
Built by Johannes Grubenmann, this covered wooden bridge over the Rhine River had a single span of 240 feet (73 meters). The bridge was destroyed by Napoleonic troops in 1799.

Additional Information: Steinman, David B. *Famous Bridges of the World*. London: Dover Publications, 1953, 1961.

Reservoir Bridge Southeast (Central Park, New York City; 1865)
Designed by Calvert Vaux and Jacob Wrey Mould, the Reservoir Bridge Southeast is one of five **cast-iron** bridges in Central Park. The bridge's span of 33 feet (10 meters) features delicate iron castings on the side.

Additional Information: DeLony, Eric. *Landmark American Bridges*. Boston: Bullfinch Press, 1993.

Reservoir Bridge Southwest (Central Park, New York City; 1864)
Designed by Calvert Vaux and Jacob Wrey Mould, the Reservoir Bridge Southwest is one of five **cast-iron** Vaux and Mould bridges in Central Park. It has a long span of 72 feet (22 meters).

Additional Information: DeLony, Eric. *Landmark American Bridges*. Boston: Bullfinch Press, 1993.

Rhine Bridge at Cologne (Cologne, Germany; 1915)
This medium-span **suspension bridge** crosses the Rhine River at Cologne in western Germany. The bridge is 605 feet (185 meters) long. It used a new technique of hanging the chains in which the chains terminate at the end of the stiffening truss rather than continuing onto a ground anchorage.

Additional Information: Brown, David J. *Bridges*. New York: Macmillan, 1993.

Rhine Bridge at Tavanasa. *See* Tavanasa Bridge.

Rialto Bridge (Venice, Italy; 1588)
Designed by Antonio da Ponte, the Rialto Bridge is a low 89-foot (27-meter) circular arch supporting a 75-foot (23-meter) roadway across the GRAND CANAL in Venice in northeastern Italy. Still in use, this pedestrian bridge is the most famous sixteenth-century bridge in the world. A line of shops is built in the center of the roadway, with steps leading up to the shops at either end of the bridge. Because the soil of Venice is soft, 6,000 wood piles were driven to a depth of 11 feet (3.4 meters) beneath each masonry-block **abutment** into the bed of the canal. The piles are set so as to lean against the current of the canal water.

Additional Information: *Academic American Encyclopedia*, (electronic version). Danbury, CT: Grolier Electronic Publishing, 1991; Steinman, David B. *Famous Bridges of the World*. London: Dover Publications, 1953, 1961.

The sixteenth-century Rialto Bridge crosses the Grand Canal in Venice, Italy.

Rideau Canal (Ottawa, Canada, to Kingston, New York; 1832)
Built by Lieutenant Colonel John By, the Rideau Canal links Ottawa in Canada with Kingston on the Hudson River in New York. Begun in 1826, the canal required 52 dams of varying sizes. The largest is the dam at Jones Falls, 35 miles (56 kilometers) northwest of Kingston. This structure was the first arched dam in North America, and also the highest dam in North America at the time. The air face is a masonry wall, 61.5 feet (19 meters) high and 27.5 feet (8.4 meters) thick at the base, tapering to a crest of 21.5 feet (6.6 meters) thick. The 350-foot-long (107-meter) crest is curved to a radius of about 175 feet (53 meters). On the water face, the dam is backed to its full height by an earth embankment that is 127.5 feet (39 meters) thick at the base. Considered a splendid example of early North American dam-building, the dam at Jones Falls is still in fine condition, and supplies water to a nearby hydroelectric plant.

Additional Information: Smith, Norman. *A History of Dams*. Secaucus, NJ: Citadel Press, 1972.

Rio-Niteroi Bridge. *See* Ponte Presedente Costa e Silva Bridge.

Riverside Avenue Bridge (Greenwich, Connecticut; 1871)
Originally, this bridge crossed the Housatonic River at Stratford in western Connecticut, but was relocated in 1890 to Greenwich in southwestern Connecticut, where it now runs over an Amtrak railroad line. In this double intersection **Pratt truss**, the vertical posts have openings to allow the diagonals to pass through them, rather than being attached on the outsides of the posts. The portal and the upper horizontal struts have decorative castings.

Additional Information: DeLony, Eric. *Landmark American Bridges*. Boston: Bullfinch Press, 1993.

Rockefeller Center (New York City; 1931-1939)

Rockefeller Center is a complex of 14 buildings in central Manhattan that runs between 48th and 51st Streets and Fifth Avenue and Sixth Avenue (Avenue of the Americas). The various buildings in Rockefeller Center house offices, stores, restaurants, exhibition rooms, and broadcasting studios. The five western buildings in the complex, a radio, television, and entertainment section, are known as Radio City. Every Christmas, Rockefeller Center is home to one of the country's largest and best-known Christmas trees, which is set at one end of Rockefeller Center's famous outdoor ice skating rink. The complex includes the 70-story RCA (Radio Corporation of America) Building.

Additional Information: Karp, Walter. *The Center: A History and Guide to Rockefeller Center.* New York: American Heritage Publishing Company, 1982.

Rockfish Tunnel. *See* Crozet Tunnel.

Rogue River Bridge (Gold Beach, Oregon; 1932)

The Rogue River Bridge carries the Oregon Coast Highway (U.S. Route 101) over the Rogue River near Gold Beach in southwestern Oregon. Built by Conde B. McCullough, the Rogue River Bridge has a total length of 1,898 feet (579 meters). The bridge has two approach spans and seven double-ribbed, reinforced-concrete deck arches of 230 feet (70 meters) each. The bridge makes extensive use of classical forms, such as columns and arches, which are **scoured** to make the concrete look like stone. To cut costs by about 10 percent, the Rogue River Bridge was the first **reinforced-concrete** arch built in the United States using the Freyssinet system. Because the method gives arches greater strength, it allows them to be much longer than would normally be the case. The method prestresses concrete by precompressing the concrete, jacking the arches apart before the concrete is fully set, and then filling in the open spaces with additional concrete plugs. The arches cure to be permanently stressed with little chance of cracking or other structural defects.

Additional Information: DeLony, Eric. *Landmark American Bridges.* Boston: Bullfinch Press, 1993; Schodek, Daniel L. *Landmarks in American Civil Engineering.* Cambridge, MA: MIT Press, 1987.

Rolph, James "Sunny Jim" Bridge. *See* Oakland-San Francisco Bay Bridge.

Roman Catacombs. *See* Catacombs.

Roosevelt, Theodore Dam. *See* Theodore Roosevelt Dam.

Roquefavour Aqueduct Bridge. *See* Marseilles Aqueduct.

Rosetta Dam/Damietta Dam (Egypt)

These two dams were the first ever built on the Nile in Egypt. Both poorly built dams leaked; a portion of the Rosetta Dam slipped from its original location in 1867. The dams were finally repaired satisfactorily by C.C. Scott-Moncrieff, a British engineer who was made chief engineer of Egypt's irrigation works.

Additional Information: Smith, Norman. *A History of Dams.* Secaucus, NJ: Citadel Press, 1972.

Ross, Betsy Bridge. *See* Betsy Ross Bridge.

Route 66. *See* U.S. Route 66.

Rove Tunnel (Arles, France; 1927)

The Rove Tunnel is part of the Canal de Marseille au Rhône, running along the Rhône River between Marseille and Arles in southeastern France. At a length of 4.5 miles (7.2 kilometers), the Rove is the world's longest canal tunnel. To accommodate sea-going vessels, the tunnel is extremely large, with a width of 72 feet (22 meters) and a height of 17 feet (5.2 meters). Although begun at one end in 1911 and at the other in 1914, tunnel construction halted for part of World War I; it was later restarted with the labor of German prisoners of war. The tunnel was first holed through in 1916. After a section of the tunnel collapsed in 1963, it was closed to traffic.

Additional Information: Beaver, Patrick. *A History of Tunnels.* Secaucus, NJ: Citadel Press, 1973. Hawkes, Nigel. *Structures: The Way Things Are Built.* New York: Macmillan, 1990.

Royal Albert Bridge (Saltash, England; 1859)

Built by I.K. Brunel, the Royal Albert Bridge carries a broad-gauge railway over the Tamar Estuary on the border of Devon and Cornwall in southwestern England. The bridge has two spans, each 445 feet (136 meters) long. It made use of a curved upper chord of oval **wrought-iron** that formed a **lenticular truss.** The Royal Albert was Brunel's last and perhaps greatest project.

Additional Information: DeLony, Eric. *Landmark American Bridges.* Boston: Bullfinch Press, 1993.

Royal Border Bridge (Berwick, England; 1850)

Designed by Robert Stephenson, the builder of the LONDON-BIRMINGHAM RAILWAY, and son of George Stephenson, the builder of the LIVERPOOL-MANCHESTER RAILWAY, the Royal Border Bridge carries railroad tracks over the Tweed River at Berwick in northeastern England near the border with Scotland. The bridge has 28 semicircular arches constructed of masonry and brick. It got its name from its location and from the fact that Queen Victoria presided at its opening in 1850. It closed the last gap in the continuous London-Edinburgh railway line.

Additional Information: Brown, David J. *Bridges*. New York: Macmillan, 1993.

Royal Gorge Bridge in Colorado spans the Arkansas River. Photo courtesy of the Royal Gorge Bridge Company.

Royal Gorge Bridge (near Canon City, Colorado; 1929)

Constructed at a cost of $350,000, the Royal Gorge Bridge crosses the Arkansas River near Canon City in south central Colorado. The bridge is one of the highest **suspension bridge**s in the world, at a height of 1,053 feet (321 meters). The Royal Gorge Bridge is 1,260 feet (384 meters) long, with a deck width of 18 feet (5.5 meters). About an hour's drive southwest of Colorado Springs, the bridge was constructed with no fatalities and no major accidents, an unusual and significant record. The bridge was designed specifically to attract tourists and vacationers to the area. It was refurbished in the mid-1980s.

Additional Information: http:\\bechtel.colorado.edu/ Graduate_Programs/Sesm/Courses/Cven3525/royal/ royal.html.

Royal Mile. *See* Edinburgh Castle.

Royal Road (Chile to Ecuador; fifteenth century)

The major road of the Inca Empire ran from what is now Santiago in central Chile, through Cuzco in southern Peru, to Quito in what is now northern Ecuador. Some of the road is built at an altitude of 3,600 meters (over 11,800 feet). Until the late nineteenth century, the Royal Road was probably the world's longest operating road; it had **suspension bridges**, embankments, and a series of stone steps. Although steep in places, the road was no problem for Inca travelers because they had no wheeled vehicles. Widths averaged from roughly 4 to 7 meters (13 to 23 feet), although in the northern section the Royal Road sometimes widened to as much as 16 meters (52 feet).

Additional Information: Lay, M.G. *Ways of the World*. New Brunswick, NJ: Rutgers University Press, 1992.

S

Sadd el-Kafara, also known as Dam of the Pagans (near Cairo, Egypt; c. twenty-eighth century B.C.)
Sadd el-Kafara is possibly the oldest surviving dam in history; it is located 20 miles (32 kilometers) south of Cairo, Egypt. Discovered by the archeologist G. Schweinfurth in 1885, the dam was constructed between 2950 and 2750 B.C. It was at least 265 feet (81 meters) long at the base and 348 feet (106 meters) long at the crest. It reached a maximum height of 37 feet (11.3 meters). The dam was built in three parts. The sections at the ends were rubble masonry walls, 78 feet (24 meters) thick. The space in the center of the dam was filled with gravel, perhaps as much as 60,000 tons (54,000 metric tons). The crest of the dam in the center was lower than the crest at the sides to assure that overflow would occur at the center. The water face had limestone blocks set in steps on the surface to protect against erosion and water action. The capacity of the associated reservoir was more than 20 million cubic feet (over 5.1 million cubic meters), not a large amount, but reasonable considering the arid climate in which the dam is situated.

Additional Information: Smith, Norman. *A History of Dams.* Secaucus, NJ: Citadel Press, 1972.

Saint Antoine Bridge (Geneva, Switzerland; 1823)
Designed by Guilaume Henri Dufour, the Saint Antoine Bridge was the world's first permanent wire-cable **suspension bridge**.

Source: DeLony, Eric. *Landmark American Bridges.* Boston: Bullfinch Press, 1993.

Saint Basil's Cathedral (Cathedral of Saint Basil the Blessed) (Moscow, Russia; 1560)
Built in the sixteenth century under the direction of Czar Ivan the Terrible, Saint Basil's Cathedral, one of the largest in the world, was constructed to celebrate Ivan's victory over the Tartars in 1555. According to

The sixteenth-century Cathedral of Saint Basil the Blessed in Moscow, Russia.

legend, when the cathedral was completed, Ivan had the eyes of the two designing architects gouged out so that they could never again design anything so beautiful. The cathedral is made up of nine interconnected chapels, each with its typical onion-shaped Russian dome. A large dome represents Jesus; the nine smaller domes represent nine of Ivan's important victories. Although the assymetric cathedral has been criticized for being a colorful hodgepodge, it is now considered an exemplar of Russian church architecture. The cathedral stands at the southern end of Moscow's Red Square, near the KREMLIN.

Additional Information: http://www.adventure.com/library/encyclopedia/ka/rfibasil.html (Knowledge Adventure Encyclopedia).

Saint Esprit Bridge (France; 1265-1309)

The Hospitaliers Pontifes built this bridge over the Rhône River in southeastern France in the late thirteenth century. The bridge is 2,700 feet (823.5 meters) long with 26 arches; for many years, this bridge was the longest bridge in the world. The piers for the arches vary in width from 20 to 50 feet (6.1 to 15 meters).

Additional Information: Kirby, Richard Shelton et al. *Engineering in History*. New York: Dover Publications, 1990.

Saint George Street (St. Augustine, Florida; c. 1680)

This 600-meter (1,968-foot) stretch of street is the oldest extant street in the United States. It was built of a concrete made from sea shells.

Source: Lay, M.G. *Ways of the World*. New Brunswick, NJ: Rutgers University Press, 1992.

Saint George's Chapel. *See* Windsor Castle.

Saint Gotthard Pass Road Tunnel (southeastern Switzerland; 1980)

The Saint Gotthard Pass Road Tunnel carries Highway N2 to Italy through the St. Gotthard Pass. Built almost parallel to the famed SAINT GOTTHARD RAILROAD TUNNEL, the road tunnel repeated the harsh experiences of the railroad tunnel builders a century earlier, although modern equipment and amenities made the task somewhat easier. Nevertheless, 19 workers died during the excavation. The 10.5-mile (17-kilometer) tunnel is the longest vehicular tunnel ever built. Innovations in building the tunnel included the installation of a "sliding floor," which was pushed into the tunnel as it was excavated. The floor provided a good surface for workers, and allowed trucks to have access to the tunnel face for removal of debris. A parallel safety tunnel, 100 feet (30.5 meters) away, was constructed for an escape route, and four vertical tunnels were built downward for ventilation, a system that remained in use after the tunnel was built to provide ventilation for traffic moving through the tunnel. Lighting in the tunnel was powered by two different systems to prevent a disastrous loss of light during the difficult work. Progress on the tunnel was guided by a laser beam system to assure that the tunnel was dug in the correct direction.

Additional Information: Hawkes, Nigel. *Structures: The Way Things Are Built*. New York: Macmillan, 1990.

Saint Gotthard Railroad Tunnel (southeastern Switzerland; 1881)

Louis Favre built this tunnel through the Swiss Alps about 13 miles (21 kilometers) north of the Italian border. In the light of the success of the MONT CENIS TUNNEL, this long desired underground train tunnel seemed, at last, to be a realistic possibility. Depending upon improved engineering techniques, such as better drills and more extensive use of dynamite, Favre confidently bid on the tunnel construction project, and promised a large payment if he did not meet his schedule. As it turned out, the tunnel project, although eventually completed, was more costly in terms of human life and health than any other project in modern times. Favre's death from heart disease, brought on by the stress of the tunnel project, left his business heirs to pay a significant penalty for lateness. Favre's bid had been for less per meter and a faster time than had been needed at Mont Cenis; the obstacles encountered at Saint Gotthard made the bid unrealistic. Running water in the tunnel, exploding from the rocks as they were drilled, was the most common obstacle to progress. Flow from the rocks reached levels that might describe the flow from a small reservoir or dam—800 cubic meters per hour in 1873, 876 cubic meters per hour in 1874, and 1,260 cubic meters per hour in 1876. Working conditions were as bad as they could possibly get. Fresh air supplied to the tunnelers was little more than 50 percent of what was actually needed. Heat in the tunnel sometimes reached 93 degrees fahrenheit. Workers fell victim to silicosis and, toward the end of the project, died of the effects of the parasite "ancylostome," which was spread by the ever-present waters. Rock cave-ins, faulty handling of dynamite, and tunnel collapses caused additional deaths. Several deaths occurred when the management of the tunnel project ruthlessly put down a strike. Overall, 310 men died on the project and another 877 were seriously injured or permanently disabled. Still one of the most magnificent engineering projects of the past several hundred years, the Saint Gothard Tunnel is, at the same time, a monument to optimism gone wild.

Additional Information: Beaver, Patrick. *A History of Tunnels*. Secaucus, NJ: Citadel Press, 1973; Epstein, Sam and Epstein, Beryl. *Tunnels*. Boston: Little Brown & Company, 1985; Gies, Joseph. *Adventure Underground: The Story of the World's Great Tunnels*. Garden City, NY: Doubleday & Company, 1962.

Saint John's Bridge (Portland, Oregon; 1931)

Built by David B. Steinman, the Saint John's Bridge carries U.S. Route 30 over the Willamette River at Portland in northwest Oregon. An impressive 1,207-foot-long (368-meter) **suspension bridge** with Gothic details,

Saint John's has 400-foot (122-meter) towers that culminate in two pointed **finials** each. The roadbed is an unusually high 200 feet (61 meters) above the Willamette River. The builders of the bridge proposed painting it green, but city officials wanted the bridge to be yellow with black stripes. To resolve the impasse, the engineers waited until St. Patrick's Day to announce that the bridge would be painted green. It has remained green since that day.

Additional Information: DeLony, Eric. *Landmark American Bridges*. Boston: Bullfinch Press, 1993; Steinman, David B. *Famous Bridges of the World*. London: Dover Publications, 1953, 1961.

Saint Lawrence Seaway (Saint Lawrence River, eastern Canada to Great Lakes; 1959)

The Saint Lawrence Seaway is one of the most successful efforts ever undertaken at creating an international transportation path. The seaway was built by the United States and Canada to connect the Great Lakes to the Atlantic via the Saint Lawrence River. Including the Great Lakes, the seaway is 2,400 miles (3,862 kilometers) long. The original recommendation for the seaway was made by an international commission in 1921, and it was built by the U.S. Army Corps of Engineers and the Saint Lawrence Seaway Authority of Canada. Work began in 1954, and the seaway was officially opened in June 1959. Within the 44-mile-long (71-kilometer) section known as the "international development area" are two navigation dams, a power dam, and a canal with two locks. The project also makes possible the production of hydroelectric power for New York, Vermont, and Ontario.

Additional Information: Condit, Carl W. *American Building Art: The Twentieth Century*. New York: Oxford University Press, 1961.

Saint Mary's Bridge. *See* Point Pleasant Bridge.

Saint Paul's Cathedral (London, England; 1710)

Built by Sir Christopher Wren, Saint Paul's stands on London's Ludgate Hill, where tradition says a Roman temple once stood. The first church on the site was destroyed by fire in 1087, and replaced by a new cathedral completed in the late thirteenth century. By the seventeenth century, this old Saint Paul's was in need of repair, and Inigo Jones was employed upon the renovations. Wren was planning further repairs when the Great Fire of London destroyed most of the church in 1666. Wren then obtained permission from the king to demolish the damaged structure and build a whole new cathedral. Wren modifed his original design of a Greek Cross with a dome over the center to provide the long nave and choir of the traditional medieval cathedral. Wren personally laid the first foundation block in 1675,

and 35 years later set the last block in place. The interior of the building consists of a three-aisled nave and choir, of equal lengths, extending east and west from the great central space at the crossing under the dome. The great dome is second in size only to the dome of

Saint Paul's Cathedral on Ludgate Hill in London. Photo courtesy of Ray Farrar.

SAINT PETER'S CHURCH in Rome. Saint Paul's was damaged by German bombs in 1940 and 1941, but rebuilt according to Wren's original plans after the war. The great dome offers a spectacular sight to travelers passing nearby on the River Thames.

Additional Information: http://www.lonelyplanet.com.au/dest/eur/lon.htm.

Saint Peter's Church (Vatican City; 1445-1626)

Saint Peter's is the largest church in the world, and the central place of worship for Roman Catholics. Construction began in the fifteenth century and continued for 181 years. The church is built on the site of the Emperor Nero's circus, where many Christians were martyred in the 60s A.D. The first Christian church on the site was built in the fourth century A.D. over the sarcophagus of the apostle Peter, who had supposedly been executed in Rome during Nero's reign. In this wooden basilica, Charlemagne was crowned Holy Roman Emperor in A.D. 800, and various popes were enthroned over the centuries. By the fifteenth century, the basilica was in need of extensive repair, and Pope Nicholas V commissioned Bernardo Rossellino to design an entirely new structure. Actual construction began in 1450, but little was accomplished until after the accession of Julius II in 1503. Julius selected a design submitted by the architect Donato d'Agnolo Bramante as the plan for the new chruch; Bramante's design called for a Greek cross with a great dome over the center. By Bramante's death in 1514, the great piers and their arches and much of the vaulting had been completed.

Saint Peter's Church in the Vatican City.

In the following decades, many architects and artists were given charge of the work, including the painter Raphael Santi, but little progress was made until 1546, when Paul III appointed the artist Michelangelo Buonarroti architect for the project. Michelangelo, a noted sculptor, had earlier painted the SISTINE CHAPEL ceiling for Julius II. Michelangelo disregarded the design changes recommended by many of his predecessors and returned to Bramante's original plan. By his death in 1564, Michelangelo had nearly completed the construction of the great dome. Michelangelo's successors on the project completed the dome in the 1590s by following his plan with but slight variations. In 1605, Carlo Maderna altered Bramante's original Greek cross plan by lengthening the nave to form a Latin cross. Pope Urban VIII dedicated the completed church in 1626. The huge dome of the church rises 404 feet (123 meters) above the pavement and has an interior diameter of 137 feet (42 meters). Beneath the dome rises the high altar, at which only the pope may say Mass. A crypt below the altar contains the tomb of Saint Peter. The interior of the church is a magnificence of marble, sculpture, gilt, and fresco art. The great forecourt of the church, SAINT PETER'S SQUARE, was constructed in the mid-seventeenth century. Today the church is in Vatican City, an independent city-state under papal control that was recognized by Italy in the 1929 Lateran Treaty.

Additional Information: http://www.iol.it/tci.sdp/html/eo706.htm.

Saint Peter's Square (Vatican City; 1629-1667)

Saint Peter's Square, the great forecourt of SAINT PETER'S CHURCH in the Vatican City (formerly part of Rome), was designed and built in the seventeenth century by the artist Gian Lorenzo Bernini. One of the largest squares in the world, Saint Peter's is a majestic elliptical plaza bounded by quadruple colonnades. The center of the square holds the Egyptian obelisk brought to Rome from Heliopolis by the Emperor Caligula in the first century A.D.; it was moved to its present location in the square by D. Fontana in 1586. The quadruple colonnade consists of 284 columns and 88 pillars, which are topped with 140 statues and the crests of Pope Alexander VII.

Additional Information: http://www.iol.it/tci.sdp/html/eo706.htm.

Saint Peter's Square, the forecourt of Saint Peter's Church in the Vatican City.

Salazar Bridge. *See* Ponte de 25 Abril.

Salginatobel Bridge (near Schiers, Switzerland; 1930)

Salginatobel Bridge is perhaps Robert Maillart's best-known work; the bridge's fame stems from both its own magnificence and its picturesque location in the Alpine foothills of Graubünden Canton (state) in eastern Switzerland. A small bridge of 295 feet (90 meters), Salginatobel is tilted upward at a slope of 1 in 34, and meets the rock wall on the upward side almost dead-on with practically no visible support. As with many of Maillart's works, the bridge was built with hollow rectangular sections.

Additional Information: Brown, David J. *Bridges*. New York: Macmillan, 1993; Lay, M.G. *Ways of the World*. New Brunswick, NJ: Rutgers University Press, 1992.

Salginatobel Bridge, near Schiers, Switzerland, is one of Robert Maillart's best-known structures. Photo courtesy of Erwin Sigrist.

Salina and Central Square Plank Road (Syracuse to Oneida Lake, New York; 1846)

Because wood was so easily available in upstate New York (and elsewhere in the northeastern U.S. in the nineteenth century), the Salina and Central Square Plank Road Company constructed this approximately 14-mile-long (23-kilometer) road from Syracuse to Lake Oneida of wooden planking. The original road was built to be 8 feet (2.4 meters) wide; as work progressed, improvements to support the planks and solve drainage problems eventually resulted in a roadway with strong supporting planks of hemlock and other hard wood laid in a careful pattern. Between 1834 and 1850, about 500 miles (804.5 kilometers) of plank road were laid in Canada, and thousands of miles of plank road were put down in the United States. Within a decade of completion, most plank roads proved themselves incapable of standing up to the traffic they were built to handle, and the appeal of plank roads quickly faded.

Additional Information: Rose, Albert C. *Historic American Roads.* New York: Crown Publishers, 1976.

Salisbury Cathedral (Salisbury, England; 1220-1260)

The Cathedral Church of the Blessed Virgin Mary is located in Salisbury in Wiltshire, a county in southwestern England. The cathedral is an impressive example of high medieval English architecture. It was begun in 1220, when the bishopric of Salisbury was transferred from nearby Old Sarum, which had been a site of settlement in Wiltshire since Celtic times. The town of New Sarum or Salisbury was planned and built around the site of the new cathedral. The great cathedral at Old Sarum, the seat of the bishopric since 1075, was torn down and parts of it were used in the construction of the new cathedral. Rising 404 feet (123 meters) into the air, the spire of Salisbury Cathedral is the highest in England. The cathedral was undamaged during World War II because German bombers used the unmistakable spire as a navigational landmark. The thirteenth-century Chapter House is noted for its exquisitely detailed stone friezes, and as the repository of one of the still extant copies of Magna Carta.

Additional Information: http://www.eluk.co.uk/spireweb/cinfo.htm.

Salt River Project (central Arizona; begun in 1905)

The Salt River Project is the current name of the operating company (a division of the state of Arizona) and the private water corporation that operates six dams and lake recreation areas in the state. The enterprise began in 1905 as a U.S. Bureau of Reclamation project to harness the Salt River and its chief tributary, the Verde River, for irrigation and power for central Arizona. The project began by building THEODORE ROOSEVELT DAM, dedicated in 1911, on the Salt River in a canyon about 80 miles (129 kilometers) east of Phoenix. Later water storage facilities built by the project include Mormon Flat Dam (1925), HORSE MESA DAM (1927), and Stewart Mountain Dam (1930) on the Salt, and BARTLETT DAM (1939) and HORSESHOE DAM (1946) on the Verde. Originally, the project's main purpose was to provide irrigation for the fruits, vegetables, and cotton grown in the Salt River Valley; today, the Salt River Project also provides power and water to the Phoenix metropolitan area.

Additional Information: http://www.srp.gov:80/aboutsrp/water/srplakes.html.

Samos Water Tunnel (Island of Samos, Greece; c. 530 B.C.)

Over 2,000 years old, the now abandoned Samos Water Tunnel was the first tunnel to bring water into a walled city from outside its gates. Polycrates, the ruler of Samos, built the tunnel in the sixth century B.C. Eupalinus, the tunnel's engineer, had only two methods of drilling through rock available to him. One method involved burning wood under or on a rock until the rock was hot. The rock was then splashed with cold water to make it crack and allow workers to chip away the broken segments. In the other method, a hole was drilled in the rock using a bow and arrow arrangement (similar to a device used by American Indians to light a fire). Emory or some other abrasive powder coated the tip of the drill. After a hole had been drilled in the rock, wood was inserted, and the rock was heated and wetted as in the first method. Using these methods required over 15 years to hole out a 3,000-foot-long (915-meter) tunnel with a 90-degree turn. The tunnel was 8 feet (2.4 meters) wide and 8 feet high, and the water flowed through clay pipes laid on the floor of the tunnel. A walkway was constructed next to the water pipes so that they could be examined by an inspector walking through the tunnel. The Greek historian Herodotus considered the tunnel one of the three greatest engineering achievements of the Greek world.

Additional Information: Sandstrom, Gosta E. *Tunnels*. New York: Holt, Rinehart and Winston, 1963.

San Mateo Dam (Spring Valley, California; 1887-1889)

The San Mateo Dam near Spring Valley in San Mateo County was the first dam to be made entirely of concrete. Built to supply water to San Francisco, the dam was 170 feet (52 meters) high and 680 feet (207 meters) long. Although no longer in use, the dam successfully survived the 1906 San Francisco earthquake.

Additional Information: Smith, Norman. *A History of Dams*. Secaucus, NJ: Citadel Press, 1972.

Sando Bridge (Kramfors, Sweden; 1943)

Eugene Freyssinet designed this deck arch bridge with a span of 866 feet (264 meters) and a low arch with a rise of only 131 feet (40 meters) at the center. When built, this was the longest concrete-arch span in the world.

Additional Information: *Academic American Encyclopedia* (electronic version). Danbury, CT: Grolier Electronic Publishing, 1991.

Sankey Viaduct (northwestern England; c. 1828)

This bridge across the Sankey Brook Canal in northwestern England was built to support George Stephenson's LIVERPOOL-MANCHESTER RAILWAY. A brick-and-masonry structure with nine arches, the Sankey Viaduct has a deck that rises 70 feet (21 meters) above water level. The **viaduct** today carries rail traffic that is significantly heavier than anything Stephenson could have imagined.

Additional Information: Sandstrom, Gosta E. *Man the Builder*. New York: McGraw-Hill, 1970.

Santa Trinita Bridge (north central Italy; 1569)

This unusually shaped multiple-arch structure across the River Arno has three "basket-handled" arches with a rise to span ratio of 1 to 7, as opposed to the more usual ratio of 1 to 4. The bridge was built in the sixteenth century by Bartolomeo Ammannati.

Additional Information: *Academic American Encyclopedia* (electronic version). Danbury, CT: Grolier Electronic Publishing, 1991.

Santee Canal (Charleston to Columbia, South Carolina; 1800)

The Santee Canal, the first American canal to be completed, connected Charleston on the Atlantic with Columbia in the interior of the state via the Santee River. Chartered in 1786, the canal was built privately at a cost of $750,000. The canal was 5 feet (1.5 meters) wide at the surface, 4 feet (1.2 meters) deep, and 22 miles (35 kilometers) long, with 13 locks of brick and stone supporting wooden gates. Most of the work on the canal was done by slaves. Five miles of the canal were frequently out of service for lack of water, but the waterway did serve its purpose of increasing commercial traffic between the two cities and bringing increased trade through Charleston harbor. The canal was replaced by a railroad line in 1850.

Additional Information: Kirby, Richard Shelton et al. *Engineering in History*. New York: 1956; Dover Publications, 1990.

Santee-Cooper Canal. *See* Santee Canal.

Sault Ste. Marie Bascule Bridge (Sault Ste. Marie, Michigan; 1941)

At a length of 336 feet (102 meters), this bridge holds the record for a double-leaf **bascule** span. It has been rebuilt since its collapse on 7 October 1941 when the two leaves failed to latch together properly, resulting in two deaths.

Additional Information: Steinman, David B. *Famous Bridges of the World*. London: Dover Publications, 1953, 1961.

Sava Bridge (Belgrade, Yugoslavia; 1956)

Built as part of the European reconstruction after World War II, the Sava Bridge is one of the longest simple **box girders** in the world. This 856-foot-long (261-meter) bridge over the Sava River in Belgrade is also one of the longest self-supporting **girder** spans in the world.

Additional Information: Brown, David J. *Bridges*. New York: Macmillan, 1993; Overman, Michael. *Roads, Bridges and Tunnels*. Garden City, NY: Doubleday & Company, 1968.

Save River Bridge (southern Mozambique; 1970)

When built, the Save River Bridge was the longest multiple **suspension bridge** in the world. It has two spans of 328 feet (100 meters) each, and three spans of 689 feet (210 meters) each.

Additional Information: *Academic American Encyclopedia* (electronic version). Danbury, CT: Grolier Electronic Publishing, 1991.

Saveh Dam (near Saveh, Iran; c. 1281-1284)

Built in the thirteenth century, the Saveh Dam was an extremely high dam for its time. The Saveh Dam is located southeast of the town of Saveh in eastern Iran. This rubble-masonry gravity dam is 60 feet (18.3 meters) high and 150 feet (46 meters) long at its crest. The water face of the dam has a number of protrusions, the function of which is not totally known; probably they were placed there as part of a scheme to allow water to flow through the dam, despite silting. Built on limestone rock, the dam itself is well constructed, but the rock on which it sits is not solid. Soon after the dam was built, water undermined it; although the dam held, water flowed from underneath it, rendering the project useless. The limestone rock extends downward below the foundation for a significant distance. According to legend, the engineer who built the dam committed suicide upon discovering that he had miscalculated the strength of the foundation rock.

Additional Information: Smith, Norman. *A History of Dams*. Secaucus, NJ: Citadel Press, 1972.

Sayano-Shushensk Dam (Yeniseysk, Central Siberia, Russia; under construction)

The Sayano-Shushensk Dam across the Yenisey River is one of the world's highest dams; this gravity-arch dam has a height above its lowest formation of 245 meters (804 feet). The dam has a gigantic power capacity of 6,400 megawatts, and construction is underway to expand its capacity even further. Currently, its reservoir can hold more than 124,000,000 **acre-feet** of water.

Additional Information: Schodek, Daniel L. *Landmarks in American Civil Engineering*. Cambridge, MA: MIT Press, 1987.

Scarlet's Mill Bridge (Scarlet's Mill, Pennsylavania; 1881)

John L. Foreman designed the Scarlet's Mill Bridge in 1881. Originally used on a Reading Railroad crossing, the bridge was moved to its present site on the Horse-Shoe Trail over the Reading Railroad near Scarlet's Mills sometime between 1907 and 1935. The bridge is unusual because it contains both **cast-iron** and **wrought-iron**, even though it was built in the late nineteenth century when both of those materials were rarely being used in bridge construction. A **Pratt truss**, the bridge is also unusual in that it uses an ellipse rather than an arch for its upper chord.

Additional Information: DeLony, Eric. *Landmark American Bridges*. Boston: Bullfinch Press, 1993.

Schaffhausen Bridge (Schaffhausen, Switzerland; 1755)

Ulrich Grubenmann built this covered bridge across the Rhine in Schaffhausen in northern Switzerland. The bridge was 390 feet (119 meters) long. The original plan was for a single span, but village officials required two spans because they thought that 390 feet was too long. Grubenmann built the bridge as required, but made it self-supporting so that it did not rest on the center pier unless it was fully loaded. The bridge was destroyed by Napoleonic troops in 1799.

Additional Information: Steinman, David B. *Famous Bridges of the World*. London: Dover Publications, 1953, 1961.

Schoharie Bridge (Amersterdam, New York; 1956)

The Schoharie Bridge crossed Schoharie Creek near Amersterdam, New York, about 40 miles (64 kilometers) west of Albany. It collapsed on 5 April 1987. The collapse of this 540-foot-long (165-meter) bridge, which killed 10, was caused by flooding, **scour**, and, most importantly, a lack of redundancy in the bridge's construction. Scour caused a pier to tip into a scoured hole, and the entire structure collapsed.

Additional Information: Matthys, Levy and Salvadori, Mario. *Why Buildings Fall Down*. New York: W.W. Norton & Company, 1992.

Schoharie Creek Aqueduct (Fort Hunter, New York; 1841)

John B. Jervis built the Schoharie Creek Aqueduct to carry the ERIE CANAL over Schoharie Creek near Fort Hunter in eastern New York. Constructed in 1841, 16 years after the opening of the Erie Canal, the **aqueduct** is a permanent structure in stone and wood with the canal's towpath running along its arches.

Additional Information: DeLony, Eric. *Landmark American Bridges*. Boston: Bullfinch Press, 1993.

Schwandbach Bridge (Schwarzenburg, Switzerland; 1924)

Robert Maillart built this curved bridge in the Swiss canton (state) of Luzerne. The bridge is made of **reinforced concrete** that is never more than 8 inches (20 centimeters) thick. The bridge, with a span of 112 feet (34 meters) over a deep ravine, is one of Maillart's early masterpieces.

Additional Information: *Academic American Encyclopedia* (electronic version). Danbury, CT: Grolier Electronic Publishing, 1991.

Sciotoville Railway Bridge (Sciotodale, Ohio; 1917)

The Sciotoville Railway Bridge crosses the Ohio River, connecting Ohio with Kentucky. The bridge has two spans of 775 feet (236 meters) each. The bridge is continuous; it has no break in the decking at the supported point. It held the length record for its bridge type until 1943.

Additional Information: Steinman, David B. *Famous Bridges of the World*. London: Dover Publications, 1953, 1961.

Sears Tower (Chicago, Illinois; 1974)

Constructed in less than two and a half years, the Sears Tower was home to Sears, Roebuck Company, one of the largest retailers in the United States. Sears no longer owns the tower. Occupying two city blocks, the tower reaches 1,454 feet (443 meters), more than one quarter of a mile, into the sky. Its 110 stories hold some 4.5 million square feet of office and commercial space. Movement within the towers is managed by 16 escalators that serve the lower levels, and 102 high speed elevator cabs for the upper floors. The tower's status as the world's tallest building ended on 13 February 1996, when the soon-to-be-completed PETRONAS TOWERS in Kuala Lumpur, Malaysia, were topped off.

Additional Information: http://cpl.lib.uic.edu/004chicago/timeline/searstower.html; Lindgren, Hugo. "Visible City: Sears Tower." *Metropolis* (November 1994), reprinted in http://www.enews.com/magazines/metropolis/archive/1994/110194.1html; http://bechtel.colorado.edu/Gradua..Cven3525/Sears/Overview/facts.html.

Second Bosporus Bridge (Istanbul, Turkey; 1992)

This recently constructed bridge across the Bosporus, the narrow waterway dividing Europe from Asia, is possibly the finest **suspension bridge** in Europe. It has a main span of 3,580 feet (1,092 meters) and an aerodynamic deck to cut down on wind resistance. Its supporting cables are hung vertically instead of in the traditional angled mode. *See also* the BOSPORUS BRIDGE.

Additional Information: Brown, David J. *Bridges*. New York: Macmillan, 1993.

Segovia Aqueduct (Segovia, Spain; second century A.D.)

The Segovia Aqueduct is one of the oldest surviving Roman **aqueducts** in Europe. It is 800 meters (2,624 feet) long, with a double row of 16.5-foot (5-meter) arches, one on top of the other, reaching 118 feet (36 meters) at the highest point. The aqueduct was built by the Emperor Trajan as part of a system to bring water to Segovia in central Spain from the Guadarrama Mountains 62 miles (100 kilometers) away to the east. The aqueduct is also impressive for its narrowness; the bases of the piers, normally the thickest part of such a structure, are only 8 feet (2.4 meters) wide.

Additional Information: Brown, David J. *Bridges*. New York: Macmillan, 1993.

Seikan Railroad Tunnel (Honshu to Hokkaido Islands, Japan; 1985)

The 33.5-mile-long (54-kilometer) Seikan is one of the longest railroad tunnels in the world; it runs under the Tsugaru Strait to connect the main Japanese island of Honshu with the northern island of Hokkaido. Plans for the tunnel were announced in 1963, and digging began in 1971. The tunnel has a main section for trains, a smaller service tunnel underneath, and a pilot tunnel in between the two. As expected, the most difficult part of the job was the 14.5-mile (23-kilometer) section underneath Tsugaru Strait. Among the many obstacles faced by the workers was the incredible depth of the proposed tunnel. Three hundred feet (91.5 meters) below the bottom of the sea, tunnelers could only safely work for a single hour out of the work day; the rest of the time was spent in pressurizing and depressurizing them to protect against the **bends**. Underwater earthquakes and ground movements also meant frequent blowouts and floods; a chemical-concrete mixture was squirted into the muck as much as 200 feet (61 meters) ahead of the shield. When this mixture hardened, it could be drilled through in a conventional manner. At one point in the construction, four months were needed to advance a mere 40 feet (12.2 meters). Despite the best precautions, 34 workers died during the construction of the project. The main tunnel was holed through on 16 March 1985, but several more years of preparation were needed to electrify the tunnel and otherwise ready it to accommodate the railroad.

Additional Information: Epstein, Sam and Epstein, Beryl. *Tunnels*. Boston: Little Brown & Company, 1985.

Selkirk Lighthouse (Selkirk, New York; 1838)

Joseph Gibbs, Abner French, and Jabez Meacham built the famous Selkirk Lighthouse near the hamlet of Selkirk at the mouth of the Salmon River along the eastern shore of Lake Ontario. In the 1830s, the area was the center of a thriving fishing and ship-building industry. The land for the lighthouse was purchased by the federal government in 1837. The local contractors completed the lighthouse by August 1838, as demanded by the federal government. Selkirk is one of only four remaining lighthouses to have an unmodified "birdcage" lantern structure. Originally, the eight lamps and reflectors in the lighthouse used whale oil; the light was upgraded in 1855 with a mineral oil lamp and a 270-degree, 6th-order **Fresnel lens**, 18 inches (46 centimeters) high and about 12 inches (30 centimeters) in diameter. In 1858, after the area's boat-building industry died away, the lighthouse was deactivated; in 1877, the lighthouse began a short period as a life saving station for Lake Ontario. After deactivation, the lighthouse was used as a home by several people, and in 1899 owner Leopold Joh began a successful resort and hotel complex, with the lighthouse at the center. The current owner, Jim Walker, bought the lighthouse in July 1987, and is restoring it with an updated lighting system. Selkirk Lighthouse is now officially listed as a "Class II navigation aid" by the Coast Guard, and Walker may upgrade it even further.

Additional Information: http://www.maine.com/lights.

Seoul Tower (Seoul, South Korea; 1980)

Designed to be the symbol of the new, industrial South Korea, this 236-meter-high (774-foot) tower stands in the heart of the city.

Source: http://www.centrepoint.com.au/tower/index.html.

Sepulcrum Antoninorum. *See* Castel Sant'Angelo.

Sergeantsville Bridge. *See* Green Sergeant's Bridge.

Seven Wonders of the Ancient World

In the fifth century B.C., the Greek historian Herodotus mentioned a list of seven wonders, as did Callimachus of Cyrene, the librarian at Alexandria, in the third century B.C. The "canonical" or accepted list of wonders of the ancient world seems to have been finalized in Europe during the Middle Ages. From oldest to youngest, the seven wonders were the GREAT PYRAMID OF GIZA in Egypt; the HANGING GARDENS OF BABYLON or the Walls of Babylon in what is today Iraq; the OLYMPIAN ZEUS, a statue of the god by Phidias at Olympia in Greece; the TEMPLE OF ARTEMIS or Artemision, a temple of the goddess Artemis at Ephesus in Asia Minor (modern Turkey); the MAUSOLEUM, the tomb of Mausolus at Halicarnassus in Asia Minor; the COLOSSUS OF RHODES, a statue of the sun god Helios at the entrance of Rhodes harbor; and the LIGHTHOUSE AT ALEXANDRIA in Egypt. The Great Pyramid of Giza is the only one of the seven still in existence.

Additional Information: http://pharos.bu.edu/Egypt/Wonders/list.html.

Severin Bridge (Cologne, Germany; 1959)

The Severin Bridge over the Rhine in western Germany is the longest "bridle chord" **box girder** ever built. The style is economical because a box girder, partly supported by cables passing over a tower, is cheaper than either a plain box girder or a **suspension bridge** in the 200- to 300-meter (656- to 984-foot) range. This asymmetrical bridge spans 297 meters (974 feet) and has three sets of paired cables.

Additional Information: Overman, Michael. *Roads, Bridges and Tunnels.* Garden City, NY: Doubleday & Company, 1968.

Severn Bridge (western England; 1966)

The Severn Bridge is the most economical giant **suspension bridge** ever built. With a span of 972 meters (3,188 feet), it carries four automobile lanes, and has a **live load** capacity of 1,620 kilograms (3,572 pounds) per meter run. Its economy can be illustrated by a comparison with the VERRAZANO NARROWS BRIDGE in New York City. The Severn Bridge has an average **dead weight** per meter run of only 4,050 kilograms (8,930 pounds), compared with the Verrazano's 50,000 kilograms (110,250 pounds). The towers each weigh 1 million kilograms (over 2.2 million pounds), compared with the Verrazano's towers of 27 million kilograms (over 59.5 million pounds). The cost of the Severn Bridge was about $33 million dollars, approximately one-tenth the cost of the Verrazano.

Additional Information: Overman, Michael. *Roads, Bridges and Tunnels.* Garden City, NY: Doubleday & Company, 1968.

Shanghai World Financial Center (Shanghai, China; expected completion 2001)

When completed, this 1,509-foot-high (460-meter), 59-story building will be the tallest in the world. An elegant, upended chisel shape, the building's height is a result of the Chinese government's construction specifications. One of several skyscrapers being built or planned in Asia, the Shanghai World Financial Center is intended to impress. According to a press release of the primary building contractor, Mori Building Company, the tower is supposed to demonstrate "the prosperity of the city of Shanghai and to embody the hopes and lofty ideals of the people of China. . . ."

Additional Information: Pacelle, Mitchell. "U.S. Architects in Asia: Only Way to Go Is Up," *Wall Street Journal*, 21 March 1996, B:1.

Shasta Dam (near Redding, California; 1938-1945)

This major northern California dam is located 12 miles (19 kilometers) northwest of Redding. Shasta Dam is a vital part of the California water system. Originally constructed between 1938 and 1945, the dam was modified between 1995 and 1996. A concrete curved gravity dam, Shasta has a total storage capacity of 4,555,000 **acre-feet**. Its crest is 3,460 feet (1,055 meters) long and 30 feet (9 meters) wide. The dam is 602 feet (184 meters) high. Shasta Dam controls flood waters and stores water for irrigation in the Sacramento and San Joaquin valleys. It also maintains navigation flows in the Sacramento River, provides water for industrial use, and generates hydroelectric power.

Additional Information: http://donews.do.usbr.gov/dams/Shasta.html.

Sheboygan Bascule Bridge (Sheboygan, Wisconsin; 1995)

The Sheboygan Bascule Bridge is the first fully unbalanced **bascule bridge** of its size in the United States. The bridge crosses the Sheboygan River in the eastern Wisconsin town of Sheboygan. Conventional bascule

A view of the Sheboygan Bascule Bridge in Sheboygan, Wisconsin. Photo courtesy of Donna Bronski.

bridges are lifted from either side; this bridge lifts upward from only one side. The design of the structure was mandated by the Environmental Protection Agency rehabilitation site below the bridge; contaminated sediment from the river below made a conventional double base too expensive to construct.

Additional Information: "News: Unconventional Bascule Bridge Crosses Superfund Site." *Civil Engineering* 65:12 (December 1995), pp. 12-14.

Shoshone Dam. *See* Buffalo Bill Dam.

Shrine of Democracy. *See* Mount Rushmore National Monument.

Shubenacadie Canal (Dartmouth-Halifax, Nova Scotia, Canada; 1858)
The Shubenacadie Canal opened to traffic in 1858, and continued in service until 1870. The entrance to the canal was via an inclined plane at Halifax harbor. The canal, using inclined planes throughout, ran in a westerly direction across Nova Scotia from Halifax to the Bay of Fundy.

Additional Information: http://ttg.sba.dal.ca/nstour/halifax/points/downdart.html.

Silver Bridge. *See* Point Pleasant Bridge.

Simplon Pass Road (Switzerland to Italy; 1805)
This road through the Simplon Pass was built by Napoleon to allow his armies access to Italy; Simplon was the first major European road project in 1,500 years. Road conditions throughout Europe were abominable in the eighteenth century when Napoleon determined to build a new road over the Alps at Simplon Pass. Because Napoleon insisted that work progress as quickly as possible, the initial construction cost several hundred lives. A new manager took over the project in 1802, and the number of deaths and injuries dropped. By 1805, the road was completed. Practically all the work was done by hand, and gunpowder, set off at close proximity, was the only explosive used. The roadwork included more than 600 small bridges of stone or wood. When completed, the roadway was 16 to 20 feet (4.9 to 6.1 meters) wide, sufficient for two-way traffic to travel comfortably. The road passed the village of Simplon at a height of 6,589 feet (2,010 meters).

Additional Information: Schreiber, Herman. *The History of Roads: From Amber Route to Motorway.* London: Barrie and Rockliff, 1961.

Simplon Tunnel (Switzerland to Italy; 1906-1908)
The Simplon Tunnel runs under Simplon Pass (see SIMPLON PASS ROAD) in the Alps between Switzerland and Italy. For many years, the Simplon Tunnel was the deep-

est mainline railroad tunnel in the world; at one point, the Simplon runs 2,100 meters (6,888 feet) below the crags of Monte Leone. It may have been the most difficult tunneling work ever undertaken. Until overtaken by the DAI-SHIMUZU TUNNEL in Japan in 1979, the Simplon was the longest railroad tunnel in the world. On the Italian side, workers blasted through 4.4 kilometers (2.7 miles) before hitting soft decomposed rock (calcerous mica-schist) that was so "fluid" the conventional timbering of the small pilot gallery was slowly crushed. Roof timbers were replaced with 40-centimeter (15.6-inch) steel joints alternating with 50-centimeter (19.5-centimeter) balks of pitch pine, but even these failed to hold. Eventually, rapid-hardening cement was forced into every gap. The tunnel next had to be opened out to full size and permanently lined. This task was done in a step-by-step method, with every cut shored up immediately with heavy timber. Temporary brickwork was added between the pilot gallery and the inner line of the finished tunnel lining to give added strength while workers laid the 2-meter-thick (6.6-foot) masonry of the permanent lining. A second obstacle was water, faced first by the Italian workers who struck what was called "the Great Spring" from which water gushed at 48,000 liters (over 12,000 gallons) per minute, originally at a pressure of about 40 kilograms per centimeter. Because of delays on the Italian side, the Swiss completed their half of the tunnel much more rapidly. To save time, the Swiss agreed to push beyond the halfway mark. After digging an additional kilometer further south, the Swiss also ran into a flood of hot water. As a safety precaution against the uncontrolled flow of hot water, iron safety doors were fitted on the Swiss side, and later the Swiss actually came across an exceptionally hot spring releasing over 2,250 liters (596 gallons) per minute; work was abandoned and the doors sealed. The Italians were left to push on, encountering more and hotter water. One spring produced 6,300 liters (1,670 gallons) per minute at a temperature of 47 degrees centigrade. The heat was so great that an additional 4,500 liters (1,193 gallons) per minute of cold water had to be pumped in and released as a spray at the working face to reduce the air temperature sufficiently for work to continue. When the Italians finally holed through to the Swiss **heading**, the hot water that was released in a torrent took half an hour to escape. The last 180 meters (590 feet) took the Italians an additional six months to finish. A second, parallel tunnel opened in 1922.

Additional Information: Epstein, Sam and Epstein, Beryl. *Tunnels.* Boston: Little Brown & Company, 1985.

Sistine Chapel (Vatican City; 1481)
The Sistine Chapel, the papal chapel in the Vatican Palace, was designed and built between 1473 and 1481 by Giovanni dei Dolci for Pope Sixtus IV, for whom the chapel is named. The Sistine is a rectangular brick structure with a barrel-vaulted ceiling and six arched win-

dows running along each of the two side walls. The outside of the chapel is unadorned and undistinguished, but the interior is decorated with some of the most magnificent and famous frescoes in Western art. The frescoes on the side walls were painted between 1481 and

The fifteenth-century Sistine Chapel in the Vatican City is best known for its ceiling frescoes, painted in the early sixteenth century by Michelangelo.

1483 by such Renaissance masters as Perugino, Pinturicchio, and Botticelli. The six frescoes on the north wall depict events from the life of Christ; the six on the south wall portray the life of Moses. Above these scenes are smaller frescoes between the windows depicting various popes. For important ceremonial occasions, the lower portions of the side walls are covered with a series of tapestries designed by Raphael and woven in Brussels (now in Belgium) between 1515 and 1519. The tapestries illustrate events from the Gospels and the Acts of the Apostles. The most famous and most important frescoes in the Sistine were painted on the ceiling by Michelangelo between 1508 and 1512 and on the west wall above the altar between 1534 and 1541. Pope Julius II commissioned the ceiling frescoes, which depict people and events from the Old Testament. Pope Paul III commissioned the "Last Judgment" fresco above the altar. A 10-year project to clean and restore the ceiling frescoes was completed in 1990. As the pope's personal chapel, the Sistine is the site of many important papal ceremonies, and is used by the College of Cardinals as a meeting place for the election of new popes.

Additional Information: Pietrangeli, Carlo et al. *The Sistine Chapel: A Glorious Restoration.* New York: H.N. Abrams, 1994.

Siuslaw River Bridge (Florence, Oregon; 1936)

The Siuslaw River Bridge carries the Oregon Coast Highway (U.S. Route 101) over the Siuslaw River near Florence in western Oregon. Built by Conde B. McCullough, this 140-foot (43-meter) double-leaf bascule **drawbridge** spans a narrow river channel. On either side of the bascules are two tied concrete arches, each 154 feet (47 meters) high. Control mechanisms for the **bascules** are sheltered in four houses at the peaks of the two piers at either side of the bascules. Twenty concrete deck **girders** approach the spans, 8 on the north side, and 12 on the south.

Additional Information: DeLony, Eric. *Landmark American Bridges.* Boston: Bullfinch Press, 1993.

Skarnsunder Bridge (Norway; 1991)

The Skarnsunder Bridge crosses Norway's Skarnsunde Strait, connecting the counties of Mosvik and Inderoya. At 3,314 feet (1,011 meters), the Skarnsunder broke the length record for a **cable-stayed bridge**. Although long, the bridge is an extraordinarily narrow 43 feet (13 meters) in width; the structure's two lanes are sufficient to handle the volume of traffic crossing the bridge. The Skarnsunder is a self-anchored cable-stayed bridge; the cables are anchored in the bridge itself. The center 1,739 feet (530 meters) of the bridge are suspended, and the sides of the bridge are supported on A-shaped pylons set in concrete piers.

Additional Information: Brown, David J. *Bridges.* New York: Macmillan, 1993; Erik Syversen at http://www.mti.unit.no/~eriks/pros...e/skarnsundet/skarnsundbridge.html.

Slaithwaite Reservoir. *See* Huddersfield Canal.

Smithfield Street Bridge (Pittsburgh, Pennsylvania; 1884)

Designed by Gustav Lindenthal, this bridge over the Monongahela River in Pittsburgh introduced the United States to an innovative European design, the **lenticular truss**. Smithfield Street Bridge is one of the oldest extant major steel trusses in the United States and was Lindenthal's earliest major project. The lenticular truss is also called the Pauli truss after the German inventor of the truss, Friedrich August von Pauli. The Smithfield Street Bridge has accepted several modifications. In 1889, an additional set of trusses and a second deck were added to carry streetcars. In 1898, these trusses were shifted upstream and widened for electric trolleys. In 1933, the bridge had its steel floor system replaced with structural aluminum beams and prefabricated aluminum decking. The reduction of the **dead load** allowed an equal increase in the bridge's ability to carry **live loads**.

Additional Information: Schodek, Daniel L. *Landmarks in American Civil Engineering*. Cambridge, MA: MIT Press, 1987; for information on the bridge's most recent maintenance upgrades, see "News: Study Shows Aluminum Deck Resists Corrosion." *Civil Engineering* 65:9 (September 1995), p. 10.

Snowden Lift Bridge (Snowden, Montana; 1913)

At the time of its construction, the Snowden Lift Bridge over the Missouri River in Montana was one of the longest **vertical lift bridges** in the world. The bridge consists of three riveted Parker **through-trusses**, each 275 feet (84 meters) long, and a 296-foot-long (90-meter) vertical lift span. The Burlington Northern Railway abandoned the bridge in the 1920s when river traffic decreased because of silting. The bridge was modified for highway use in 1926.

Additional Information: DeLony, Eric. *Landmark American Bridges*. Boston: Bullfinch Press, 1993.

Soldier's National Monument (Gettysburg National Battlefield Park, Gettysburg, Pennsylvania; 1869)

This highly ornate obelisk supposedly marks the site where Abraham Lincoln delivered the Gettysburg Address on 19 November 1863. In actuality, the address was delivered close by, but not on the monument site. The monument was rededicated in 1936 by President Franklin Roosevelt. Gettysburg is a small town in southeastern Pennsylvania near the Maryland border; one of the largest battles of the Civil War was fought here during the first three days of July 1863.

Additional Information: Eicher, David. *Civil War Battlefields: A Touring Guide*. Dallas: Taylor Publishing Company, 1995.

South Fork Dam, also known as the Johnstown Dam (Johnstown, Pennsylvania; 1839)

The South Fork Dam, one of the earliest dams in the United States, was to become one of the most infamous in May 1889 when its collapse caused the devastating Johnstown Flood. Built originally to supply a canal, the dam was abandoned for this use in 1857. The dam functioned well, and was properly made. It was 72 feet (22 meters) high and 840 feet (256 meters) long, with a thickness of 10 feet (3 meters) at the crest that grew to 200 feet (61 meters) at the base. The inner portion of the dam had a core of slate, and the water face was covered with loose stone. Downstream, the face had a **rockfill** pile of rubble masonry. A masonry-lined tunnel served as a low-level outlet. After 1857, the dam was used for fishing, and between 1862 and 1872, it provided water for locomotive boilers. In 1875, the dam was taken over by a hunting and fishing club,

which raised the height of the dam to 75 feet (23 meters) and partially screened off the dam's spillway. As a result of these changes, the dam could not withstand the heavy rains that fell in May 1889. With the spillway overloaded, the water soon cut into the embankment, and the entire dam collapsed. Further down the valley, 2,000 people were killed by the sudden and overwhelming rush of water. Changes that seemed innocuous were responsible for one of the worst dam disasters in history.

Additional Information: Matthys, Levy and Salvadori, Mario. *Why Buildings Fall Down*. New York: W.W. Norton & Company, 1992; Smith, Norman. *A History of Dams*. Secaucus, NJ: Citadel Press, 1972.

South Fork Newaukum River Bridge (near Onalaska, Lewis County, Washington; 1930)

The South Fork Newaukum River Bridge carries State Route 508 across the South Fork of the Newaukum River in southwestern Washington. This metal pony **Warren truss** remains in excellent condition, and is a classic example of the Warren truss bridge type.

Source: http://wsdot.wa.gov/eesc/environmental/Bridge-WA-112.htm.

South Halsted Street Bridge (Chicago, Illinois; 1894)

Built by J.A.L. Waddell, the South Halsted Street Bridge is the world's first **vertical-lift bridge**.

Additional Information: Steinman, David B. *Famous Bridges of the World*. London: Dover Publications, 1953, 1961.

South Saskatchewan Dam (Saskatchewan, Canada; 1966)

This impressive earth dam is 223 feet (68 meters) high and 16,700 feet (5,094 meters) long. The dam's volume is 86,328,000 cubic yards (66,041,000 cubic meters), with an additional reservoir capacity of 8,000,000 **acre-feet**.

Additional Information: Smith, Norman. *A History of Dams*. Secaucus, NJ: Citadel Press, 1972.

Southwark Bridge (London, England; 1819)

John Rennie built this early **cast-iron** arch over the Thames River to connect London with Southwark, its southern suburb. The bridge had a central span of 240 feet (73 meters), and used plates of solid cast-iron. The strength of iron was not completely understood in the early nineteenth century; the ribs of the structure could have been open-work **voussoirs** and they still would have been strong enough to support the bridge.

Additional Information: Kirby, Richard Shelton et al. *Engineering in History*. New York: Dover Publications, 1990.

Space Needle (Seattle, Washington; 1962)

Conceived by Edward E. Carlson and designed by John Graham, the Space Needle was built for the Seattle World's Fair of 1962. The 605-foot-high (185-meter) Space Needle dominates the area around it with its futuristic look. It has a viewing deck, a revolving restau-

A nighttime view of the Space Needle in Seattle, Washington.

rant, and numerous stores. Despite its commercial orientation, the Needle is an amazing structure. The revolving restaurant on top of the Needle is balanced so well that it moves with just the force of a one horse-power motor (with the highest gear ration in the world of 360,000 to 1). Its foundation hole, 30 feet by 120 feet (9 by 37 meters), took 467 truck-loads of cement to fill, the largest continuous concrete pour ever performed in the United States. At completion, the foundation weighed almost 6,000 tons (5,400 metric tons), as much as the structure itself.

Additional Information: http://spaceneedle.com/whatis.html; http://spaceneedle.com/histfaq.html

Spassky Tower. *See* Kremlin.

Sphinx. *See* Great Sphinx.

Spotsylvania Confederate Cemetery. *See* Fredericksburg and Spotsylvania National Military Park.

Standedge Tunnel (near Huddersfield, England; 1811)

This 3-mile (4.8-kilometer) tunnel was part of the HUDDERSFIELD CANAL, which ran through northwestern England northeast of the city of Manchester. Gunpowder was used to blast away hard rock. During the 17 years of construction, many injuries and deaths occurred, but the completed tunnel became a vital part of England's economic growth during the later stages of the Industrial Revolution. Like many canal tunnels, the only way to navigate the Standedge Tunnel was for bargemen to lie on top of their barges and push the barge through by pushing with their feet against the top or the sides of the tunnel (a process called "legging"). The trip through the Standedge usually took about four hours, although a speed record was set by David Whiteman, who once moved his barge through the tunnel in one hour and 25 minutes.

Additional Information: Epstein, Sam and Epstein, Beryl. *Tunnels.* Boston: Little Brown & Company, 1985.

Staple Bend Tunnel (near Johnstown, Pennsylvania; 1835)

The first railroad tunnel built in the United States, the Staple Bend Tunnel was constructed on the Allegheny Portage Railroad in southwestern Pennsylvania to replace canal locks through the area's low hills. The railroad is gone but its single tunnel, the 700-foot-long (213.5-meter) Staple Bend, still exists.

Additional Information: Epstein, Sam and Epstein, Beryl. *Tunnels.* Boston: Little Brown & Company, 1985.

Starrucca Viaduct (near Lanesboro, Pennsylvania; 1848)

Designed by Julius Adams and James P. Kirkwood, the Starrucca Viaduct carried the Erie Railroad over Starrucca Creek. Pressure from investors and financiers pushed the **viaduct** to completion in just two years. Although one of the most expensive building projects of its time, the Starrucca Viaduct was both beautiful and financially successful. The structure has 17 arches, each 51 feet (15.5 meters) long, and a total length of 1,040 feet (317 meters). The viaduct is 110 feet (34 meters) above the stream bed. The main piers constituted one of the first significant uses of concrete in the United States. Although built to carry 50-ton (45-metric-ton) locomotives, the viaduct is still in use today; it is able to carry four 100-ton (90-metric-ton) trains with no difficulty.

Additional Information: DeLony, Eric. *Landmark American Bridges.* Boston: Bullfinch Press, 1993; Schodek, Daniel L. *Landmarks in American Civil Engineering.* Cambridge, MA: MIT Press, 1987.

Statue of Liberty (New York City; 1884)

The colossal Statue of Liberty, one of the best-known of American national monuments, stands on Liberty Island in New York harbor. Known originally as "Lib-

A view of the Statue of Liberty in New York harbor.

erty Enlightening the World," the staute was proposed by Edouard Laboulaye as a commemoration of the American and French revolutions. The Franco-American Union was organized in 1875 to raise funds for the project. F. A. Bartholdi designed the statue in the shape

of a woman with arm uplifted holding a torch. The 152-foot (46-meter) statue was constructed of copper sheets from Bartholdi's 9-foot (2.7-meter) model. The Franco-American Union presented the statue to the United States in 1884, and it was dedicated in 1886. Emma Lazarus's 1886 sonnet to the statue, "The New Colossus," was engraved on the statue's pedestal. The base of the statue is an 11-pointed star, part of Fort Wood, which originally stood on the island to protect New York harbor. The 150-foot-high (46-meter) pedestal is made of concrete faced with granite. The statue was one of the first sights seen by turn-of-the-century immigrants as they came into New York harbor bound for processing on nearby Ellis Island. In 1924, the Statue of Liberty became a national monument.

Additional Information: Trachtenberg, Marvin. *Statue of Liberty.* New York: Viking Press, 1976.

Steel Bridge (Portland, Oregon; 1912)

The Steel Bridge carries the Union Pacific and Southern Pacific Railroads over the Willamette River at Portland in northwestern Oregon. This unique double-deck steel structure has a 211-foot (64-meter) double-deck **vertical-lift** truss that can telescope. The engineers both invented and patented the design. The lower railway deck can be raised by itself or, if necessary, both the lower deck and the upper vehicular deck can also be raised to allow a tall ship to pass.

Additional Information: DeLony, Eric. *Landmark American Bridges.* Boston: Bullfinch Press, 1993.

Step Pyramid of Djoser (Zoser) (Saqqara, Egypt; twenty-seventh century B.C.)

The step pyramid of the Third Dynasty pharaoh Djoser predates the GREAT PYRAMID OF GIZA, which, along with its companion pyramids, was constructed about a century later by rulers of the Fourth Dynasty. Djoser's step pyramid and its surrounding mortuary complex were the first major buildings in the world to be constructed

The Step Pyramid of Djoser at Saqqara, Egypt, predates the Great Pyramid of Giza.

entirely of stone. Construction of the pyramid was directed by the pharaoh's talented vizier (i.e., first minister) Imhotep. The pyramid consisted of six unequal steps rising to a height of 204 feet (62 meters). The base of the pyramid is 358 by 411 feet (109 by 125 meters). The substructure of the pyramid is a honeycomb of shafts and tunnels, some of which were dug later by grave robbers. Vast quantites of stone vases, many of great artistry, were found beneath the pyramid, which, like the larger structures at Giza, was meant as a tomb for the pharaoh. A series of smaller tombs and temples comprise the mortuary complex, which was enclosed with the pyramid within a rectangular colonnade. Djoser's family and chief officials, including Imhotep, were likely buried in the tombs surrounding the pyramid.

Additional Information: http://ccat.sas.upenn.edu/arth/zoser.html.

Stewart Mountain Dam. *See* Salt River Project.

Stewartstown Railroad Bridge (Stewartsville, Pennsylvania; 1870)

This bridge carries the Stewartstown Railroad over the Valley Road in Stewartsville, Pennsylvania. Jacob H. Linville originally constructed the bridge in Denver, Colorado, for the Keystone Bridge Company of Pittsburgh, a predecessor of the U.S. Steel Corporation. The bridge's features include decorative end-post castings and bosses on the column spacers. Keystone bridges used distinctive split **compression** members.

Additional Information: DeLony, Eric. *Landmark American Bridges*. Boston: Bulfinch Press, 1993.

Stone Arch Bridge (Viaduct) (Minneapolis, Minnesota; 1883)

Colonel Charles C. Smith built this bridge for the Burlington Northern Railway at a time when masonry structures were generally unknown in the area. The **viaduct** carried the main transcontinental passenger line across the Mississippi and was an important link in the development of the American Northwest. Previous bridges on the same site included an 1855 **suspension bridge** with a 550-foot (168-meter) span, which was rebuilt in 1876 and removed in 1889. The site for the Stone Arch Bridge was chosen for its closeness to Minneapolis and to nearby rail routes. Because of the terrain, this **arch bridge** had to be built at an angle to the Falls of Saint Anthony. The unique curve in the design caused high building costs, with the project coming in at $690,000.

Because of the curve, a train passenger can see the bridge at the same time that it is being crossed. The bridge is a double-track structure, 2,100 feet (640.5 meters) long, 76 feet (23 meters) high, and 26 to 28 feet (8 to 8.5 meters) wide, with 23 circular stone arches spanning lengths of between 40 and 100 feet (12.2 and 30.5 meters). The bridge required 30,554 cubic yards (23,374 cubic meters) of masonry, and 18,000 cubic yards (13,770 cubic meters) of filling for construction. It was repaired and rebuilt several times. Modifications included reinforcement and new drains between 1907 and 1911, wider tracks in 1925, and repair of flood damage in 1965. Also in 1965, three piers were found to be weakened by flooding and **scour**. To prevent vibrations that might bring the bridge down, preventive measures were taken, although no power tools were used during the repairs in an effort to imitate the type of work that would have been done at the time of the bridge's original construction.

Additional Information: Schodek, Daniel L. *Landmarks in American Civil Engineering*. Cambridge, MA: MIT Press, 1987.

Stonehenge (Salisbury Plain, Wiltshire, England; c. 2500 B.C.)

Stonehenge, on Salisbury Plain in Wiltshire in southern England, is the best known of the many megalithic monuments found in the British Isles. Stonehenge is a

Stonehenge is a ring of standing stones erected on Salisbury Plain in southwestern England probably in the third millennium B.C.

group of large standing stones arranged in four series. The two outermost series form circles, while the third series takes a horseshoe shape, and the fourth an ovoid shape. Within the ovoid lies the stone altar. The whole structure is enclosed by a circular ditch that measures 300 feet (91 meters) in diameter. The outer series is made of sarsen stones that stand about 13.5 feet (4.1 meters) in height and form a circle about 100 feet (30.5 meters) in diameter. About half of the 30 original uprights are still standing. The uprights were originally topped by a continuous line of dove-tailed lintels held by **mortise**-and-tenon fastening to two uprights. The exact purpose of the structure is unclear, though some kind of religious function seems likely. Recent theories hold that the structure served as a huge astronomical instrument, for the stones seem to be arranged in such a way as to point directly at the rising sun on the summer solstice. Estimates of the age of the structure have varied widely, but recent evidence from excavations at nearby Neolithic and Bronze Age sites has placed construction of the circle in the middle of the third millenium B.C.

Additional Information: Chippindale, Christopher. *Stonehenge Complete*. London: Thames and Hudson, 1983.

Storm Cliff Tunnel. *See* Mitchell's Point Tunnel.

Storstrom Bridge (Sjaelland and Falstar Islands, Denmark; 1937)

This massive structure uses 50 steel spans to link the Danish island of Sjaelland, on which stands Copenhagen, with the island of Falstar. A **beam bridge**, Storstrom's 47 side spans are plate **girders** varying in length from 53 to 61 meters (174 to 200 feet) and weighing up to 500,000 kilograms each. The final placement of the beams was on piers 30 meters (98 feet) above water level using a huge 610,000-kilogram floating crane. The main navigation spans are **tied arches**, two of 101 meters (331 feet) and one of 134 meters (440 feet); the arches were placed in parts—first the two bottom chords supported at their centers by trestles; then a smaller crane was placed across these girders and used to erect the arch ribs and hangers.

Additional Information: Overman, Michael. *Roads, Bridges and Tunnels*. Garden City, NY: Doubleday & Company, 1968.

Suez Canal (northeastern Egypt; 1869)

The desirability of a canal to connect the Red Sea with the Mediterranean was long recognized. In the nineteenth or twentieth century B.C., the Egyptians built a canal in the area to connect the two bodies of water. In the fifth century B.C., when the Red Sea had receded, the Persian king Xerxes I, who controlled Egypt, ordered the canal extended. The waterway was restored several times, notably by the Egyptian ruler Ptolemy II in the third century B.C., and by the Roman emperor Trajan at the start of the second century A.D. After the eighth century A.D., the canal was closed and fell into complete disrepair. Planning for the modern canal began in 1844. Ferdinand de Lesseps, a French engineer and entrepreneur, began construction on 25 April 1859. De Lesseps bowed out of the project after 1863, but construction continued until 1869, when the canal opened with a performance of Verdi's opera *Aida,* which was written specifically for the opening. The canal extends from Port Said on the Mediterranean southward in a straight line to Lake Timsah, where a new cutting takes the waterway to the Bitter Lakes (now one body of water) and then to the Gulf of Suez, which leads into the Red Sea. Port Tewfik near the town of Suez is the southern terminus of the canal. The canal is over 100 miles (161 kilometers) long, with a maximum bottom width of 196 feet (60 meters) and a depth of over 42 feet (13 meters). The canal has no locks and can accommodate ships of almost any draft. The original canal that opened in 1869 was not the efficient transport system the canal is now. Great Britain acquired control of the canal in 1875 by purchasing the Egyptian ruler's interest in the waterway. The British, who were anxious for better and faster communications with their important colony in India, assumed a protectorate over Egypt in 1885 and undertook to widen and improve the canal. The British were guarantors of the neutrality and accessibility of the canal until 1956, when Egypt nationalized the waterway. The Egyptian action led to a joint British-French-Israeli invasion of the canal zone, but the invaders quickly withdrew under Soviet and American pressure and the canal remained under Egyptian control. The Suez Canal was closed to Israeli shipping from 1948 to 1975 and to all shipping from 1967 to 1975, following the Arab-Israeli War of 1967. The waterway reopened in 1975, after being cleared of mines and war wreckage, and was enlarged in 1976-80 to accommodate oil supertankers.

Additional Information: Kinross, Lord John Patrick Douglas Balfour. *Between Two Seas: The Creation of the Suez Canal*. New York: William Morrow and Company, 1969; Sandstrom, Gosta E. *Man the Builder*. New York: McGraw-Hill, 1970.

Sunderland Bridge (Sunderland, England; 1796)

Thomas Wilson built this bridge across the River Wear near Sunderland in northeastern England in 1796. The Sunderland Bridge was the next major **arch bridge** built after COALBROOKDALE BRIDGE was constructed across the Severn in western England in 1779. The Sunderland Bridge had an unprecedented span of 236 feet (72 meters). For years, the Sunderland was thought to contain parts adapted from Thomas Paine's design for the WEAR BRIDGE, which had been exhibited in London in 1788.

Source: DeLony, Eric. *Landmark American Bridges*. Boston: Bullfinch Press, 1993.

Sunshine Skyway Bridge (Tampa Bay, Florida; 1986)

A replacement for an earlier bridge destroyed by a passing freighter, the Sunshine Skyway Bridge crosses the central navigation channel of Tampa Bay at a height of 175 feet (53 meters). The bridge is an attractive **cable-stayed** design with a length of 21,877 feet (6,673 meters). It is one of the largest concrete spans in the United States.

Additional Information: Brown, David J. *Bridges*. New York: Macmillan, 1993.

Sweet Track (near Glastonbury, England; c. 3300 B.C.)

Sweet Track, in western England near Glastonbury, is likely the world's oldest extant manufactured track or way. It linked settlements in the hills with an island in a swamp. Using longitudinal logs supporting V-shaped pegs, with planks placed between the Vs to create a footpath, the 2-kilometer (1.2-mile) way could have been constructed in one day if the wood for its construction were available on site.

Additional Information: Lay, M.G. *Ways of the World*. New Brunswick, NJ: Rutgers University Press, 1992.

Sydney Harbour Bridge (Sydney, Australia; 1932)

This trussed steel **arch bridge** crossing the harbor of Sydney in southeastern Australia has a span of 1,650 feet (503 meters). The bridge has two hinged arches,

Sydney Harbour Bridge in Sydney, Australia. Photo courtesy of Lawrence D. Savoy.

and is a half-through type with a 56-foot-wide (17-meter) roadway running between the top and bottom of the arch. The bridge also accommodates four sets of railroad tracks and two footpaths. Although shorter than the BAYONNE BRIDGE, the Sydney Harbour Bridge is designed to carry much heavier loads than the Bayonne. The thrust (pressure) exerted by one of the arch ribs against the **abutment** is about 22,000 tons. Dealing with heavy thrusts is a serious problem at some sites. In this bridge, each bearing (where the foundation meets the ground or rock underneath) is capable of withstanding a pressure of 20 million kilograms (over 44 million pounds) at an angle of 45 degrees. The surface rock was excavated to between 10 and 12 meters (33 to 39 feet). Trenches were cut into the excavated rock and filled with solid concrete. To avoid development of concrete planes that would permit shifting, the concrete was poured in individual hexagonal cells. Beneath the concrete are 23.75-centimeter-thick (9.3-inch) high-tensile bearing plates.

Additional Information: Steinman, David B. *Famous Bridges of the World*. London: Dover Publications, 1953, 1961; Overman, Michael. *Roads, Bridges and Tunnels*. Garden City, NY: Doubleday & Company, 1968.

Sydney Opera House (Sydney, Australia; 1973)

Actually a performing arts center rather than a single opera house, this gigantic and attractive complex contains nearly 1,000 rooms, including four auditoriums, five rehearsal studios, four restaurants, 60 dressing rooms and suites, and a library and artists' lounge. Built along Bennelong Point in the harbor of Sydney in southeastern Australia, the Sydney Opera House was designed by the famed Danish architect Jørn Utzon. Promotion of the project began in the early 1950s, and an appeal fund was set up to raise the estimated construction cost of $7 million. When less than $1 million was raised, the Opera House Lotteries were established; the lotteries eventually raised more than $100 million for the project. After an international competition, the design contract was awarded to Utzon. Construction began in March 1959 at Bennelong Point, which was named after the first Aborigine to speak English, who was born on the site. Although Utzon's design was recognized as a masterpiece, it proved structurally impossible to construct, and plans were modified after several years; the most important change was to give the roof vaults a spherical shape, allowing the roofs to be prefabricated off site, thereby reducing both time and cost. Utzon resigned from the project in February 1966, and a team of Australian architects completed the work. The first performance at the Opera House occurred on 28 September 1973—a production of Prokofiev's *War and Peace*. The Opera House was officially opened by Queen Elizabeth II on 20 October 1973. The building covers an area of about

Sydney Opera House on Bennelong Point in the harbor of Sydney, Australia. Photo courtesy of Lawrence D. Savoy.

4.5 acres (1.8 **hectares**) on a site of 5.5 acres (2.2 hectares). Inside, the building has about 11 acres (4.5 hectares) of usable floor space. The Opera House is 6,112 feet (1,864 meters) long and 380 feet (116 meters) wide at its widest point. Set on 580 concrete piers sunk to 82 feet (25 meters) below sea level, the entire building weighs 161,000 tons (144,900 metric tons), including a weight of 27,230 tons (24,507 metric tons) for the roofs alone. The construction includes 67,000 square feet of French glass, 400 miles (644 kilometers) of electrical cable, 12 miles (19 kilometers) of air-conditioning duct work moving 1 million cubic feet (255,000 cubic meters) of air per minute, and 400 miles of electrical cable, supplying enough power to support a town of 25,000 people. The Concert Hall, which seats 2,690 people, includes the largest mechanical tracker action organ in the world. The organ, designed and built by Ronald Sharp between 1969 and 1979, has 10,500 pipes, six keyboards (five manual and one pedal), 127 stops, and 205 ranks.

Additional Information: http://www.anzac.com/aust/nsw/soh.html.

Sydney Tower (Sydney, Australia; 1981)

One of the highest towers in the world, and the highest building in the southern hemisphere, the Sydney Tower rises 1,000 feet (305 meters) above the nearby streets, and 1,065 feet (325 meters) above the harbor. Near the top of the tower is an observation deck where one can see a full 360 degrees around the tower across the harbor, the city, and the surrounding area.

Additional Information: http://www.centrepoint.com.au/tower/index.html.

T

Tacoma Narrows Bridge (Galloping Gertie)

(Tacoma to Bremerton, Washington; 1940)

Opened to traffic on 12 July 1940, the Tacoma Narrows Bridge immediately acquired the nickname of "Galloping Gertie" for the undulating motions it made

The reconstructed Tacoma Narrows Bridge opened to traffic in 1950. Photo courtesy of the Bethlehem Steel Corporation.

in the wind. The bridge, designed by noted engineer Leon Moisseiff, crossed the Narrows of Puget Sound in western Washington to connect Tacoma with Bremerton. Over a mile long, the structure was a combination of a cable-supported **suspension bridge** and two steel-plate **girder** bridges for the approaches. The two towers of the bridge were 2,800 feet (854 meters) apart, with cables anchored another 1,000 feet (305 meters) away from center. Up and down sway was a recognized problem soon after the bridge opened; various attempts were made to minimize sway, including the use of tie-down cables anchored to 50-ton (45-metric-ton) concrete blocks, and the connection of main cables to stiffening girders. However, the problem was never considered to be more than an annoyance; most engineers believed the bridge was safe. But the flexibility of the bridge was extreme, more than three times the flexibility of the GEORGE WASHINGTON BRIDGE (the only longer bridge in the world at that time). The width to length ratio for the Tacoma Narrows Bridge was a high 1 to 72 compared with 1 to 33 for the George Washington. Undulating movements on the bridge became pronounced during a storm, and the Tacoma Narrows Bridge shook itself apart in November 1940 amidst 42-mile-per-hour (68-kilometer) winds. There was enough notice of the pending collapse for people to get off the bridge. Although a pet dog trapped in a car on the bridge died, there were no human fatalities. Much subsequent aerodynamic research was based on the film record of the vibrations that preceded the bridge's collapse. The design error in the Tacoma Narrows Bridge was almost universally ascribed to a lack of general knowledge in the profession rather than to any fault of Moisseiff's. A new Tacoma Narrows Bridge opened in 1950.

Additional Information: Matthys, Levy and Salvadori, Mario. *Why Buildings Fall Down.* New York: W.W. Norton & Company, 1992.

Taft Memorial Bridge (Washington, DC; 1907)

George S. Morison built this bridge to carry Washington's Connecticut Avenue over Rock Creek. Designed in the "classic Roman style," the Taft Memorial Bridge has five semicircular arches of 150 feet (46 meters),

A view (c.1908) of the Taft Memorial Bridge in Washington, DC. Photo courtesy of the Library of Congress, World's Transportation Commission Photograph Collection.

and two of 82 feet (25 meters). The bridge has a length between **abutments** of 1,036 feet (316 meters) and a total length of 1,314 feet (401 meters). The main arches are hingeless (without reinforcing).

Additional Information: DeLony, Eric. *Landmark American Bridges.* Boston: Bullfinch Press, 1993.

Tagus Bridge. *See* Ponte de 25 Abril.

Tain-Tournon Bridge (Lyons, France; 1825)

Marc Seguin built the Tain-Tournon Bridge across the Rhône River in southern France in 1825. The Tain-Tournon was the first wire **suspension bridge** to carry both vehicles and general traf-

fic. It had two spans, each 275 feet (84 meters) long. The design was highly innovative for its time; it used "festoons," multiple cables (three per side), rather than single cables to carry the deck. Other innovations were Howe-like trusses to stiffen the deck, adjustable cables, and hydraulic cement foundations. The bridge was restricted to foot traffic in 1847 because it was considered too weak for vehicles.

Additional Information: Kirby, Richard Shelton et al. *Engineering in History.* New York:1956; Dover Publications, 1990.

Taj Mahal (Agra, India; 1648)

In 1629, Mumtaz Mahal ("Chosen One of the Palace"), wife of the Mogul emperor Shah Jahan, died in childbirth. She had been the emperor's inseparable companion since their marriage in 1612. To give expression to his grief, Shah Jahan ordered a mausoleum to be built for his late wife on a site along the Yamuna River outside the city of Agra. Work on the structure, which was designed by a Turkish architect, began in 1630 and was completed about 1648. The Taj Mahal is set in a walled garden, which is adorned with fountains and marble pavements. The building is reflected in an oblong pool, and dark cypress trees surround it on three sides. The white marble exterior bears Arabic inscriptions and is inlaid with semiprecious stones and floral and other designs. The building rises from a 313-square-foot platform and bears a slender minaret at each corner. The interior designs are accented with agate, jasper, and colored marbles. On the inside, the roof dome is 80 feet (24 meters) high and 50 feet (15 meters) in diameter; outside, the dome is bulb-shaped, tapering to a spire topped by a crescent. The octagonal tomb chamber contains two sarcophagi, although Shah Jahan and his wife are buried in a vault be-

The Taj Mahal in Agra, India.

neath the floor. The chamber is softly illuminated by light that passes through intricately carved marble screens set high on the walls. More than 20,000 workers labored daily on the Taj Mahal, which is today considered one of the most beautiful buildings in the world and the finest example of Indian Muslim architecture.

Additional Information: http://www.bucknell.edu/departments/library/multimedia/new/taj_mahal.html.

Tancarville Suspension Bridge (LeHavre, France; 1959)

The Tancarville Suspension Bridge crosses the River Seine near Le Havre in northwestern France. This 610-meter (2,001-foot) **suspension bridge** is notable because its cables are made of coiled steel-wire ropes, rather than spun cables, an unusual but not unheard-of method of making cables for suspending long bridges.

Additional Information: Overman, Michael. *Roads, Bridges and Tunnels*. Garden City, NY: Doubleday & Company, Inc. 1968.

Tanna Railway Tunnel (near Tokyo, Japan; 1934)

The Tanna Railway Tunnel runs under Takiji Peak on the main Japanese island of Honshu. Work began on this 5-mile-long (8-kilometer) tunnel on 1 April 1918. During 16 years of construction, the project was beset by an unprecedented number of hazards. In 1921, on the third anniversary of the start of the tunnel, 200 feet (61 meters) of rock fell in, killing 16 workers and burying another 17 workers for seven days before they were brought to safety. In February 1924, a sudden sea of mud suffocated another 16 workers. The roof collapsed twice on the tunneling shield; the crew of the shield was killed in one of those incidents. Continual inundations of water reached as high as 65,000 gallons per minute, greater than the capacity of the drainage system installed in the tunnel floor. A second tunnel, built specifically for drainage, could not totally handle the flooding either, and additional tunnels were needed. Before construction was completed, the mountain was pierced through with several tunnels; in one location, five different parallel **headings** were being worked at the same time. Toward the end of construction in November 1930, an earthquake hit the area, collapsing a portion of the main tunnel and killing three more workers. The tunnel was finally holed through in 1934.

Additional Information: Beaver, Patrick. *A History of Tunnels*. Secaucus, NJ: Citadel Press, 1973.

Tavanasa Bridge (northern Switzerland; 1905)

An early **reinforced concrete** three-hinged arch, the Tavanasa Bridge was the first of Robert Maillart's three-span arches, and one of his first mature works. The bridge crossed the Rhine River in northern Switzerland. Many areas not under stress were omitted in this 167-foot (51-meter) span and replaced with triangular cut-outs. The concrete itself tapered upwards to thin crowns. The bridge was destroyed by a landslide in 1927.

Additional Information: *Academic American Encyclopedia* (electronic version). Danbury, CT: Grolier Electronic Publishing, 1991; Brown, David J. *Bridges*. New York: Macmillan, 1993.

Tay Bridge (Dundee, Scotland; 1878)

Thomas Bouch designed this long-needed **railroad bridge** to carry a single-track line across the 2-mile-wide (3.2-kilometer) River Tay near Dundee in central Scotland. Bouch was knighted for his achievement by Queen Victoria in 1879. However, on 28 December 1879, the **girders** at the middle of the bridge fell during a great gale. A mail train crossing the bridge fell with the central portion of the structure; all 75 people on the train died. Bouch died the following year, tremendously depressed by the failure of his bridge. In 1887, W.H. Barlow and his son, Crawford, completed a new two-track bridge on the site. The Barlows' bridge remains standing.

Additional Information: Ellis, Hamilton. *The Pictorial Encyclopedia of Railways*. Prague, Czechoslovakia: Paul Hamlyn, 1968.

Taylorsville Dam. *See* Miami Conservancy District.

The Tearaght. *See* Inistearaght Lighthouse.

Tecolote Tunnel (southern California; 1955)

This 6.5-mile-long (10.5-kilometer) tunnel moves water from the Cachuma Reservoir under the Santa Ynez Mountains to Glen Annie Canyon. The tunnel's builders claimed that Tecolote was the most difficult tunnel ever built in the Americas. Work began in 1950 and proceeded until January 1951 without any particular problem. After 8,949 feet (2,729 meters) had been driven on the reservoir side, a sudden water break occurred, nearly drowning the work crew, who were forced to drop their tools and run for their lives. As the men were running, a methane explosion occurred at the work face, burning the faces of 11 workers and seriously singeing the rest of the men. Sand and water filled 200 feet (61 meters) of the **heading**. It took several months to stop the inflow with bulkheads and high pressure grouting. When work continued, grouting and the placing of bulkheads was done continuously until the tunnel had progressed a distance of 14,919 feet (4,550 meters). At this point, tunneling from the inlet side was stopped while a tunnel lining was installed, and another heading opened from the opposite side of the tunnel. Water was also a problem on the Glen Annie Canyon side, and unsuccessful attempts were made to grout out the 3,500 gallons of water per minute that were making progress impossible. Smaller tunnels on either side were built especially to relieve the flow of water. As soon as the flow decreased, the main tunnel was enlarged and lined with concrete, with a drainage pipe set underneath. Intense heat, especially under the mountain, was another tremendous obstacle. At 14,000 feet (4,270 meters), air temperature reached 112 degrees. The heat from water flow sometimes softened or even melted the cartridges for the charges. Workers rode to and from their work in mining skips filled with cold water—the skips were left on site so that workers could cool off during their shifts. In July 1953, with less than a mile to go, the contractors were forced to give up work because of the intense heat. Work restarted under new contractors in January 1954. Hot water was flowing into the tunnel at a rate of nearly 10,000 gallons a minute. Workers also encountered large pockets of hydrogen sulfide that burned workers' eyes. Progress was made, however, and the tunnel was finally holed through on 15 January 1955.

Additional Information: Beaver, Patrick. *A History of Tunnels*. Secaucus, NJ: Citadel Press, 1973.

Tehri High Dam (near Delhi, India; under construction)

The Tehri High Dam is being built on the Bhagirathi River about 200 miles (322 kilometers) northeast of Delhi, India. Construction began in 1978 and is expected to finish in 1999. The Tehri will be one of the world's highest dams with a height above its lowest formation of 261 meters (856 feet). When completed, the dam will deliver 2,400 megawatts of electricity to northern India. Its attached reservoir will be so huge that more than 100 villages with a combined population of 70,000 people will be displaced. The base of the dam will be more than a half mile wide, tapering up to 65 feet (20 meters) wide at the top. Because the dam is sited along the earthquake fault where India pushes against Eurasia, concern has arisen about the structure's safety, not only for the surrounding area, but even for areas hundreds of miles away. To meet this concern, the dam will be "earthquaked" so it can resist a quake of some force, and is being built as a **rockfill** dam, which is considered more resistant to quakes than other types of dams. Still, some geologists believe that the area is overdue for a powerful quake that will destroy the dam and devaste the region with floods. The social effects of relocating so many people and villages is another matter of concern.

Additional Information: Svitil, Kathy A. "The Coming Himalayan Catastrophe." *Discover* 16:7 (July 1995), pp. 80-84.

Temple of Artemis (Diana), also known as the Artemision (Ephesus, Asia Minor [modern Turkey]; c. 550 B.C.)

One of the SEVEN WONDERS OF THE ANCIENT WORLD, the Temple of Artemis in Ephesus, a Greek city in southwestern Asia Minor about 50 kilometers (31 miles) south of the present-day Turkish city of Izmir, was built in the sixth century B.C. by Chersiphron for Croesus, the king of Lydia. The structure stood on an earlier temple foundation that dated back to the seventh century B.C. Both a temple and a marketplace, it was considered by those who saw it to be the most beautiful building in the world. The temple was decorated with bronze statues by famous sculptors, and probably contained superb paintings inside. The building was made of marble with a decorative facade that overlooked a large courtyard. The temple's 127 columns stood 20 meters (66 feet) high and were aligned around a high terrace. The temple was burned down on the night of 21 July 356 B.C. by Herstratus, who desired to immortalize his name. Coincidentally, this was the same night that Alexander the Great was born; Plutarch speculated that the goddess Diana was too occupied with Alexander's birth to save her temple. The temple was restored over the next several years, and totally rebuilt when Alexander the Great conquered Asia Minor in the 330s B.C. St. Paul must have seen the reconstructed temple when he visited Ephesus in the first century A.D. In A.D. 401, Christians, at the direction of Saint John Chrysostom, tore the temple down.

Additional Information: Ashmawy, Alaa K., ed., "The Seven Wonders of the Ancient World" at http://pharos.bu.edu/Egypt/Wonder.Home.html.

Tennessee Valley Authority (TVA) (Tennessee River Valley; 1933)

The TVA is a U.S. government-owned corporation created by Congress in 1933 to stimulate and integrate development in the Tennessee River Valley. Headquartered in the region, and directed initially by David Lilienthal, the project undertook to construct a series of multipurpose dams and steam-generating plants to produce relatively cheap electric power for residents of parts of Tennessee, Kentucky, Virginia, North Carolina, Georgia, Alabama, and Mississippi. The TVA built 30 dams, created 175,000 acres (70,875 **hectares**) of lakes, built 1,300 miles (2,092 kilometers) of highways and 200 miles (322 kilometers) of railroads, installed 9.4 million kilowatts of power-generating capacity, planted 234 million trees, wiped out malaria in the region, rejuvenated 3 million acres (over 1.2 million hectares) of farmland, and gave jobs and job training to over 200,000 workers. Since the 1940s, the availability of low-cost power provided by the TVA has attracted new businesses to the region, creating additional jobs and improving the overall economic climate. The TVA also conducted programs in land and wildlife conservation, and in health and education. Today, the area uses about 65 times as much power as it did in 1933; in the 1970s, the TVA began building nuclear power facilities to meet the increased demand. Although opposed at the time of its inception by private companies that feared government intrusion into power generation, the TVA has become an example of government efficiency and strength at its best. Despite this achievement, a project like the TVA has never again been attempted anywhere in the world.

Additional Information: Sandstrom, Gosta E. *Man the Builder*. New York: McGraw-Hill, 1970.

Ternavasso Dam (near Turin, Italy; c. 1600)

Ternavasso Dam created the Lago di Ternavasso 18 miles (29 kilometers) southeast of Turin in northern Italy. An earth dam, the Ternavasso has a maximum height of 25 feet (7.6 meters). At its crest, it is between 17 and 22 feet (5.2 and 6.7 meters) thick. Its base measurement is unknown but is probably between 50 and 60 feet (15 and 18 meters) thick. Because the topology of the site prevented the construction of a straight dam, Ternavasso is shaped like a huge Z, with legs in lengths of 690 feet (210 meters), 230 feet (70 meters), and 150 feet (46 meters). The central section is perpendicular to each leg. The water face is covered with a vertical brick wall that is 2 feet (0.6 meters) thick, and covered, in turn, by a layer of mortar. The brick wall is supported by tapering buttresses. At one end of the dam is an overflow spillway that is 12 feet (3.7 meters) wide and about one foot deep. The dam was also built with three outlets at varying heights for the drainage of excess water. However, only the lower, central outlet could have ever been used to draw water away on a regular basis. Built to irrigate the lands of the estate around it, the dam is the only remaining reservoir structure in Italy built before 1800, except for some Roman ruins.

Additional Information: Smith, Norman. *A History of Dams*. Secaucus, NJ: Citadel Press, 1972.

Ternay Dam (Annonay, France; 1868)

The Ternay Dam is a masonry-gravity dam that was intentionally built as a replica of the FURENS DAM. The dam, 112 feet (34 meters) high, provides flood protection and supplies water for Annonay in southeastern France.

Additional Information: Smith, Norman. *A History of Dams*. Secaucus, NJ: Citadel Press, 1972.

Thames Tunnel (London, England; 1842)

Marc Brunel dug this tunnel under the Thames over an 18-year period by using his brilliant invention, the tunneling shield. Brunel drove both carriageways of this 353-meter-long (1,158-foot) twin tunnel simultaneously; both were done full size from the start without a pilot tunnel and without forced ventilation. Every bit of sand, clay, and mud was removed by hand from the 7-meter-high (23-foot) and 19-meter-wide (62-foot) face of the tunnel. The brick lining of the tunnel, built before the invention of Portland cement, has remained entirely waterproof for a century and a half. Brunel used Roman cement and had every brick hammer tested before it was allowed into the shaft. He fired every brick-

layer found to have laid even a single loose brick. The side walls of the two finished vaulted archways are 1.5 meters (5 feet) thick, and the floor and arch crown are one meter thick. In addition to the seepage of dirty, sewage-laden Thames water through the working face, tunnelers had to face the inflow of pockets of methane gas, which burst into flame upon coming in contact with the naked gas lamps used for illumination. The tunnel flooded several times during construction. The usual method of repairing breakthroughs of river water was for Brunel's son to first survey the situation in a diving bell from a raft moored in the Thames. After the leak was located, it was then plugged by dumping thousands of sacks of clay over the opening, pinning a huge tarpaulin over the mend, and then pumping the tunnel dry with steam pumps. Although built as a road tunnel, Brunel's structure never had approach roads built to it, and in 1865 it became part of the Underground, and is still in use by the Underground today. The use of Brunel's shield made possible the construction of the first modern subaqueous tunnel; without the shield, the muck that was being dug through would have collapsed before a tunnel lining could be constructed. The Brunel shield consisted of 12 vertical **cast-iron** sections or frames set into one gigantic face. Each section pivoted on its foot; it protected workers behind it from mud, clay, and sand by means of a large number of stout horizontal oak timbers called "poling boards." Each poling board could be removed independently from inside to enable a worker to dig out the tunnel directly ahead of it before forcing it forward into the void by means of its own small screw jacks acting as struts pushing against the frame. When the tunnel face in front of an entire frame had been excavated and all the poling boards pushed forward, the frame was again jacked forward by simultaneously slackening off all the poling board jacks and pushing with two huge master jacks fitted between the foot and head of the frame, and between the invert and arch crown of the tunnel brickwork. By moving one frame at a time, less jacking force was required than is used with modern one-piece shields. The height of the tunnel's working face and of each shield frame was 7 meters (23 feet), and each frame was divided into three cells, one above the other, with each cell accommodating one laborer. The entire shield thus had 36 cells, and required 36 workers at the face in each shift.

Additional Information: Epstein, Sam and Epstein, Beryl. *Tunnels*. Boston: Little Brown & Company, 1985; Overman, Michael. *Roads, Bridges and Tunnels*. Garden City, NY: Doubleday & Company, 1968.

Theodore Roosevelt Dam (near Phoenix, Arizona; 1903-1911)

The U.S. Reclamation Service built Theodore Roosevelt Dam as part of the SALT RIVER PROJECT to bring water and power to the Salt River Valley (now the Phoenix metropolitan area) in central Arizona. Roosevelt Dam is one of only seven dams in the United States that have been designated national historic engineering landmarks. The

Theodore Roosevelt Dam in Arizona. Photo courtesy of Salt River Project.

dam is the largest cyclopean (one opening) masonry arch dam in the world. It is a composite thick arch made of concrete overlaying masonry. Located 76 miles northeast of Phoenix, the dam is built across the Salt River; its reservoir is Theodore Roosevelt Lake. The dam has been modified several times, most recently in 1987 and 1996, when the dam's height was raised 77 feet (24 meters), to provide additional storage and drainage outlet capacity. Modifications were also made to help the dam withstand potential flooding and earthquake. Presently, the dam is 356 feet (109 meters) high, with a crest elevation of 2,218 feet (676 meters) and crest a length of 723 feet (221 meters).

Additional Information: "News: Roosevelt Dam Reaches a New Height of Safety." *Civil Engineering* 66:5 (May 1996), p. 10; http://donews.do.usbr.gov/dams/TheodoreRoosevelt.html.

Third New York City Water Tunnel. *See* City Tunnel No. 3.

Thomas Mill Bridge (Philadelphia, Pennsylvania; 1855)

The Thomas Mill Bridge is the only historic covered bridge remaining within the city of Philadelphia. A **Howe truss** crossing Wissahickon Creek, the bridge is 86.5 feet (26 meters) long, with a width of almost 19 feet (5.8 meters). An unusual feature of the bridge is a sawtooth decoration at the roof line of the portals.

Additional Information: http://william-king.www.drexel.edu/top/bridge/CBTmill.html

Thomas Viaduct (Washington, DC; 1835)

The Thomas Viaduct crosses the Patapsco River near Washington; it was built on the Washington Branch of the BALTIMORE AND OHIO (B&O) RAILROAD. The **viaduct** is built on a four-degree curve, 62 feet (19 meters) above the river. It has eight arches and a total length of 612 feet (187 meters).

Additional Information: DeLony, Eric. *Landmark American Bridges*. Boston: Bullfinch Press, 1993.

Three Gorges Dam Project (Yangtze River, China; under construction)

This project is a gigantic undertaking to dam the Yangtze River in southeastern China. If successful, the project will control flooding, to which the area has been subject throughout history, and generate an immense amount of electricity. As planned, the $18 billion project includes 26 hydroelectric turbines, each with 700 megawatts of capacity. Begun in 1993, the Three Gorges enterprise is probably the largest river dam project in the world. It was highly debated because its size could cause tremendous changes in the environment and in the society of the local population. Environmentalists fear that the dam could cause landslides and tidal waves as water levels fluctuate. Human rights groups are concerned about the forced relocation of over 1 million people. In October 1995, the White House advised the U.S. Export-Import Bank that it did not support loans to companies participating in the project. The bank is not under legal obligation to follow White House recommendations, but usually gives great weight to presidential positions.

Additional Information: Greenberger, Robert S. "U.S. Opposes Aid for Firms to Help Build China Dam." *Wall Street Journal*, 13 October 1995, A3.

Tien An Men Square (Beijing, China)

Beijing's ancient Tien An Men Square is the largest square in the world, covering an area of 98 acres (40 **hectares**). Tien An Men ("Heavenly Peace") is a central part of life in Beijing. In 1989, the square was the scene of student protests and riots for greater political freedom in China. The square is surrounded with impressive structures; the Great Hall of the People, the home of the Chinese legislature, is on the west side, and the Museum of the Chinese Revolution and Chinese History is on the east. In ancient times, edicts were read from a balcony on the Tien An Men Gate, on the north side of the square. Tien An Men is also where Chinese leaders gather on holidays to review parades and festivities.

Additional Information: http://www.ihep.ac.cn/tour/bj.html.

Todd Brook Dam. *See* Peak Forest Canal.

Tokyo Millennium Tower. *See* Millenium Tower.

Tien An Men Square in Beijing, China.

Tokyo Tower (Tokyo, Japan; 1958)

At 333 meters (1,092 feet), Tokyo Tower is one of the highest towers in the world. The tower is used as a common carrier by eight Tokyo television and radio stations. Its observation deck at 250 meters (820 feet) gives

The Tokyo Tower in Tokyo, Japan.

a view of Mount Fuji. Another observation deck is at a height of 150 meters (492 feet).

Additional Information: http://www.centrepoint.com.au/tower/index.html.

Tom Miller Dam. *See* Austin Dam.

Tower Bridge. *See* Tower of London Bridge.

Tower of London (London, England; 1078)

The White Tower, the high, square, four-turreted central structure in the tower of London complex was built in the 1070s on the orders of William the Conqueror by Gundulf, bishop of Rochester. Constructed on the north bank of the Thames at what was then the eastern end of London, William intended to use the fortress to overawe the city. Various other towers and surrounding structures were built throughout medieval and early modern times. Sir Christopher Wren, the designer of SAINT PAUL'S CATHEDRAL, restored part of the exterior in the seventeenth century. The tower has served the Crown and the government as a royal residence, a prison, an armory, a treasury, a mint, and an archive. At one time, it also contained the royal zoo. Today the tower houses the Crown jewels and is an important London tourist attraction. During the Middle Ages, the tower was a royal prison. Many famous and important figures in English history passed into the tower through Traitors' Gate, which gave access to the fortress by water from the river, or were held in the Bloody Tower. Henry VI was murdered in the tower in 1471, and the two young sons of Edward IV disappeared in the tower in 1483, presumably murdered by their uncle, Richard III. Many state prisoners, condemned for treason, were executed in the tower, including Sir Thomas More and two of Henry VIII's wives, Anne Boleyn and Catherine Howard. These queens are just two of hundreds of beheaded persons who lie buried beneath the floor of the Chapel of Saint Peter ad Vincula ("in chains"), which stands within the tower walls near Tower Green where most of these executions took place. Other famous victims of the tower include George, duke of Clarence, who was supposedly drowned there in a butt of malmsey wine in the fifteenth century, and Thomas Cromwell, Sir Walter Raleigh, the earl of Essex, Lady Jane Grey, the earl of Strafford, and the duke of Monmouth, all of whom were beheaded there in the sixteenth or seventeenth centuries. The White Tower is enclosed by double castellated walls which are in turn surrounded by a dry moat. The entire complex covers about 13 acres (5.3 **hectares**). The tower is still guarded by the Yeoman of the Guard (the

"Beefeaters"), who dress in Tudor garb. The tower suffered some damage from air raids in 1940.

Additional Information: Stiles, Benjamin A. "Origins of the Tower of London" at http://www.millersv.edu/~english/homepage/duncan/medfern/tower.html.

William the Conqueror began construction of the Tower of London along the Thames east of London in the 1070s.

Tower of London Bridge, also known as Tower Bridge (London, England; 1894)

A famous double-leaf **bascule**, the Tower of London Bridge crosses the Thames near the TOWER OF LONDON. The bridge has a footwalk at the top of the towers which allows the bridge to be crossed even when the span is open. The bridge was designed by Sir Horace Jones and built by John Wolfe-Barry.

Additional Information: Brown, David J. *Bridges*. New York: Macmillan, 1993.

Tower of Pisa. *See* Leaning Tower of Pisa.

Trans-Canada Highway (Newfoundland to British Columbia; 1950-1962)

A large portion of this 4,859-mile-long (7,818-kilometer) road was paid for by the Canadian government. Such funding is unusual in Canada where road construction is typically the responsibility of the provinces. The Trans-Canada Highway runs from St. John's in Newfoundland in the east to Victoria in British Columbia in the west.

Additional Information: *Academic American Encyclopedia* (electronic version). Danbury, CT: Grolier Electronic Publishing, 1991.

Transco Tower (Houston, Texas; 1982)

The Transco Tower was designed by Philip Cortelyou Johnson and John Burger. At a height of 277 meters (909 feet), the Transco Tower is one of the highest towers in the world. Headquarters of the Transco Corp., it is one of the most impressive office buildings in the world.

Source: http://www.centrepoint.com.au/tower/index.html.

Trezzo Bridge (Trezzo, Italy; c. 1375)

For many years, the 230-foot (70-meter) Trezzo Bridge was the longest single arch in the world. It crosses the River Adda at Trezzo.

Additional Information: Kirby, Richard Shelton et al. *Engineering in History*. New York: Dover Publications, 1990.

Tucurui Dam (Tocantins River, Brazil; mid-1980s)

This gigantic dam, built at a cost of $4 billion, converted the Tocantins River into a series of lakes. This 4,000-megawatt dam in the Brazilian Amazon region is 1,180 feet (360 meters) long. It is among the world's most powerful hydroelectric dams. Located near the city of Tucurui in the Para region of northeastern Brazil, this earth and rockfill gravity dam reaches a height of 93 meters (305 feet).

Additional Information: Hawkes, Nigel. *Structures: The Way Things Are Built*. New York: Macmillan, 1990. Additional details were provided by Scott K. Nelson, The National Performance of Dams Program (NPDP), Stanford, California.

Tower Bridge crosses the Thames near the Tower of London. Photo courtesy of Ray Farrar.

Tunkhannock Viaduct (Nicholson, Pennsylvania; 1915)

Tunkhannock Viaduct carries the Delaware, Lackawanna & Western Railroad over Tunkhannock Creek near Nicholson in northeastern Pennsylvania. Tunkhannock Viaduct is the largest reinforced-concrete structure of its kind in the world. Construction involved moving massive amounts of land and actual hills. The

viaduct is also unique because it carries the railroad transversely to the regional drainage pattern. The scale of the project is huge. One embankment is 140 feet (43 meters) high and 2,000 feet (610 meters) long, and contains 1 million cubic yards (765,000 cubic meters) of fill and a 3,630-foot (1,107-meter) double-track tunnel. The viaduct has 13 piers, some as much as 95 feet (29 meters) below ground level, all carried to bedrock. The end towers are 3,028 feet (924 meters) apart, with a central tower in between. Ten 180-foot (55-meter) arches have five sets of three-hinged steel arch centers. Each center is composed of four ribs weighing 47 tons (42.3 metric tons) each, spaced almost 4 feet (1.2 meters) to center, providing support for the concrete ribs that compose each span. Overall, the viaduct is 245 feet (75 meters) high and 2,375 feet (724 meters) long.

Additional Information: DeLony, Eric. *Landmark American Bridges*. Boston: Bullfinch Press, 1993; Schodek, Daniel L. *Landmarks in American Civil Engineering*. Cambridge, MA: MIT Press, 1987.

Turn of River Bridge (Stamford, Connecticut; 1892)

The Berlin Iron Bridge Company built this bridge across the Rippowam River in Stamford in western Connecti-cut. This parabola-shaped **lenticular pony truss** is one of several iron bridges made by the historic Berlin Iron Bridge Company at the end of the nineteenth century. The bridge once lay on Turn of River Road, a thorough-fare that was rendered almost useless by the building of the Merritt Parkway in the 1930s. Fifty-three feet (16 meters) long, with geometric and floral patterns cast into the iron, the bridge now blocks the entrance to a proposed housing development. Although pedestrians can climb onto the bridge, road traffic is prevented by a tree on one side and some boulders on the other. The housing developer has volunteered to remove the bridge and reassemble it at some other spot, but, so far, the bridge has had no takers. Two other Berlin Iron Bridge Company structures in Stamford are the WEST MAIN STREET BRIDGE and the disassembled PULASKI STREET BRIDGE.

Additional Information: Sierpina, Diane Sentementes, "Stamford; A (Sort of Purple) Iron Bridge Gathers Champions." *New York Times*, 4 June 1995, XIIICN:2.

Tuscarora Tunnel. *See* Pennsylvania Turnpike.

TVA. See Tennessee Valley Authority.

U

Umpqua River Bridge (Reedsport, Oregon; 1936)
This bridge carries the Oregon Coast Highway (U.S. Route 101) over the Umpqua River near Reedsport in western Oregon. Built by Conde B. McCullough, this 430-foot (131-meter) steel Parker truss swing span has two **reinforced concrete** through-tier arches that are 154 feet (47 meters) in length. The attractive Umpqua River Bridge makes use of strong classical forms combined with **Art Deco** and Modern styles, and Gothic arches on its underside.

Additional Information: DeLony, Eric. *Landmark American Bridges*. Boston: Bullfinch Press, 1993.

Union Bridge, also known as the Waterford Union Bridge (Waterford, New York; 1804)
Theodore Burr built this bridge across the Hudson River at Waterford, New York. The Union Bridge was built with the first system of trusses to be widely used in the United States, the Burr truss. The system's four reinforced trusses were a combination of arch and parallel chord trusses. Considered Burr's masterpiece, the Union Bridge burned in 1909.

Additional Information: DeLony, Eric. *Landmark American Bridges*. Boston: Bullfinch Press, 1993.

Union Bridge (border of Scotland and England; 1820)
Built by Captain Samuel Brown, Union Bridge crosses the Tweed River to connect England and Scotland. Union Bridge was the earliest major **suspension bridge**; it used a suspension chain with a circular **eyebar**. The deck was supported by vertical rods suspended from three pairs of iron chains. The chains are still intact, although modern steel cables have been added to assist them.

Additional Information: Brown, David J. *Bridges*. New York: Macmillan, 1993.

Union Canal (southeastern Pennsylvania; 1827)
For a time, the Union Canal was the only commercial link between east and west Pennsylvania. The Union Canal Tunnel connected the Schuylkill and Susquehanna rivers near Lebanon in southeastern Pennsylvania. William Penn suggested a canal for the area as early as 1690. Two companies were chartered in 1791 to build the canal, but financial problems and scandals impeded progress. Work eventually began in 1815 under the "guidance" of Loammi Baldwin, son of the builder of the MIDDLESEX CANAL. He left the project when his plans for dams and feeders for water supply were opposed. Canvass White, who later worked on the ERIE CANAL, took over the leadership of the project in 1824. The eastern portion of the Union Canal was completed in 1826, a 108-mile (174-kilometer) stretch along the Schuylkill River that terminated in Philadelphia. Sixty-two miles (100 kilometers) were canal, and the rest of the stretch consisted of slack water pools and portions of river made navigable by dams, which were 80 feet (24 meters) long and 17 feet (5.2 meters) wide. Part of this eastern section was the Union Canal Tunnel, which was built in 1827 by three brothers—Job, Samson, and Solomon Fudge. The tunnel is a 400-foot (122-meter) drive through a ridge. Because a small shift in the canal plan would have avoided the tunnel, it may have been built to provide publicity for the canal. The western portion of the canal, 81 miles (130 kilometers) long, was completed in 1827, and connected the Schuylkill

at Reading with the Susquehanna at Middletown. A 729-foot-long (222-meter) tunnel had to be driven through the watershed ridge about a mile west of Lebanon, Pennsylvania. This tunnel is 18 feet (5.5 meters) wide and 14 feet (4.3 meters) high; it was built by Simeon Guilford who excavated it between 1824 and 1826. The canal involved 93 lift locks, 43 waste **weirs**, 135 bridges, and 14 **aqueducts** to supply water. The canal was popular and certainly useful, but its profits were used up by construction problems and mistakes. The channel and locks were too narrow for most vessels, and the locks on the Schuylkill were 15 feet (4.6 meters) longer and the channel twice as wide as those on the western section. The canal finally went out of operation in 1862 after flooding wiped out one of the dams. The canal was then abandoned. Further repairs or rebuilding made no sense because the canal could not compete with the railroads.

Additional Information: Kirby, Richard Shelton et al. *Engineering in History*. New York: Dover Publications, 1990; Schodek, Daniel L. *Landmarks in American Civil Engineering*. Cambridge, MA: MIT Press, 1987.

Union Canal Tunnel. *See* Union Canal.

United States Capitol Building (Washington, DC; 1800)

Pierre Charles L'Enfant, the designer of the city of Washington, DC, was to have built the capitol building of the new nation. However, President George Washington reluctantly removed L'Enfant from office in 1792 when the designer refused to produce drawings of his

late plan that was considered eminently acceptable, and was formally approved by President Washington on 25 July 1793. Thornton's design was carried out by three architects in succession; most of the early work was performed by James Hoban, the architect of the White House. Thornton's original design had called for a central section with a low dome, flanked by rectangular wings on the north and south sides for the two bodies of Congress. Although the building has undegone some modifications over the years, the Capitol remains essentially true to its original design. On 14 August 1814, during the War of 1812, the British briefly occupied Washington and set the building on fire; only a sudden rainstorm saved the building from complete destruction. Repairs were begun under the stewardship of Benjamin Henry Latrobe, widely recognized as the first professional engineer in the United States. In 1863, the "Statue of Freedom" was lifted, in portions, to the top of the Capitol. Designed by Thomas Crawford, the 19.5-foot-high (5.9-meter) statue weighed a massive 14,985 pounds (6,743 kilograms), and depicted a woman holding a sword and a shield. In December, the statue was assembled at the top of the dome while forts around the city fired gun salutes. Restoration of the statue was required in 1993, when it was lifted from its pedestal by helicopter. Recent modifications include construction of a subway terminal under the Senate steps, the addition of 90 new rooms, and the installation of improved lighting. George M. White, the present architect of the Capitol, faces such current renovation projects as conservation of the Rotunda canopy and the Statue of Freedom, replacement of tiles in the Senate corridors, repair and restoration of stairways, and installation of a visitor's center with exhibits, theaters, and all the other necessary educational paraphernalia. As it stands today, the Capitol covers an area of 175,170 square feet, and has a floor area of about 16.5 acres (6.7 **hectares**). Measured north to south, the building has a length of slightly over 751 feet (229 meters); its greatest width, including approaches, is 350 feet (107 meters). It has about 540 rooms, 850 doorways, and 658 windows, including 108 windows in the dome itself.

The United States Capitol, meeting place of Congress, in Washington, DC.

proposed buildings, claiming that he carried the design "in his head." A public competition for the project was held, and none of the 17 designs submitted was judged appropriate. Dr. William Thornton, a Scottish physician living in Tortola, British West Indies, submitted a

Additional Information: http://www.acc.gov/history/cap_hist.html (this website from the Office of the Curator of the United States Capitol is an extremely detailed and fascinating account of the Capitol's history); Ditzel, Paul C. *How They Build Our National Monuments*. New York: Bobbs-Merrill, 1976; Cagley, James R. "Restoring Freedom at the Capitol Dome." *Civil Engineering* 64:6 (June 1994), pp. 57-59.

University Heights Bridge (New York City; 1895)
George W. Birdsall of the New York City Department of Public Works designed this bridge in consultation with William H. Burr and Alfred T. Boller. The bridge crosses the Harlem River, linking the New York boroughs of Manhattan and the Bronx. The bridge was floated downstream to its current site from its original location on the river in 1908. The 270-foot-long (82-meter) **swing bridge** is designed as a bridge on a turntable. The span is a **through-truss** with four Gothic-style **finial**s built on rock-faced granite piers with ornamental, iron openwork on its sides. After the 1908 move, two stone gatehouses and four pedestrian shelters were added to its original design. In the bridge's most current rebuilding in 1990, the gatehouses and shelters were moved to one side, and the footpath on the other side was eliminated.

Additional Information: Gray, Christopher. "Streetscapes: The University Heights Bridge; a Polite Swing to Renovation for a Landmark Span." *New York Times*, 25 February 1990, section X, p. 9.

Upper Otay Dam (near San Diego, California; 1900)
Upper Otay Dam was built to supply water to San Diego, California. It is the first dam to use concrete at its base, which was reinforced with iron wires.

Additional Information: Smith, Norman. A History of Dams. Secaucus, NJ: Citadel Press, 1972.

Upper Pacific Mills Bridge (Merrimack College, North Andover, Massachusetts; 1864)
Upper Pacific Mills Bridge originally spanned the North Canal in Lawrence, Massachusetts, but is now located at Merrimack College in North Andover, Massachusetts. This early iron bridge was designed, built, and patented by Thomas Moseley, who brought the **wrought-iron** bridge to the United States. Two counterbraces serve as reverse curves to the main arch. The bridge was being cut up for scrap when Francis Griggs, a civil engineering professor at Merrimack happened by and negotiated on-the-spot for removal of the bridge to the college.

Additional Information: DeLony, Eric. *Landmark American Bridges*. Boston: Bullfinch Press, 1993.

Urnerloch Tunnel (Italy to Switzerland; 1707)
Petro Morettini built this tunnel to replace the thriteenth-century DEVIL'S BRIDGE between Italy and Switzerland. The tunnel was driven 240 feet (73 meters) through a mountain to become the first Alpine tunnel. Urnerloch was a narrow 12 feet (3.7 meters) by 10 feet (3 meters) until widened in 1830 to accommodate larger carriages.

Additional Information: Hawkes, Nigel. *Structures: The Way Things Are Built*. New York: Macmillan, 1990.

U.S. Interstate Route 1. *See* Boston Post Road.

U.S. Interstate Route 5 (Mexico to Canada; 1953-1979)
Running through the west coast states of California, Oregon, and Washington, I-5 eventually meets the TRANS-CANADA HIGHWAY in Vancouver, British Columbia. In California, I-5 runs 797 miles (1,282 kilometers), mostly through the central San Joaquin Valley; I-5 is the longest freeway in the state. Built in stages between 1953 and 1979, I-5 is the only continuous highway connecting Mexico and Canada.

Additional Information: "Mexico, U.S. and Canada Linked by 4.6 Mile Road Completing I-5." *New York Times*, 13 October 1979.

U.S. Interstate Route 70 (Denver to Vail, Colorado; 1975)
Cutting through the Rocky Mountains, this 100-mile (161-kilometer) stretch of I-70, which was built especially to accommodate winter skiers, reaches the highest level of any interstate highway, 11,165 feet (3,405 meters). The highway contains the two longest land vehicular tunnels in the United States, the 8,959-foot-long (2,733-meter) E. Johnson Memorial Tunnel, and the 8,941-foot-long (2,727-meter) Eisenhower Memorial Tunnel. At the Eisenhower Tunnel near Boulder, traffic on the highway in 1995 hit 8.5 million vehicles. To keep the traffic flowing in bad winter weather, Colorado Transportation Department authorities use computerized message signs, highway radio stations, and emergency roadside telephones. A fax system sends traffic and weather updates to ski areas so that proprietors can properly advise their customers of the weather on the roads below the slopes.

Additional Information: Brooke, James. "Tackling a Crucial Highway to Keep Skiers Happy." *New York Times*, 23 December 1995, p. 7.

U.S. Interstate Route 80 (New York City to San Francisco, California; 1975)
I-80 is the long-desired transcontinental route between the east and west coasts. At a length of 2,899 miles (4,665 kilometers), I-80 is the longest highway in the world. Built in portions, sometimes by local authorities to federal specifications, I-80 is the premiere example of an interstate highway. From west to east, the road runs through 11 states: California, Nevada, Utah, Wyoming, Nebraska, Iowa, Illinois, Indiana, Ohio, Pennsylvania, and New Jersey. Such highways have a special appeal to some folks, but are repugnant to others. The late Charles Kuralt (of "On the Road" fame) observed that "Thanks to the Interstate Highway System, it is now possible to travel across the country from coast to coast without seeing anything."

Additional Information: Kost, Mary Lu. *Milepost I-80: San Francisco to New York*. Milepost Publications, 1993.

U.S. National Museum. *See* Arts and Industries Building.

U.S. Route 30. *See* Lincoln Highway.

U.S. Route 40. *See* Cumberland Road.

U.S. Route 66 (Chicago, Illinois, to Santa Monica, California; 1920s)

Built in the 1920s, this 2,400-mile (3,862-kilometer) stretch of road, winding from the Midwest to California, proudly carries more historical weight than any other American road built in the twentieth century. Long before Martin Milner and George Maharis traveled the road in the 1960s seeking adventures on behalf of a vicarious television audience, the road had a major impact on the lives of Americans. It provided the way west for citizens fleeing the Dust Bowl of the 1930s, and was, in many ways, what historian and geographer Arthur Krim has termed the "Main Street of America." Throughout its length, the route was always built to improve transportation between important areas. In Oklahoma in 1925, for example, the road was constructed to link Oklahoma with the two end points in Illinois and California. The makers of Phillips 66 gasoline specifically used the "66" and the standard U.S. route marker shape to identify their product with the road. Today there are Route 66 museums, souvenirs, fan clubs, websites, and hundreds of entrepreneurs trying to cash in on the fame of the road.

Additional Information: Brown, Patricia Leigh. "Driving, Buying, Reading and Remembering Route 66." *New York Times*, 24 August 1995, A1; Wallis, Michael. *Route 66: The Mother Road*. New York: St. Martin's Press, 1990; a search of the WWW will also produce several pages dealing with Route 66.

V

Vaiont Dam, or Vajont Dam (near Venice, Italy; 1961)

The Vaiont Dam crossed a reservoir on the Piave River about 50 miles (80 kilometers) north of Venice, Italy. Built by the Adriatic Electric Society, the Vaiont Dam was the third highest concrete dam in the world, reaching a height of 858 feet (262 meters). The Vaiont was 624 feet (190 meters) long; it was 72 feet (22 meters) thick at the top and tapered at the base. Its reservoir held 137,000 **acre-feet** of water. When this arch dam collapsed on 9 October 1963, its reservoir was almost completely full. The collapse sent a wall of water racing into the Piave River Valley; the flood engulfed the entire town of Longarone and nearly destroyed at least six other nearby towns. Fatalities eventually totalled nearly 900. The dam collapse tragically underscored the necessity of understanding both dam engineering and the architecture of the soil and rock upon which a dam is built. The Vaiont was destroyed by underground streams that seeped through the unstable rock below the dam. Technically, the dam did not collapse; it was tipped by the force of the water tunneling beneath it.

Additional Information: Smith, Norman. *A History of Dams.* Secaucus, NJ: The Citadel Press, 1972; "Giant Dam Falls in Italy; Hundreds Reported Dead as Water Engulfs Towns." *New York Times*, October 10, 1963, A:1.

Vajont Dam. *See* Vaiont Dam.

Verrazano Narrows Bridge (New York City; 1964)

The Verrazano Narrows Bridge carries Interstate 278 over Verrazano Narrows in New York harbor; the bridge connects Staten Island and Brooklyn. The Verrazano is the longest span in the United States, and one of the longest **suspension bridge**s in the world. The bridge spans 4,260 feet (1,300 meters) from tower to tower; its total length is 7,200 feet (2,196 meters). The four main cables of this massive bridge are each 3 feet (0.9 meters) in diameter, and contain 142,200 miles (228,800 kilometers) of wire. Its double-deck roadway

The Verrazano Narrows Bridge spans the entrance to New York Harbor. Photo courtesy of the Bethlehem Steel Corporation.

carries 12 lanes of traffic. The vertical clearance for ships is 216 feet (66 meters). The Verrazano is a gravity anchorage bridge; its western anchorage is a massive 216 feet wide and 315 feet (96 meters) long, and is founded 75 feet (23 meters) below ground level on compact sand. The bridge's other anchorage is even larger, but only extends downward 52 feet (16 meters). The footings on the anchorages are spread to distribute the downward force and protect against settling. As with other giant bridges, the Verrazano's deck had to be curved slightly in construction to accommodate the curve of the earth. The bridge opened on 21 November 1964, the eighty-fifth birthday of its designer, Othmar Amman.

Additional Information: DeLony, Eric. *Landmark American Bridges*. Boston: Bullfinch Press, 1993; Overman, Michael. *Roads, Bridges and Tunnels*. Garden City, NY: Doubleday & Company, 1968; Young, Edward M. *The Great Bridge: The Verrazano Narrows Bridge*. New York: Ariel/Farrar, Strauss and Giroux, Inc., 1965.

Versailles Palace (Versailles, France; late seventeenth century)

Louis XIII of France built a royal hunting lodge at Versailles, about 14 miles (23 kilometers) southwest of Paris, in 1624. In 1661, the architect Louis Le Vau began construction on the site of a new palace ordered by Louis XIV. The original palace, which took more than

Versailles Palace was built in the seventeenth century by Louis XIV of France.

40 years to complete, was subsequently enlarged by later kings. The palace is more than a quarter mile (0.4 kilometer) long and contains over 1,300 rooms. The richly decorated interior contains many famous rooms, including the private apartments of the king and queen, the Room of Hercules, and the Hall of Mirrors, designed by Charles Le Brun. The ceiling of Le Brun's long mirror-lined hallway is painted with scenes glorifying the

achievements of Louis XIV. The palace also contains a magnificent royal chapel and a private theater. The palace gardens, originally laid out by Andre Le Notre in the 1660s, have been enlarged over time to cover almost 250 acres (101 **hectares**). Laid out in grand geometric patterns, the gardens contain many fountains and exotic plants and much statuary. The palace park also contains two small palaces—the Grand Trianon and the Petit Trianon, as well as the stables, the orangery (a greenhouse for growing orange trees), and the *hameau*, a miniature farm designed in the late eighteenth century by Marie Antoinette, the wife of Louis XVI. Both a place of entertainment and a center of government, Versailles in the seventeenth and eighteenth centuries housed some 1,000 courtiers and over 4,000 attendants. Over 14,000 soldiers and servants lived in the town of Versailles, which was founded in 1671 and grew to 30,000 inhabitants by the death of Louis XIV in 1715. In the eighteenth century, the palace became a symbol of royal extravagance; mobs invaded the palace during the French Revolution in the 1790s, and much of the furniture and art was stolen or destroyed. In 1837, Louis-Philippe restored the palace and turned it into a national museum. In 1870, the Prussian army besieging Paris used Versailles as its headquarters, and the German Empire was proclaimed in the Hall of Mirrors in 1871. In the 1870s, the palace was the seat of the French Parliament, and the constitution of the Third Republic was proclaimed there in 1875. In 1919, Versailles was the site of the peace conference ending World War I; the Treaty of Versailles between Germany and the Allies was signed there. President Charles de Gaulle restored and modernized the palace in the mid-twentieth century, making Versailles one of the most visited tourist sites in France.

Additional Information: Walton, Guy. *Louis XIV's Versailles*. Chicago: University of Chicago Press, 1986.

Veurdre Bridge (near Moulines, France; 1910)

In building this bridge across the River Allier, designer Eugene Freyssinet anticipated that the structure might, over a short period of time, have a problem with "concrete creep," that is the concrete might grow slightly larger as it cured. Freyssinet left an opening at the top of the crown of each arch of the bridge, and then, one year after completion, jacked the sections apart and refilled the openings.

Additional Information: Brown, David J. *Bridges*. New York: Macmillan, 1993.

Via Appia, also known as the Appian Way (Rome, Italy; 312 B.C.)

Appius Claudius Caecus, the Roman censor (a city magistrate) who was responsible for the construction of the AQUA APPIA aqueduct, also built this road. Appius skillfully played off the plebeians (the non-noble classes in Rome) against the aristocrats to win Senate approval for the two projects. A majority of Roman citizens had opposed both projects as too expensive. Neither the road nor the **aqueduct** were complete on the expiration of Appius's term of office. Fearing that his successors would refuse to complete the projects, Appius convinced the Senate to extend his term until both projects had been finished. Appius then secured his place in history by naming the two projects after himself. The most famous of Roman roads, the Via Appia originally connected Rome with Capua; it was later extended southward to Beneventum (Benevento), Tarentum (Taranto), and Brundisium (Brindisi). Branch roads led to Neapolis (Naples) and other Italian ports. The 350-mile (563-kilometer) road became the chief highway to Greece and the Roman East. Built of extremely large polygonal blocks of lava, the road is still in existence today. The section leaving Rome was lined with Roman tombs and temples and Christian CATACOMBS.

Additional Information: Hawkes, Nigel. *Structures: The Way Things Are Built*. New York: Macmillan, 1990; O'Connor, Colin. *Roman Bridges*. Cambridge, England: Cambridge University Press, 1993; Schreiber, Herman. *The History of Roads: From Amber Route to Motorway*. London: Barrie and Rockliff, 1961.

Vietnam Veterans Memorial Wall

(Washington, DC; 1982, 1984)

The Vietnam Veterans Memorial Wall was designed by Maya Ying Lin. The concept of a memorial to veterans of the Vietnam War was conceived in 1979 by Jan Scruggs, a former infantry corporal who had served in Vietnam. As fund raising for the monument progressed, a national design competition for the monument proceeded. Of the 1,421 entries, Lin's sober, simple, and stately monument was the unanimous choice of the organization. Lin, from Athens, Ohio, was at the time a 21-year-old student at Yale University. The monument is made of walls of polished black granite inscribed, in chronological order, with the names of the 58,191 dead of the Vietnam War. The wall is starkly effective; Lin was absolutely correct in believing that the names would eventually "become the memorial." The walls are 246.75 feet (over 75 meters) long. To support the memorial, 140 pilings go down approximately 35 feet (10.7 meters) to bedrock. The equally moving sculpture at the entrance to the wall was designed by Frederick Hart. The sculpture depicts young Vietnam soldiers wearing the uniforms and carrying the equip-

ment of their war. Hart sought to underscore "the poignancy of their sacrifice" by illustrating the contrast between the soldiers' innocence and youth and the

The Vietnam Veterans Memorial in Washington, DC.

weapons of war they carried. The memorial was dedicated on 23 November 1982. The entire $7 million cost of establishing the memorial was paid for by contributions from individuals, businesses, and organizations.

Additional Information: http://www.nps.gov/nps/ncro/nacc/vvm.html.

Vioreau Dam. See Nantes-Brest Canal

Volta Bridge (near Accra, Ghana)

Volta Bridge is located 100 kilometers (62 miles) north of Accra, Ghana. This 242-meter-long (794-foot) **arch bridge** was built by using a new variation on the **cantilever** method. The arch, designed as a series of high-tensile shop-welded steel sections, was butted and bolted at each junction by four high-tensile steel bolts. Bolted joints made the **girder**s less vulnerable during the **tension** stage of cantilever erection when steel rivets are inclined to change shape permanently (creep) under the temporary tensile load. The new method saved both time and expense.

Additional Information: Overman, Michael. *Roads, Bridges and Tunnels*. Garden City, NY: Doubleday & Company, 1968.

Volta Dam (near Accra, Ghana; 1965)

Volta Dam crosses the Volta River northeast of Accra in eastern Ghana. This large rock dam is 370 feet (113 meters) high and 2,100 feet (640.5 meters) long. The dam's volume is 10,350,000 cubic yards (7,918,000 cubic meters), with an additional capacity in the associated reservoir of 120,000,000 **acre-feet**.

Additional Information: Smith, Norman. *A History of Dams*. Secaucus, NJ: Citadel Press, 1972.

Vyrnwy Dam (Powys County, Wales; 1892)

Vyrnwy Dam was the first large masonry dam built in Great Britain, and the first use of a dam in Wales to retain water for a city in England (Liverpool). The dam is 1,350 feet (412 meters) long and 136 feet (42 meters) high, and is much thicker than any of its contemporaries. It was also the first dam constructed with a network of drainage tunnels, a safety measure designed to prevent pressure on the water face from becoming too severe. The building of what was then the largest dam in Europe involved the flooding of an entire village to create the artificial Lake Vyrnwy in the Welsh county of Powys.

Additional Information: Smith, Norman. *A History of Dams*. Secaucus, NJ: Citadel Press, 1972.

W

Wabasha Street Bridge River Crossing (St. Paul, Minnesota; 1997)

Four successive bridges have been constructed at this site on the Mississippi River, and a fifth was completed in the fall of 1997. The planning for the first bridge began in 1854, on the same day that the city of Saint Paul was incorporated. The first bridge, a wooden **Howe truss**, was completed in 1859. This bridge was replaced by a second Howe truss, and then by an all-iron **Pratt truss**. All three bridges were designed to allow traffic to flow between the trusses, leaving room below for navigation. The fourth bridge, built in 1889, was an iron **cantilever** deck-truss; until it was removed in February 1996, it was the only surviving cantilever truss bridge in Minnesota and was listed in the National Register of Historic Places. Because of deterioration of the old bridge, work began in early 1996 on a new structure to replace it. With an estimated cost of $24 million, the new bridge is a concrete **box girder** with a split deck. It has a bicycle lane in each direction, and 12-foot-wide (3.6-meter) walkways. A small bridge leads from the south side of the new Wabasha Street Bridge to Raspberry Island.

Additional Information: http://bridges.sppaul.gov/ WabINFO.html.

Wachusett Dam (Clinton, Massachusetts; 1905)

Frederic P. Stearns designed and built this dam across the south branch of the Nashua River near Clinton, Massachusetts, 40 miles (64 kilometers) west of Boston. The all-masonry Wachusett Dam was part of an urgently needed project to deliver water to Boston. Still an integral part of Boston's water system, the dam delivers water first through a 12-mile-long (19-kilometer) tunnel that empties into the Sudbury Reservoir, which was built at the same time as the dam. The dam is 944 feet (288 meters) long with an overflow spillway that is more than 450 feet (137 meters) long. The maximum water height is 250 feet (76 meters).

Additional Information: Jackson, Donald C. *Great American Bridges and Dams*. Washington, DC: The Preservation Press, 1988.

Waco Suspension Bridge (Waco, Texas; 1869)

Thomas M. Griffith built this **suspension bridge** across the Brazos River in the eastern Texas city of Waco. Considered an engineering marvel when it first opened, the bridge is 475 feet (145 meters) long; cables are suspended from two Gothic-type brick towers. The cables and other steel for the building were provided by John A. Roebling & Son of New York; Roebling became the master bridge builder of the nineteenth century with his later construction of the BROOKLYN BRIDGE.

Additional Information: http://ohare.easy.com/waco/ bridge.html.

Walnut Lane Bridge (Philadelphia, Pennsylvania; 1949)

Gustav Magnel designed the Walnut Lane Bridge, the first major prestressed and precast concrete **girder** bridge in the United States. The bridge crosses Wissahickon Creek in the Roxborough area of northwest Philadelphia. The bridge has 13 long girders of 160 feet (49 meters), and 14 shorter girders of 74 feet (23 meters). Each of these 27 girders weighs up to 150 tons (135 metric tons).

Additional Information: Steinman, David B. and Watson, Sara Ruth. *Bridges*. New York: Dover Publications, 1941, 1957.

Walnut Street Bridge (Chattanooga, Tennessee; 1891)

This bridge carries Walnut Street across the Tennessee River in the southeastern Tennesse city of Chattanooga. The Walnut Street Bridge was the first multi-use bridge to span the Tennessee River, with the exception of a military bridge built during the Civil War. The bridge was substantially rebuilt in 1993. Until the rebuilding, the 1891 bridge was the oldest surviving truss bridge of its size in the South, a pin-connected camel-back **through-truss** bridge. It included space for a mule-driven steetcar line and two pedestrian walkways. The rebuilt bridge is 2,370 feet (723 meters) long.

Additional Information: http://bertha.chattanooga.net/chamber/walnut.html.

Walt Whitman Bridge (Camden, New Jersey, to Philadelphia, Pennsylvania; 1957)

One of the 10 busiest bridges in North America, thc Walt Whitman Bridge crosses the Delaware River, connecting Philadelphia with Camden, New Jersey. The bridge is named for the famed nineteenth-century American poet, Walt Whitman, who lived most of his

The Walt Whitman Bridge carries traffic across the Delaware River between Philadelphia and New Jersey. Photo courtesy of the Bethlehem Steel Corporation.

life in Camden. The Walt Whitman has mile-long, gradually elevated approaches. The highest point of the bridge is 374 feet (114 meters) above the water. The bridge has a 2,000-foot (610-meter) main span, and covers a total of 6.5 miles (10.5 kilometers) from approach to approach.

Additional Information: Bishop, Gordon. *Gems of New Jersey*. Englewood Cliffs, NJ: Prentice-Hall, 1985.

Washington, George Bridge. *See* George Washington Bridge.

Washington Monument (Washington, DC; 1884)

More than 555 feet (169 meters) tall, this monument in the form of an Egyptian obelisk is the largest masonry structure in the world. Robert Mills, an eminent

The Washington Monument in Washington, DC.

architect and engineer, began planning for a monument to George Washington in 1836; Congress authorized the structure in 1848. Construction progress was slow after Mills' death in 1855, and came to a complete halt during the Civil War in the 1860s. When construction resumed in 1876, the type of marble used for the outside of the monument was no longer available, and visitors can see a difference in the marble beginning about 150 feet (46 meters) up. One of the problems when construction resumed was that the rope used to haul supplies to the top of the monument from the inside had been pulled down. This problem was solved by Lieutenant Colonel Thomas Lincoln Casey, who was appointed by President Ulysses S. Grant to complete the monument. Casey tied a thread to the leg of a pigeon and released the bird inside the monument. A gun blast was set off to scare the bird, which flew up and out the open tower. The bird was then shot down, and stronger and stronger wires were tied to the thread and hauled up and out of the monument. Mills had designed a Greek-style rotunda for the monument but that was never constructed and there is no interest in doing so now.

Additional Information: http://sc94.ameslb.gov/TOUR/ washmon.html; Ditzel, Paul C. *How They Build Our National Monuments*. New York: Bobbs-Merrill, 1976.

Washington Square (Philadelphia, Pennsylvania; 1682)

Washington Square was one of five public parks included in William Penn's 1682 drawings of Philadelphia. Known then as "Southeast Square," the park was almost immediately used as a "potter's field," a burial ground for the indigent, as well as a gathering place. In the late 1770s, the square became a place to bury the dead from the Revolutionary War battles fought near Philadelphia, or those who had been wounded further afield but then died in Philadelphia's hospitals. Facing the square at the time was the Walnut Street Jail, established by the British during their 1777 occupation of Philadelphia. When inhumane treatment led to the deaths of a large number of captured revolutionaries, their bodies were interred in mass graves in the nearby square. In 1793, the square was used as a burial place for victims of a yellow fever epidemic. Five thousand residents of Philadelphia, one-tenth the city's population, died and were buried in trenches at the square during the epidemic. By 1825, the city decided to beautify the area, which was renamed Washington Square in honor of the commander of the Revolutionary dead. More than 100 years later, in 1954, the Washington Square Planning Committee erected a "Tomb of the Unknown Soldier" on the site by locating and reburying the body of a revolutionary war soldier. A wall nearby has an eternal flame, and the tomb itself proclaims: "Beneath this stone rests a soldier of Washington's army who died to give you liberty."

Additional Information: http://www.libertynet.org/iha/ _tomb.html.

Washington Street Tunnel (Chicago, Illinois; 1869)

J.K. Lake constructed the Washington Street Tunnel under the Chicago River in 1869. Chicago's first traffic tunnel, the Washington Street Tunnel was built to allow the river bridges to be raised for long periods without slowing up traffic across the river. Built at a cost of $517,000, the tunnel was 1,605 feet (490 meters) long, and remained in service until 1953.

Additional Information: http://cpl.lib.uic.edu/004chicago/ timeline/tunneltrffc.html.

Waterford Union Bridge. *See* Union Bridge.

Wear Bridge (Sunderland, England; 1796)

Created for a design competition by Thomas Paine, the American patriot and philosopher, the Wear Bridge design makes use of **cast-iron voussoirs** to form a bridge with a 20 to 1 span to rise ratio. Paine lost interest in the bridge (as well as financial backing) when the French Revolution broke out in 1789. The bridge was redesigned somewhat and erected across the Wear River at Sunderland (see entry for SUNDERLAND BRIDGE) in northeastern England in 1796, where it remained in service until 1858.

Additional Information: Lay, M.G. *Ways of the World*. New Brunswick, NJ: Rutgers University Press, 1992.

Wertz's Bridge (near Reading, Pennsylvania; 1867)

Wertz's Bridge crosses Tulpehocken Creek in Berks County Heritage Park in southeastern Pennsylvania. Built by Amandas Knerr, this Burr truss covered bridge, known to locals as the "Bridge to Nowhere" because it is closed to traffic, is 15 feet (4.6 meters) wide and 204 feet (62 meters) long.

Additional Information: http://william-king.www.drexel.edu/ top/bridge/CBWertz.html.

West Cornwall Covered Bridge (West Cornwall, Connecticut; 1864)

One of Connecticut's few remaining covered bridges, the West Cornwall Bridge crosses the Housatonic River near West Cornwall in the northwestern part of the state. The wooden truss bridge was threatened by destruction and replacement in 1961. A huge ice jam began to build up around the bridge's **abutments**, and the state had to dynamite the ice to save the bridge and its deck. Shortly after, the state suggested replacing the bridge with a more modern structure. When residents protested, the state worked out a method of dealing with the ice problem that was satisfactory to everyone involved. Rather than do away with the bridge, the state raised the bridge 2 feet (0.6 meters), and replaced only the wooden deck with a steel one. When the project was completed in 1973, it won an award from the Federal Highway Administration for an outstanding example of historic preservation.

Additional Information: Failla, Kathleen Saluk. "A Decade After Disaster State Rebuilds at Rapid Pace." *New York Times*, 29 June 1993, 13CN:1.

West Gate Bridge (Melbourne, Australia; collapsed 1970)

The West Gate Bridge across the River Yarra at Melbourne in southeastern Australia collapsed in 1970 while under construction. Failure of a large steel **box girder**, which split at the central joint between its two box sections, led to the deaths of 35 men. Some evidence suggests that the failure was partly due to human error; one account of the failure states that engineers removed too many connection bolts at midspan while attempting to make a repair.

Additional Information: "Disaster in the Making: An Engineering Case Study [of the Quebec Bridge]," http:// www.civeng.carleton.ca/ECL/reports/ and related pages; Brown, David J. *Bridges*. New York: Macmillan, 1993.

West Main Street Bridge (Clinton, New Jersey; 1870)

The West Main Street Bridge, built by William and Charles Cowin, carries West Main Street over the South Branch Raritan River in the western New Jersey town of Clinton. This structure is one of only three similar bridges in New Jersey built by the Cowins. (The other two are in New Hampton [1868] across the Musconetcong River, and in Glen Gardner [1970] across Spruce Run.) Made of composite **cast-iron** and **wrought-iron Pratt trusses**, the three bridges all include a "**tension** adjuster," which was patented by William Johnson in 1870. The upper chord member is octagonally shaped, and the railing on the walkway is carefully detailed. Since its construction, the West Main Street Bridge has only had its deck beams and deck surface replaced.

Additional Information: DeLony, Eric. *Landmark American Bridges*. Boston: Bullfinch Press, 1993.

West Main Street Bridge, also known as the Purple Bridge (Stamford, Connecticut; 1888)

The Berlin Iron Bridge Company built this classic iron bridge, a parabola-shaped lenticular **pony truss**, across the Rippowam River in downtown Stamford in western Connecticut. The 124-foot-long (38-meter) West Main Street Bridge has intricate detailing in its iron-work. It was built to service woolen mills in the area, but is now surrounded by apartment houses and small stores. Its pedestrian walkways, once made of wood, have been replaced with metal in spots where the wood has broken or rotted away. In the late 1980s, the bridge was mistakenly painted purple, and it is almost always referred to now by people in Stamford as the Purple Bridge. Renovations began in the winter of 1995-1996; what its color will be after the renovation has not been determined. Two other Berlin Iron Bridge Company structures in Stamford are the TURN OF RIVER BRIDGE and the disassembled PULASKI STREET BRIDGE.

Additional Information: Sierpina, Diane Sentementes. "Stamford; A (Sort of Purple) Iron Bridge Gathers Champions." *New York Times*, 4 June 1995, XIIICN:2.

Westminster Abbey (London, England; begun 1245)

Westminster Abbey, officially the Collegiate Church of Saint Peter in Westminster, is England's national shrine and one of the most important examples of Gothic architecture in the country. King Sebert of Essex is re-

Westminster Abbey, built in the eleventh century and rebuilt in the thirteenth, stands along the Thames in London, England.

ported to have built the first church on this site along the Thames in 616. About 1050, King Edward the Confessor began to build a new church on the site, near an existing Benedictine abbey. The Confessor's new church at Westminster, a village west of London, was consecrated in 1065. In 1245, Henry III demolished the Confessor's church and initiated a long period of work on a new, larger structure that was to serve as the abbey church of the Westminster Benedictine community until 1539, when Henry VIII dissolved the monastery. Henry of Reims, Henry III's architect, completed a good portion of the church building by 1258. The abbey exhibits the High Gothic character of CHARTRES CATHEDRAL in France, but deviates with several important and artistically successful English modifications, such as the unvaulted gallery on the second level and the use of delicate patterning on the walls. Work continued for centuries on the monastic structures accompanying the church. The octagonal chapter house was completed in 1250, and the cloisters, abbot's house, and other major monastic buildings were added in the fourteenth century. The nave of the church was completed in the fifteenth century, and Henry VIII finished the Lady Chapel, and dedicated it to his father, Henry VII, in the early sixteenth century. The Henry VII chapel, with its superb fan vaulting, is an excellent example of the Perpendicular style. Sir Christopher Wren and Nicholas Hawksmoor built the two western towers of the church in the eighteenth century. The church is cruciform in plan, and both the nave, which is the highest in England, and the transept have side aisles. Like NOTRE DAME in Paris, Westminster Abbey has strongly emphasized flying buttresses. All English monarchs since William the Conqueror in the eleventh century have been crowned in the abbey, and more than a dozen monarchs, including Edward the Confessor, Henry III, Edward III, Richard II, Henry V, Henry VII, Mary I, and Elizabeth I, are buried there. Noted political figures, such as David Lloyd George, and other distinguished English citizens have been buried in the abbey since the fourteenth century. The Poets' Corner in the south transept contains the remains of such important literary figures as Chaucer, Browning, and Tennyson. The abbey was somewhat damaged by German air raids in 1941, but it remained structurally sound and was repaired after the war.

Additional Information: *Academic American Encyclopedia* (electronic version). Danbury, CT: Grolier Electronic Publishing, 1991.

Westminster Hall. *See* Westminster Palace.

Westminster Palace, or the Houses of Parliament (London, England; 1840-1860)

The present Houses of Parliament were built in the mid-nineteenth century on a site along the north side of the Thames River to the west of London in Westminster; the site had been occupied by an aggregation of ancient buildings that included the old royal palace of Westminster. WESTMINSTER ABBEY stands immediately northwest of the Houses of Parliament. The original palace and most of the surrounding structures were destroyed by fire in 1834. King Edward the Confessor built the original palace at Westminster in the mid eleventh century; the structure served as a royal residence until the sixteenth century, when the palace buildings became the assembly place for the Houses of Parliament, with the Commons meeting usually in St.

Stephen's Chapel. Westminster Hall, built in the late eleventh century by William II, became the fixed meeting place of the chief courts of the common law and the

"Big Ben" clock, in the eastern tower of Westminster Palace, the meeting place of the British Houses of Parliament. Photo courtesy of Ray Farrar.

site of some of the most dramatic events in English history, including the deposition of Richard II, the trial of Sir Thomas More, and the trial and condemnation of Charles I. Richard II rebuilt the hall in the late fourteenth century, giving the structure its magnificent 68-foot-wide (21-meter) hammer-beam roof. The only part of the palace to survive the fire of 1834, Westminster Hall was severely damaged by incendiary bombs in May 1941. Sir Charles Barry built the present perpendicular Gothic structure, incorporating Westminster Hall as an entrance for the new building. Both the chambers of the Commons and the Lords were heavily damaged by bombing in 1940 and 1941. For a time, the House of Commons was forced to meet in the Lords chamber until repairs were completed on the Commons chamber. The best-known portion of the new building is the tall, ornate clock tower on the eastern end of the structure. The huge clock in the tower is often known as "Big Ben," although that name actually applies to the large bell that was installed in the tower in 1856 and named for Sir Benjamin Hall, who was commissioner of works when the bell was put in place.

Additional Information: http://www.parliament.the-stationery-office.co.uk.

Wheeling Suspension Bridge (Wheeling, West Virginia; 1849, 1854, 1859, 1872)

Designed and built by Charles Ellet, Jr., the massive and still somehow delicate Wheeling Suspension Bridge established the United States as the leader in the design and construction of long **suspension bridge** spans. The Wheeling Suspension Bridge crosses the Ohio River, connecting Wheeling, West Virginia, with eastern Ohio. Completed only a few months after John Roebling's DELAWARE AQUEDUCT, the bridge was the longest span in the world. Like the GRAND PONT SUSPENDU in Switzerland, the Wheeling's major technological feat was the use of individual wires bundled into cables. On the Wheeling bridge, Ellet used cable garlands, multiple cables to carry the deck. Later, after several cables were blown away by storms, and a near-total collapse in 1854, the multiple cables were bound into a single cable. In 1859, an associate of Ellet's widened the rebuilt 1854 structure. In 1872, an engineer with Roebling's company, Wilhelm Hildenbrand, rebuilt the deck structure and added diagonal stay wires between the towers and the deck to help withstand winds. The bridge was rehabilitated again in the early 1980s by the West Virginia Department of Highways. Still in use, the bridge has retained more than a passing resemblance to Ellet's original construction.

Additional Information: Brown, David J. *Bridges*. New York: Macmillan, 1993; DeLony, Eric. *Landmark American Bridges*. Boston: Bullfinch Press, 1993; Hopkins, H.J. *A Span of Bridges: An Illustrated History*. New York: Praeger, 1970; Jackson, Donald C. *Great American Bridges and Dams*. Washington, DC: The Preservation Press, 1988.

Whinhill Dam (Whinhill, England; 1796)

Whinhill Dam was built to supply water to nearby towns. Although it failed several times in the twentieth century, Whinhill is now functioning well; its height of 40 feet (12.2 meters) is distributed along an 800-foot (244-meter) length, with a gently sloping masonry-lined water face.

Additional Information: Smith, Norman. *A History of Dams*. Secaucus, NJ: Citadel Press, 1972.

Whipple Cast-Iron & Wrought-Iron Bowstring Truss Bridge (near Albany, New York; 1869)

Squire Whipple patented his truss design in 1841, and he and Simon DeGraff built this first example of the design over the ERIE CANAL near Norman's Kill Ravine in eastern New York near Albany. Whipple's mathematical analysis of stresses helped turn bridge building into a practice of professional engineers, rather than an activity of carpenters, ironworkers, or other types of craftsmen. The bowstring truss also helped move American bridges away from wood toward steel and iron. The sides of the bridge are relatively open except for the iron parallelograms and their diagonals, which support the arch.

Additional Information: DeLony, Eric. *Landmark American Bridges*. Boston: Bullfinch Press, 1993.

White Bowstring Arch-Truss Bridge (Poland, Ohio; 1877)

The White Bowstring Arch-Truss Bridge carries Cemetery Drive over Yellow Creek in the eastern Ohio town of Poland. William Rezner, John Glass, and George Schneider planned this bridge as an oval **wrought-iron** tubular arch, a bridge design they patented in 1867. Across the top of the arch is a sizable reinforcement structure, similar to the structure supporting the sides of the bridge. This upper chord section is designed to give the bridge lateral strength to resist sideways twisting.

Additional Information: DeLony, Eric. *Landmark American Bridges*. Boston: Bullfinch Press, 1993.

White House (Washington, D.C.; begun 1792)

The White House is the official residence of the president of the United States. It is located in Washington, D.C., at 1600 Pennsylvania Avenue on a site on the south side of the street selected by George Washington.

The White House, the official residence of the president of the United States, in Washington, DC.

Designed by James Hoban, whose original plans referred to the structure as "the Palace," the White House is constructed of Virginia freestone. It is the oldest public building in Washington, its cornerstone having been laid in 1792. It was first occupied in 1800 by President John Adams and his wife Abigail. The British burned the building in 1814 when they briefly held Washington during the War of 1812. The building unofficially acquired the name "White House" when its smoke-stained walls were repainted white as part of the post-war reconstruction. The name became official in the early twentieth century when President Theodore Roosevelt had it engraved on his presidential stationery. The main building is four stories high, about 170 feet (52 meters) long, and about 85 feet (26 meters) wide. The east and west terraces, the executive office, and the east wing are later additions. Large receptions are often held in the East Room, which is 82 feet (25 meters) long and 40 feet (12.2 meters) wide; the elliptical Blue Room is also used for many diplomatic and official functions. The Red and Green Rooms are used for smaller, more informal gatherings. The White House is surrounded by 18 acres (7.3 **hectares**) of lawns and grounds originally designed by Andrew Jackson Downing.

Additional Information: http://www.whitehouse.gov.

Whitestone Bridge (Bronx to Queens, New York; 1939)

Othmar Ammann built this bridge across the East River between the Bronx and the Whitestone area in the borough of Queens. A vital link between two of the five boroughs of New York City, this 2,300-foot (701.5-meter) **suspension bridge** was opened in time to provide access to the 1939 World's Fair in Queens. Although sturdy, the bridge has a seeming lightness to it; after the failure of the TACOMA NARROWS BRIDGE in 1940, visible 14-foot (4.3-meter) trusses were added, partly to strengthen the bridge and partly to reassure drivers going over the bridge. The apparent lightness shows the skill in the early work of Ammann. The bridge was a favorite project of Robert Moses, chairman of the Triboro Bridge and Tunnel Authority, and a leader (albeit controversial) in building transportation systems in and around New York City.

Additional information: Maitland, Leslie, "Moses, 90, Nostalgic About Whitestone Bridge, 40." *New York Times*, April 30, 1979, B:5.

Whitman, Walt Bridge. *See* Walt Whitman Bridge.

Williamsburg Bridge (New York City; 1903)

The Williamsburg Bridge spans the East River, connecting Manhattan and Brooklyn. The Williamsburg stands about one mile (1.6 kilometers) north and east of the MANHATTAN BRIDGE and about 1.5 miles (2.4 kilometers) northeast of the BROOKLYN BRIDGE. The Williamsburg

Bridge spans 1,600 feet (488 meters), a little more than the Brooklyn Bridge. It was the first large **suspension bridge** to be built with steel towers rather than iron, concrete, or stone towers.

Additional Information: Brown, David J. *Bridges*. New York: Macmillan, 1993.

A view (c.1903) of the Williamsburg Bridge between Manhattan and Brooklyn under construction. Photo courtesy of the Library of Congress, World's Transportation Commission Photograph Collection.

Wilson Dam (Muscle Shoals, Alabama; 1925)

The U.S. Army Corps of Engineers built this dam across the Tennessee River near Muscle Shoals in northwestern Alabama. Planning for the dam began in 1917, when the U.S., about to enter World War I, had need of the munitions that power from the dam could help manufacture. However, the dam was not completed until well after the war ended. Wilson Dam provides a great deal of electric power for the surrounding region, with a generating capacity of 400,000 kilowatts. The water from the 4,800-foot-long (1,464-meter), 137-foot-high (42-meter) dam eases navigation on the Tennessee River. Now under the control of the TENNESSEE VALLEY AUTHORITY, the dam was the center of a major political controversy in 1937 when Henry Ford tried unsuccessfully to buy the dam from the government for a fraction of its construction cost.

Additional Information: Jackson, Donald C. *Great American Bridges and Dams*. Washington, DC: The Preservation Press, 1988.

Wiman Mek Palace (Bangkok, Thailand; c. 1868)

Surrounded by water on all sides, the Wiman Mek Palace is the world's largest building made entirely of golden teak. A three-story structure, it has 81 rooms, halls, and anterooms, as well as many terraces and verandas. Originally, King Chulalongkorn (Rama V, who reigned 1868-1910) had the building constructed on Srchang Island in the Gulf of Siam. Chulalongkorn (the model for the Siamese king in the musical "The King and I") moved the building to its present site in 1901. The king had his permanent residence on the third floor where he enjoyed what is believed to have been the first shower ever installed in Thailand (the water tank had to be filled by hand). The first and second floors were residences for his consorts and female children. After Chulalongkorn's death, the palace was briefly used by the succeeding two kings, and then closed by King Rama VII. It remained shut for more than 50 years until Queen Sirikit began to renovate the building.

Additional Information: http://www.asiatour.com/thailand/e-03bang/et-ban72.htm.

Winch Bridge (near Middleton, England; 1741)

This foot bridge across the River Tess was the first metal **suspension bridge** in England. Suspended on iron chains, it was about 70 feet (21 meters) long but only 2 feet (0.6 meters) wide. With a handrail on only one side, the swaying bridge must have been a fearful experience to cross. The bridge collapsed in 1802, probably because of deterioration of the iron chains. Its replacement, known as Wynch Bridge (the different spelling probably was to differentiate it from its predecessor) held until 1908.

Additional Information: Hopkins, H.J. *A Span of Bridges: An Illustrated History*. New York: Praeger, 1970.

Winchester Mystery House (San Jose, California; 1922)

Built by Sarah Winchester, the widow of William Wirt Winchester, who was the son of the manufacturer of the Winchester repeating rifle, this house was a work in progress for 38 years. Construction only ended with the death of Mrs. Winchester. Containing at least 160 rooms, the house has an amazingly complex organization that has never been completely or accurately mapped. Upset by the deaths of her husband and daughter, Mrs. Winchester consulted a psychic who told her that the deaths were the revenge of all those people who had been killed by the rifles manufactured by her family. The only way to escape the curse, and achieve eternal life, was by buying a house in the West and continually building

and expanding it. Mrs. Winchester moved to California and began the continual renovation of her house. While she lived, she maintained a staff of 18-20 domestic servants, 12-18 gardeners, and 10-22 carpenters. Until her death, the house was a scene of constant building, demolition, rebuilding, and remodeling. Staircases go nowhere; rooms open into other rooms or dead ends. Parts of the house may never have been seen, and it is uncertain if even Mrs. Winchester knew where everything was.

Additional Information: http://www.netview.com/svg/tourist/winchester/story.html.

Winchester-Coombs #2 Covered Bridge (Winchester-Swanzey town line, New Hampshire; c. 1837)

This 118-foot (36-meter) **Town Lattice truss** crosses the Ashuelot River in southwestern New Hampshire. The stone **abutments** are unusual in that they were built without mortar.

Additional Information: http://vintagedb.com/guides/covered1.html.

Windsor Castle (Windsor, England; c. 1090)

Windsor Castle, the official residence of the British monarch, was begun in the eleventh century by William the Conqueror. The site was chosen because it placed the castle high above the River Thames to protect the western approaches to London, which is now only about 20 miles (32 kilometers) away. The castle was improved, enlarged, and rebuilt by various monarchs throughout the Middle Ages. Edward III built the massive Round Tower in the fourteenth century, and Edward IV built St. George's Chapel, the church of the Knights of the Garter, in the fifteenth century. Henry

Windsor Castle on the Thames west of London. Photo courtesy of Ray Farrar.

VI, Edward IV, Henry VIII and his wife Jane Seymour, Charles I, and George VI are among the monarchs buried in the chapel. The castle itself stands in the Home Park, which is surrounded by the vast Great Park. The tree-lined Long Walk extends out from the castle for several miles. When the monarch is in residence, his or her official standard flies above the castle. Now almost completely open to tourists, the castle, which contains famous works of art and important artifacts of English history, is a major destination for visitors to the London area. Among the castle's attractions is Queen Mary's Doll House. The current ruling family, the House of Windsor, took their name from the castle.

Additional Information: http://www.londonmail.co.uk/windsor/default.html.

Windsor Locks Canal (Windsor Locks, Connecticut; c. 1840)

This canal was a 6-mile (10-kilometer) bypass around the Enfield Falls, which are located on the Connecticut River 12 miles (19 kilometers) above Hartford. It ran from Windsor Locks to about one mile above the town of Suffield. For a while, the canal played an essential part in the state's growing economy and was the most important canal on the Connecticut River. The opening and closing dates of the canal are not precisely known. The canal probably opened in the 1840s and closed by the end of the 1880s. It was definitely not in use in the 1890s.

Additional Information: Spangenburg, Ray and Moser, Diane K. *The Story of America's Canals*. New York: Facts on File, 1992.

Winter Palace (St. Petersburg, Russia; 1764)

The architect Bartolomeo F. Rastrelli designed this palace for Empress Elizabeth, the daughter of Peter the Great. The winter Palace was the home of the tsars for 150 years until the 1917 Russian Revolution. Rastrelli was a well-known builder, a French-educated Italian who spent much of his working life in Russia until driven out of the country by Catherine the Great. With a total perimeter of nearly 2 kilometers (1.2 miles), the massive white and gold building is an "endless array of columns and statues." Although an impressive structure, the palace has been criticized for not being a fully integrated whole, as are some of the smaller works of Rastrelli. The ornamental interior was destroyed by fire in 1837, and was rebuilt under the direction of Czar Nicholas I, who restored Rastrelli's 1,050 rooms and 117 staircases. The rebuilding also added a French Empire theme to Rastrelli's interior. The building now boasts floors of rosewood and ebony, and walls and ceilings of marble, gold, malachite, and jasper. Parts of the interior of the palace are now accessible to tourists.

Additional Information: Bartoli, Georges. *Moscow and*

Leningrad Observed. New York: Oxford University Press, 1975; Hamilton, George Heard. *The Art and Architecture of Russia*. 3rd ed. Pelican History of Art Series. New York: Penguin Books, 1983.

Woodward Avenue (Detroit, Michigan; 1909)

A famed central artery in this major city, Woodward Avenue, completed in July 1909, boasts the first rural mile of concrete highway.

Source: Patton, Phil. *Open Road: A Celebration of the American Highway*. New York: Simon and Schuster, 1986.

Woolworth Building (233 Broadway, New York City; 1913)

Sixty stories high, this building supplanted the FLATIRON BUILDING as the tallest in the world. It held the title for 17 years. Architect Cass Gilbert used glazed terra-cotta as a masonry building material, a style he, along with

A view (c.1920) of the Woolworth Building in New York City. Photo courtesy of the Library of Congress, Detroit Publishing Company Photographic Collection.

several other noted architects of the period popularized. The Woolworth Building was the first skyscraper to feature such modern advantages as fireproof steel framing, pump-driven plumbing, and telephone service. Owner and builder Frank Woolworth was the founder of the "five-and-ten-cent" store.

Additional Information: http://www.hotwired.com/rough/usa/mid.atlantic/ny/nyc/city/midtown.html.

World Trade Center (New York City; 1973)

This building complex in lower Manhattan consists of 110-floor twin towers rising to 1,350 feet (412 meters), making the structure second in height in the United States to the SEARS TOWER in Chicago. Designed by Minoru Yamasaki and Emery Roth, the structure was completed in 1973 at a cost of $750 million. In 1993, an underground parking structure at the World Trade Center was the site of a terrorist bombing that took several lives and caused millions of dollars in damage.

Additional Information: Robins, Anthony. *The World Trade Center*. Englewood, FL: Pineapple Press, 1987.

The twin towers of the World Trade Center in lower Manhattan.

Y

Yamantau Mountain Military Complex
(Beloretsk area, Russia; under construction)

In 1996, a gigantic engineering project of unknown function was reported to be under construction in the former Soviet Union. Begun under the regime of Leonid Brezhnev, the project is said to be larger than the Washington, D.C., area inside the beltway. PENTAGON officials believe that the project will serve as a "command and control center for nuclear weapons and a bunker for military leaders." Earlier characterizations of the project have included a "mining site, a repository for Russian treasures, a food storage area, a dump for nuclear materials and a bunker for Russia's leaders in case of nuclear war." Development of the project, if it goes forward under the new Russian regime, means development of a railroad, a highway, and towns for thousands of workers and their families. The project's status as of mid-1997 is unknown.

Additional Information: Gordon, Michael R. "Despite Cold War's End, Russia Keeps Building a Secret Complex." *New York Times*, 16 April 1996, A:1.

Yaquina Bay Bridge (Newport, Oregon; 1936)

The Yaquina Bay Bridge carries the Oregon Coast Highway (U.S. Route 101) over Yaquina Bay near Newport in western Oregon. Built by Conde B. McCullough, the Yaquina Bay Bridge is a combination of steel and **reinforced concrete** arches that stretches 3,260 feet (994 meters). The main channel span is 600 feet (183 meters) long and 226 feet (69 meters) high, and suspends the roadway from hangers. Two steel deck arches on either side are each 350 feet (107 meters) long. On either side of these are a total of 10 ribbed open-**spandrel** deck

arches, ranging in length from 160 feet (49 meters) to 265 feet (81 meters) long.

Additional Information: DeLony, Eric. *Landmark American Bridges*. Boston: Bullfinch Press, 1993.

Yarmuk Dam Project (Baquora, Jordan; proposed)

Agreed upon in 1995, this $535 million project will construct two dams at the Yarmuk and Jordan Rivers to send water into Jordan for at least five months of the year. Built on territory that Israel ceded to Jordan in 1994, and which Jordan then leased back to Israel in 1995, the agreement to build the dams was signed by Jordan's King Hussein, Israeli Prime Minister Yitzhak Rabin, and Helmut Kohl, chancellor of Germany. Germany will contribute at least 28 percent of the cost of the project, which is being funded by the European Union.

Yaxchilan Suspension Bridge (across the Usumacinta River between modern Mexico and Guatemala; c. eighth century)

The evidence for the existence of a rope **suspension bridge** over the Usumacinta is highly convincing. James O'Kon, an engineer and amateur archeologist, suggests that a 600-foot-long (183-meter) rope structure supported by two stone and masonry towers and stone **abutments** once spanned the river. His evidence includes the remnants of one tower and an abutment, as well as a possible rock guideway for the ropes of the bridge. The bridge must have been high enough to allow people to pass over the river even in times of flooding, which was a common occurrence in this rain-forest region. A bridge longer than the Yaxchilan was not constructed in Europe until the fourteenth century.

Additional Information: O'Kon, James A. "Bridge to the Past." *Civil Engineering* 65:1 (January 1995), pp. 62-65.

Yemen Dam. *See* Marib Dam.

Yerba Buena Tunnel (underneath Yerba Buena Island in San Francisco Bay between San Francisco and Oakland; 1936)

Constructed on U.S. INTERSTATE ROUTE 80 as part of the OAKLAND-SAN FRANCISCO BAY BRIDGE system, the Yerba Buena Tunnel is the world's largest bore tunnel with a width of 76 feet (23 meters) and a height of 58 feet (18 meters). The tunnel is 1,700 feet (519 meters) long. Yerba Buena ("good herb") was the original name for the city of San Francisco; the island had previously been called "Goat Island."

Additional Information: http:// www.dot.ca.gov/dist4/calbridgs.htm#ab.

York Minster (York, England; thirteenth century)

York Minster, the Cathedral of Saint Peter in York in northeast England, occupies the site of a wooden church in which King Edwin of Northumbria was baptised into Christianity by the Roman missionary Paulinus on Easter Day in A.D. 627. The present cathedral was begun by the Anglo-Saxons and enlarged and renovated in

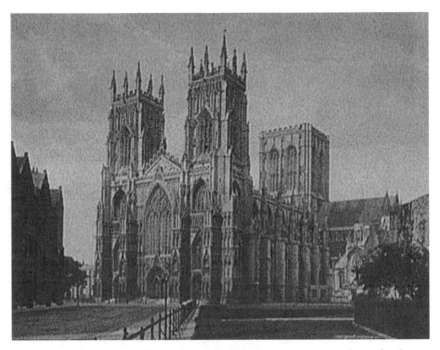

York Minster, the thirteenth-century Cathedral of Saint Peter in York, England. Photo courtesy of the Library of Congress, Detroit Publishing Company Photographic Collection.

Norman times. Completed in 1225, York Minster may be the largest cathedral in the world; it is certainly the largest Gothic structure north of the Alps. Its most incredible features are its magnificent stained glass windows depicting biblical scenes; the most magnificent is the Great East Window, which is 77 feet (24 meters) high and 32 feet (10 meters) wide.

Additional Information: Clucas, Philip. *England's Churches.* New York: Crescent Books, 1984.

Z

Zambezi Gorge Bridge (Zambia-Zimbabwe border; 1907)

Ralph Freeman designed this 500-foot (152.5-meter) steel arch across Victoria Falls on the Zambesi River, which forms the border between Zimbabwe and Zambia. In Freeman's time, the Zambesi flowed through the British colony of Rhodesia. The bridge was built out from the sides of the gorge. The sections were tied back by cables until the two halves met. The bridge was originally built to handle two railroad tracks, but one track was replaced with a roadway and footpath in 1932.

Additional Information: Brown, David J. *Bridges*. New York: Macmillan, 1993.

Zhaozhou Bridge. *See* An Ji Bridge.

Zola Dam (east of Aix-en-Provence, France; before 1850)

The Zola Dam on the Infernet River is the only arch dam built in France before 1850; it was the highest dam in the world until 1887. It was built by François Zola, the father of the famed French novelist, Emile Zola. The dam is 118 feet (36 meters) high, and the base extends another 21.5 feet (6.6 meters) down into its foundation trench. It is curved at a mean radius at the crest of 168 feet (51 meters). The water face is vertical except for three horizontal steps, each one foot wide. The air face is sloped in three different sections—at a slope of 1 in 5 for the upper 56 feet (17 meters), at a slope of 1 in 20 for the center 33 feet (10 meters), and at a slope of 1 in 4 for the bottom 30 feet (9 meters). The base is 43 feet (13 meters) thick. The Zola Dam is made of rubble masonry and is faced with cut stone blocks; it has stood up well over time. The dam has three outlets, two for drainage (one of which is now unused) and a 26-foot-wide (8-meter) overflow spillway cut onto one side of the dam. The reservoir capacity is 650,000 **acre-feet**. The dam has no **scouring** gallery.

Additional Information: Smith, Norman. *A History of Dams.* Secaucus, NJ: Citadel Press, 1972.

Zook's Mill Bridge (Lancaster County, Pennsylvania; 1849)

Built by Henry Zook, this Burr truss bridge crosses Cocalico Creek in southeastern Pennsylvania. This 89-foot-long (27-meter) covered bridge has stood up to the elements extremely well. It supposedly needed no major repairs from the time of its construction until it was flooded by Hurricane Agnes in 1972. Its long health may be partly due to its location at a broad, flat spot near the creek. In 1972, water covered the bridge to a height of 6.5 feet (2 meters); after the flood waters receded, some rotting timbers were replaced, and the horizontal clapboard siding was replaced with vertical board.

Additional Information: http://william-king.www.drexel.edu/top/bridge/CBZook.html.

Zuos Bridge (near Zuos, Switzerland; 1901)

When Robert Maillart designed this bridge across the River Inn in eastern Switzerland at the end of the nineteenth century, the temper of the times did not trust concrete bridges. Thus, when concrete bridges were built at all, they were camouflaged with a reassuring covering of stone or timber. Maillart, who was to become the acknowledged master of the concrete bridge, went against tradition and taste in allowing the bridge to be obviously made of concrete rather than clad or covered to look like another material. The 38-meter-long (125-foot) Zuos Bridge made use of a revolutionary, hollow-rectangular member made of **reinforced concrete**.

Additional Information: Brown, David J. *Bridges*. New York: Macmillan, 1993.

Zuos Bridge in Switzerland is one of Robert Maillart's early bridges. Photo courtesy of Erwin Sigrist.

Appendix A
Bridge and Truss Designs

Shown on the following pages are illustrations of some of the most important bridge truss designs developed in the United States in the nineteenth century. Descriptions of most of these various truss designs can be found in the "Glossary" section.

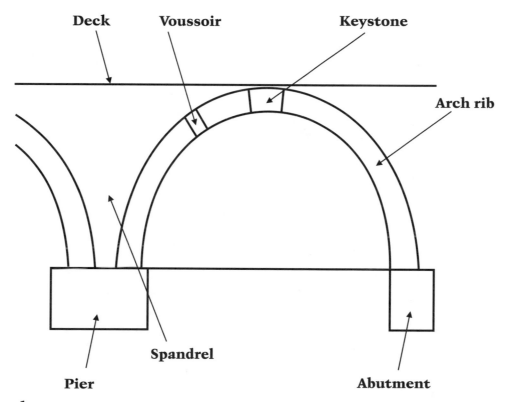

Parts of an arch.

Based on a drawing in Colin O'Connor, *Roman Bridges*, Cambridge: Cambridge University Press, 1993, p. 164.

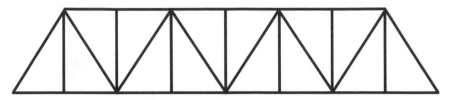

Warren Truss

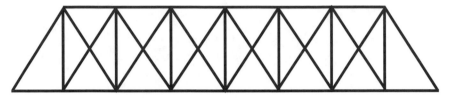

Quadrangular Warren Truss

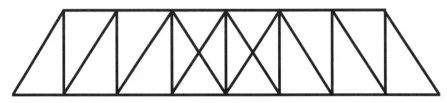

Howe Truss

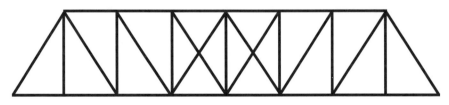

Pratt Truss

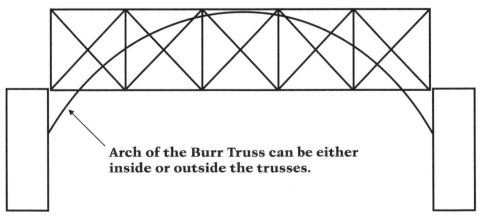

Arch of the Burr Truss can be either inside or outside the trusses.

Burr Truss

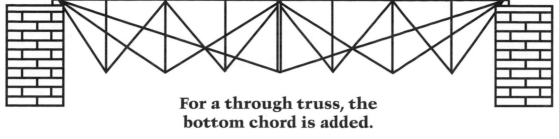

**For a through truss, the
bottom chord is added.**

Fink Deck Truss

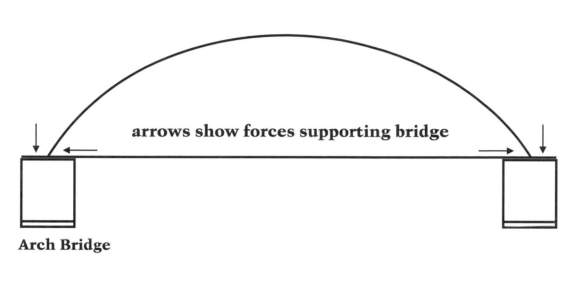

arrows show forces supporting bridge

Arch Bridge

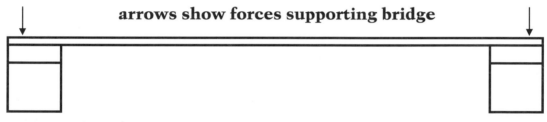

arrows show forces supporting bridge

Beam Bridge

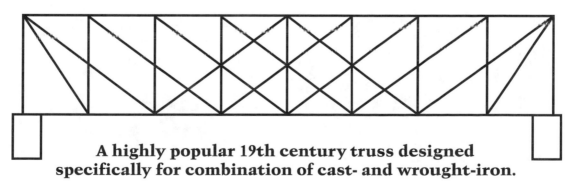

**A highly popular 19th century truss designed
specifically for combination of cast- and wrought-iron.**

Whipple Truss

APPENDIX B
A PORTFOLIO OF OHIO COVERED BRIDGES

Although covered bridges exist all over the world, the craft of making these wooden truss structures flourished in the nineteenth-century United States. Covered bridges, many still in working order or in the process of being refurbished, are part of the romance of the United States because they represent the ability and innovative talent of American craftspeople. These bridges are also an important aspect of civil engineering; covered bridges bear, on a smaller scale, the same sense of design elegance associated with the largest and most famous civil engineering landmarks of the world. Even a modern engineer, fresh from the construction of a miles-long structure, can admire the skill, knowledge, and thought that went into the building of nineteenth-century covered bridges.

Numerous covered bridge societies exist in the United States; these organizations are dedicated to understanding, preserving, and restoring these wonders of history. The Ohio covered bridges pictured here come from the files of Miriam F. Wood of the Ohio Covered Bridge Historical Association, and are reproduced with the permission of the association.

The Harshman Covered Bridge, Pueblo County, Ohio (1894), a view of the north portal (left) and of the interior (right). Built by Everett S. Sherman, this bridge is an example of a rare "Childs Truss," a wooden bridge that makes use of steel tension rods as counter-braces. Still able to carry traffic, the bridge has a respectable load capability of at least 4 tons.

A view of the portal (left) and interior (right) of the Roberts Covered Bridge, Eaton, Ohio (1829). Ohio's last surviving Burr Truss and last surviving two-lane covered bridge, the Roberts, built by Orlistus Roberts, was seriously damaged in a 1986 fire. In 1990, it was relocated to Eaton where it was rebuilt.

The Bergstresser Bridge, Franklin County, Ohio (1887). In 1991, the bridge's roof was removed and the bridge underwent complete renovation (left). This Partridge Truss was originally built by the Columbus Bridge Company of Columbus, Ohio. In the photo on the right, taken in 1992, the bridge was ready to go back into service.

Interior of Salt Creek Covered Bridge, Muskingum County, Ohio (1876). This fine example of a wooden Warren Truss was constructed by Thomas Fisher.

The Germantown Covered Bridge, Montgomery County, Ohio (1870),was built by D.H. Morrison. The Germantown is a unique covered suspension-truss bridge. As indicated by the barrier in front of the bridge, it has been removed from everyday use.

A 1992 interior view (left) and a 1985 exterior view (right) of the Vinton County, Ohio, Humpback Bridge, built in 1874 by Martin McGrath and Lyman Wells. Photos by Henry Davidson.

The Buckeye Furnace Bridge, Jackson County, Ohio (1871). This Robert Smith Truss bridge was constructed by Dancy, McCurdy & Co. The Robert Smith truss had a well-earned reputation for lightness and durability.

The east side of the north portal of the Hills Covered Bridge, Washington County, Ohio (1878). This excellent example of a Howe Truss was built by the Hocking Valley Bridge Works.

The Creek Road Bridge, Ashtabula County, Ohio, was renovated in 1994; this picture was taken before the sides of the bridge were replaced. Probably erected sometime in the 1870s, the name of the builder is unknown. The renovation was done under the supervision of John Smolen, Ashtabula County Engineer. Photo by Brian McKee.

The Harperfield Bridge (1868) is Ohio's longest covered bridge. Although its original builder is unknown, this Howe Truss was renovated under the direction of John Smolen, Ashtabula County Engineer.

An exterior (upper left) and interior (upper right) view of the Upper Darby Bridge, Union County, Ohio (1868). The bridge is a Partridge Truss constructed by Reuben Partridge. In some cases, as in the 1973 Axe Handle Road Bridge, Union County, Ohio (lower), the Partridge Truss has been augmented with a laminated arch.

APPENDIX C
THE TALLEST, LONGEST, AND HIGHEST FOR 1997

Trying to determine if a bridge, dam, or building is the largest, longest, tallest, biggest, or widest of its type is a difficult issue. New projects are announced almost weekly and, depending on the varying interest of the media, several years can pass before someone notes, almost in passing, that an important record has been achieved. The following lists reflect available information as of 1996-1997. Because civil engineering is moving ahead so rapidly, the lists given below will change quickly.

20 Tallest Buildings in the World in 1997

Rank	Building	Location	Date	Height (feet/meters)
1	Petronas Tower 1	Kuala Lumpur, Malaysia	1996	1476/450
2	Petronas Tower 2	Kuala Lumpur, Malaysia	1996	1476/450
3	Sears Tower	Chicago, Illinois	1974	1454/443
4	Jin Mao Building	Shanghai, China	est. 1998	1379/420
5	One World Trade Center	New York City	1972	1368/417
6	Two World Trade Center	New York City	1973	1362/415
7	Empire State Building	New York City	1931	1250/381
8	Central Plaza Building	Hong Kong, China	1992	1227/374
9	Bank of China Tower	Hong Kong, China	1989	1209/369
10	T&C Tower	Kaoshiung, China	est. 1997	1140/347
11	Amoco Building	Chicago, Illinois	1973	1136/346
12	John Hancock Center	Chicago, Illinois	1969	1127/344
13	Shun Hing Square	Shenzen, China	1996	1066/325
14	Sky Central Plaza	Guangzhou, China	1996	1056/322
15	Baiyoke Tower II	Bangkok, Thailand	1997	1050/320
16	Chrysler Building	New York City	1930	1046/319
17	Nationsbank Plaza	Atlanta, Georgia	1992	1023/312
18	First Interstate World Center	Los Angeles, California	1990	1018/310
19	Texas Commerce Tower	Houston, Texas	1982	1000/305
20	Ryugoyong Hotel	Pyongyang, North Korea	1995	984/300

Source: Dupre, Judith. *Skyscrapers.* Black Dog & Leventhal Publishers, Inc., abstracted by Filiz Bermek, January 1997, at "100 Tallest Buildings in the World," http://www.civeng.rutgers.edu/asce/buildings.html

20 Longest Suspension Bridges in the World in 1997

Rank	Bridge	Location	Year Open	Length (feet/meters)
1	Akashi-Kaikyo	Japan	est. 1998	6066/1990
2	Great Belt Link	Denmark	1996	5328/1624
3	Humber River	England	1981	4626/1410
4	Verrazano Narrows	New York City	1964	4260/1298
5	Golden Gate	San Francisco, California	1937	4200/1280
6	Mackinac Straits	Michigan	1957	3800/1158
7	Minami Bian-Seto	Japan	1988	3668/1118
8	Second Bosporus	Turkey	1992	3576/1090
9	First Bosporus	Turkey	1973	3524/1074
10	George Washington	New York City-New Jersey	1931	3500/1067
11	Tagus River	Portugal	1966	3323/1013
12	Forth Road	Scotland	1964	3300/1006
13	Kita Bisan-Seto	Japan	1988	3300/1006
14	Severn	England	1966	3240/988
15	Shimotsui Straits	Japan	1988	3136/956
16	Ohnaruto	Japan	1985	2874/876
17	Tacoma Narrows	Tacoma, Washington	1950	2800/853
18	Innoshima	Japan	1983	2526/770
19	Kanmon Straits	Japan	1973	2336/712
20	Angostura	Venezuela	1967	2336/712

Source: Brockenbough, Roger L. and Frederick S. Meritt. *Structural Steel Designer's Handbook*. McGraw Hill, Inc., abstracted by Filiz Bermek, January 1997, at "100 Longest Suspension Bridges in the World," http://www-civeng.rutgers.edu/asce/suspension.html.

10 Highest Dams in the World in 1997

Rank	Dam	River	Country	Type	Height (meters)	Year Completed
1	Rogun	Vakhah	Russia	Earth-Rockfill	335	Under construction
2	Nurek	Vakhah	Tajikistan	Earth	300	1980
3	Grand Dixence	Dixence	Switzerland	Gravity	285	1961
4	Inguri	Inguri	Georgia	Arch	272	1980
5	Chicoasen	Grijalva	Mexico	Rockfill	261	1980
6	Tehri	Bhagirathi	India	Earth-Rockfill	261	Under construction
7	Kishau	Toas	India	Earth-Rockfill	253	Under construction
8	Ertan	Yalong-Jiang	China	Arch	245	Under construction
9	Sayano-Shushensk	Yenisei	Russia	Arch	245	Under construction
10	Guavio	Guavio	Colombia	Rockfill	243	Under construction

Source: USCOLD Register of Dams, as published by the National Performance of Dams Program at http://blume.stanford.edu/~npdp/npdp/damhigh.html.

BIOGRAPHIES

Ammann, Othmar (1879-1965)

A Swiss-born engineer, Ammann had, early in his career, worked with the great Gustav Lindenthal. Ammann was well-known and highly respected for his work on a number of different projects, including heading up an investigation into the collapse of a large cantilever bridge in Quebec in 1907. In 1923, Ammann proposed constructing a bridge between New York and New Jersey; in 1925, the New York Port Authority appointed him as bridge engineer. Work on the George Washington Bridge began in 1927 and was completed in 1931. The bridge, which spans the Hudson to connect Manhattan with New Jersey, was the accomplishment of which Ammann was proudest, although some of his later works, such as the Verrazano Narrows Bridge, are considered by many to be even greater achievements.

Beach, Alfred Ely (1826-1876)

Editor and co-founder of *Scientific American* magazine, Beach began in 1869 to construct a tunnel for a pneumatic railroad under lower Manhattan. The tunnel was only about 1,000 feet (305 meters) long and featured a propulsion method that still makes technological sense today. Beach had devised and patented a shield that used hydraulic rams to push a railroad car through a tight space. The system worked well, and for a while he sold rides on this first New York subway. But the political machine of Boss Tweed prevented him from securing approval to build a city subway system. With a less corrupt city government, Beach might have been the progenitor of a successful underground system in New York, but his tunnel was eventually sealed and the railroad closed off. In the twentieth century, workers digging underground broke into Beach's nearly forgotten tunnel station.

Additional Information: Jacobs, David and Neville, Antony E. *Bridges, Canals and Tunnels*. New York: American Heritage Publishing Company, 1968.

Saint Bénézet (1165–1183)

The patron saint of bridge builders, and probably the only bridge builder in history to achieve canonization. Originally a shepherd, Bénézet supposedly constructed the Pont d'Avignon under divine instruction and assistance.

Additional Information: Lay, M.G. *Ways of the World*. New Brunswick, NJ: Rutgers University Press, 1992.

Bessemer, Sir Henry (1813-1898)

Sir Henry Bessemer was the inventor of the process for making modern steel; the "Bessemer furnace" blows air through molten pig iron, burning out the iron's impurities. The new steel, which was lighter and stronger than either cast-iron or wrought-iron, soon became the standard material for bridges and other architectural structures. Although the process has been greatly improved since the nineteenth century, Bessemer's furnace is still the basis of the modern steel industry.

Bollman, Wendel (1814-1884)

A self-taught engineer from Baltimore, Wendel Bollman patented his design for the Bollman truss in 1852 as the "suspension and trussed bridge." His truss arrangement used all iron instead of a combination of iron and timbers, and also provided a large element of what is now called "redundancy," i.e., the truss would still support a bridge despite the failure of any of its members. Because the weight of trains, and especially engines, has increased since the Bollman railroad bridges were originally built, many Bollman bridges were rebuilt or replaced, and few are left today. However, Bollman's bridges spearheaded the railroads' drive to replace timber bridges with iron ones, and his design helped make possible the growth of railroads in the United States in the mid-nineteenth century.

Additional Information: Schodek, Daniel L. *Landmarks in American Civil Engineering.* Cambridge, MA: MIT Press, 1987.

Brindley, James (1716-1772)

A famed early canal builder, Brindley was apprenticed at an early age to a millwright before becoming an engineer. His success at several projects, including the building of a water engine to drain a coal mine, led to employment with the duke of Bridgewater, for whom he built the Bridgewater Canal. He had many subsequent successes, among them the Trent and Mersey Canal. Illiterate throughout his life, he solved his problems without writing or drawing. He had a well-known and lifelong habit of thinking through a major problem while in bed, remaining there until he arrived at a solution.

Source: Magnusson, Magnus, ed. *Cambridge Biographical Dictionary.* Cambridge, England: Cambridge University Press, 1990.

Brunel, Isambard Kingdom (1806-1859)

Son of the famed tunnel engineer, Sir Marc Isambard Brunel, the younger Brunel shared much of the work on the arduous and important Thames Tunnel. He engineered a number of canals and bridges, including the Clifton Suspension Bridge, which was completed after his death and which made use of chains from one of his earlier projects. In 1833, he was appointed engineer to the Great Western Railroad, which linked London and Bristol. Brunel was responsible for approving all and designing many of Great Western's projects, an amazing number of tunnels, bridges, viaducts, and decks.

Source: Magnusson, Magnus, ed. *Cambridge Biographical Dictionary.* Cambridge, England: Cambridge University Press, 1990.

Brunel, Sir Marc Isambard (1769-1849)

Beginning in 1825, Brunel built a tunnel under the Thames River in England. Despite numerous construction difficulties, he persevered for 18 years, completing the project at a total cost of more that $3 million in today's money. His son, Isambard Kingdom Brunel, worked with him on the project. During construction, Brunel developed the Brunel Shield, a device that provided protection to diggers as they worked. The shield had 12 narrow frames, 22 feet (6.7 meters) high, which were placed side by side; each frame had three openings at different heights to allow workers to dig. The shield, which was jacked forward as work progressed, also protected the workers from flooding and tunnel collapse. As the shield was pushed forward, other workers built a brick support lining for the tunnel. The entire shield weighed 80 tons and was 38 feet (12 meters) wide. Prior to the shield's invention, tunnels were dug by hand in whatever opening was available, and supports were added to a tunnel as work opened it up, thus increasing the chances for collapse and loss of life. The concept of the Brunel Shield was a striking innovation at the time. The shield has been greatly developed and improved, and many modern boring machines are actually combinations of shield and borer. The most recent of these devices is the "Excalibore," built by the Lovet Tunnel Equipment Company of Etobicoke, Ontario, Canada; this device moves forward at the incredible rate of approximately 8 meters (26 feet) per day. Unlike the bricks that the Brunels used, Excalibore uses 40-ton precast concrete liners to support the portion of the tunnel already excavated.

Burr, Theodore (1771-1822)

Burr was a famous bridge builder who constructed large covered bridges, most of them in Pennsylvania in the early nineteenth century. He also invented the Burr arch truss, which he patented in 1817. The Burr truss superimposes an arch on a king post truss. After building a number of successively larger timber bridges, Burr attempted to span the Susquehanna River in Pennsylvania at McCall's Ferry. Although he successfully completed the 360-foot (110-meter) structure, it was swept away by ice in the winter of 1814-1815. Burr built many other bridges and builders continued to use his design through the end of the century.

Additional Information: Condit, Carl W. *American Building Art: The Nineteenth Century.* New York: Oxford University Press, 1960.

DeLesseps, Ferdinand Marie (1805-1894)

A French builder and diplomat, DeLesseps, assisted by his son Charles, built the Suez Canal at the invitation of Said Pasha, ruler of Egypt. Work began on the canal in 1859 and was completed in 1869. Although the canal is a superb engineering achievement, DeLesseps' eagerness to build it may have led to a willingness to tolerate the use of unwilling labor, if not outright slave labor. DeLesseps' ability and charisma enabled him to secure funding in Western Europe for the project in a country ruled by a repressive and undemocratic regime. Certainly a P.T. Barnum of engineers, DeLesseps was able to deliver the canal that he promised, affecting the tide of modern politics as well as the movement of modern civil engineering.

Additional Information: Sandstrom, Gosta E. *Man the Builder*. New York: McGraw-Hill, 1970.

Eads, James Buchanan (1820-1887)

One of the great American entrepreneurs, Eads was a true genius. Although he never finished high school, having left at age 13 for a job on a Mississippi steamboat, he was successful in a number of fields, including engineering. In 1842, he started his first business venture, salvaging boat wrecks in the Mississippi River at St. Louis. Due to Eads's engineering brilliance, the business was highly successful. He later expanded his operation to include the construction of steel-hulled boats, eventually manufacturing them for the government. After the Civil War, his knowledge of the Mississippi River, and of the steel business, won him the contract to bridge the Mississippi, which he did in 1874. This bridge, a magnificent three-arch structure, set a record for length. At the beginning of the project, Eads lost several workers to the "bends," a disease afflicting people coming up too quickly from highly pressurized work areas. Eads took action to control the affliction by instituting rules that required slow decompression following work in the caissons. He also provided a floating hospital and food for his workers. His efforts at control were almost totally successful. At the east pier, where work began first, 80 cases of the bends left 12 workers dead and 2 permanently crippled. For the later east abutment, after Eads took action to prevent the bends, 27 of 28 cases were successfully treated. (The one death was reportedly caused by the worker's refusal to abide by decompression rules.) This bridge, the only one Eads ever built, was the first steel bridge in the United States made of the new Bessemer steel. The Eads Bridge still stands proudly, a testament to the genius of Eads.

Additional Information: Hopkins, H.J. *A Span of Bridges: An Illustrated History*. New York: Praeger, 1970; Jacobs, David and Neville, Antony E. *Bridges, Canals and Tunnels*. New York: American Heritage Publishing Company, 1968.

Eiffel, Alexandre Gustave (1832-1893)

Popularly known as the creator of the Eiffel Tower in Paris, Eiffel was an engineering genius whose skills also extended to bridges, railroad viaducts, harbors, churches, steelworks and other structures. Some of his major achievements were the construction of an iron bridge over the Garonne River in France, the building of the railway bridge over the Douro River at Oporto in Portugal, the construction of the well-known Bon Marche store, and the design of lock gates for the Panama Canal. At Auteil in 1912, Eiffel founded the first laboratory for the study of aerodynamics. Although its main focus was the effects of air currents on airplanes, his work was applicable to the effects of wind on bridges. Eiffel's college studies were in chemistry. His interest in civil engineering stemmed from an early job at a railway equipment company.

Ellet, Charles (1810-1862)

Known as "the Brunel of America," Ellet was a flamboyant but highly competent engineer and bridge builder. His notable works include suspension bridges over the Schuylkill River at Fairmount, Pennsylvania (1842), and the Ohio River at Wheeling, West Virginia (1849). In 1841, Ellet was invited to build a bridge over the Niagara River, about 2 miles (3.2 kilometers) below the falls. In a brilliant strategic combination of public relations and practicality, Ellet offered a prize to the first child who flew a kite over the gorge. The prize was won by a young man named Homan Walsh. Ellet then followed the path of the kite string using ever-thicker wires until he had a cable strong enough to support his weight. Ellet used a bucket to haul himself across. Not long after, with the cable supporting a narrow service road, Ellet actually rode a horse across the span. The much desired bridge, with the attendant publicity, became a money-maker. Although it cost $30,000 to build, it brought in $5,000 in its first year of service. Ellet soon devised a plan to carry a four-passenger iron car across the cable, a scheme that paid for itself within a year. Ellet resigned as administrator in 1848, following a dispute on the distribution of the profits from the path, but his reputation had been made. In 1849, Ellet completed the bridge at Wheeling. The collapse of this bridge five years later ended Ellet's career as a bridge builder, but he continued to work in the engineering field. Ellet was the designer of the ram ships used by the Union during the Civil War; he died in 1862 after being wounded during a naval battle.

Additional Information: Jacobs, David and Neville, Antony E. *Bridges, Canals and Tunnels*. New York: American Heritage Publishing Company, 1968.

Fink, Albert (1827-1897)

A railway engineer and, later, an executive, Fink designed and built bridges following his emigration from Germany to the United States in 1849. He developed the Fink through-truss bridge, in which the roadway is suspended from the lower chords of the supporting trusses and the train passes between, or through, the parallel trusses. He first used this innovative technique in 1852 for a bridge over the Monongahela River at Fairmont, Virginia (now West Virginia), the longest iron railroad bridge in the United States at the time. He surpassed his own record in 1857, when he built a longer bridge over the Green River in Kentucky, and surpassed it yet again after the Civil War with a one-mile-long bridge over the Ohio River at Louisville, Kentucky.

Source: *Who Was Who in America, Historical Volume 1607-1896*. rev. ed. Chicago: Marquis Who's Who, 1967, p. 494.

Finley, James (1762–1828)

American designer of the first suspension bridge, Finley published a book describing how to build such a bridge, and even built one himself near Uniontown, Pennsylvania. His designs showed a sharp awareness of the characteristics of suspension bridges, including the cables and towers, and the method of hanging a deck from the cables with suspenders. It is uncertain whether Ellet, Roebling, and others knew about Finley's work, although it seems likely.

Freyssinet, Eugene (1879-1962)

From the beginning of his career, Freyssinet, a French engineer, promoted the use of reinforced concrete in bridges. His masterpiece, the Plougastel Bridge across the Elorn River at Brest in western France, has three 612-foot (187-meter) spans and was the largest reinforced-concrete bridge then constructed. In 1930, Freyssinet determined that a high-strength steel at a high level of stress could be used to best achieve a permanent pre-stress in concrete. In 1938, he developed a tool to apply that stress properly, and that enabled prestressed concrete to be used as a common construction material.

Source: *Encyclopaedia Britannica*. 15th ed. Chicago: Encyclopaedia Britannica, Inc., 1992.

Goethals, George Washington (1858-1928)

George Washington Goethals, the man who built the Panama Canal, was a uniquely skilled person for a unique job. Goethals entered the College of the City of New York at the age of 14, a rare achievement then as now. He left three years later after winning an appointment to the U.S. Military Academy at West Point, from which he graduated second in the class in 1880. His army career was marked by steady and well-deserved promotions. When he retired from the army in 1917 as a major general, he had gained canal lock and dam expertise, and experience in the construction of harbor facilities and fortifications. Goethals returned to active service that same year, with the outbreak of World War I, serving first as general manager of the Emergency Fleet Corporation and, the next year, as acting quartermaster general for all U.S. armed forces. He was awarded the Distinguished Service Medal in 1919. After his second retirement in 1919, he formed his own consulting firm and worked on several major projects in the New York-New Jersey area, including the Holland Tunnel, the George Washington Bridge, and the bridge between Elizabeth, New Jersey, and Staten Island, New York, that now bears his name. The original engineers for the Panama Canal soon found that there was more than an engineering challenge involved in constructing the massive waterway. The job called for someone who could deal with the intricacies of a bureaucracy as well as with the finer points of engineering. After President

Theodore Roosevelt put him in charge of the canal project in 1907, Goethals dealt successfully with a board of overseers meeting thousands of miles away from the construction site, excessive red tape, and massive logistical problems.

Grubenmann, Hans Ulrich (1709-1783) and Grubenmann, Johannes (1707-1771)

Swiss carpenters from the village of Teufen, the brothers, working sometimes together and sometimes separately, built three covered timber bridges, as well as area churches. The Limmat River Bridge, which they built together at Wettingen (1766), was the first timber bridge to incorporate a true arch in its design. Well ahead of their times, the Grubenmanns had a major influence on bridge design as their fame spread in Europe. Unfortunately, their three bridges were destroyed during the Napoleonic Wars.

Source: Steinman, David B. *Famous Bridges of the World*. London: Dover Publications, 1953, 1961. *Encyclopaedia Britannica*. 15th ed. Chicago: Encyclopaedia Britannica, Inc., 1992.

Haupt, Herman (1817-1905)

Upon his graduation from the United States Military Academy at West Point in 1835, Herman Haupt had a bright future in engineering. After serving in the army for three months, Haupt resigned his commission, went to work in the railroad industry, and moved ahead professionally at a respectable pace. Among his interests was the construction of railroad bridges, and Haupt published (anonymously) a pamphlet called *Hints on Bridge Construction*, which generated both interest and controversy. In 1845, he was appointed to the position of professor of mathematics at Pennsylvania College in Gettysburg, a post he held until 1847. While there, he wrote his *General Theory of Bridge Construction*, which is widely acknowledged to be an important and authoritative work in the field. By 1856, Haupt had risen through a series of important positions with the Pennsylvania Railroad. In 1855, he conducted what is now termed a "feasibility study" for a railroad tunnel through the Hoosac Mountains in Massachusetts, and concluded that the tunnel was indeed possible to build. Resigning his position with the Pennsylvania Railroad, Haupt took a financial interest in the tunnel and set about raising funds for its construction, as well as actually building the tunnel. Haupt was eventually squeezed out of the tunnel project, which was a political hot potato, and spent years in litigation before recovering even a portion of his investment. While constructing the tunnel, Haupt had introduced the use of his new pneumatic drill, a major advance in tunneling technology. Haupt left the unfinished tunnel project and served with great distinction in the Civil War until 1863, when he returned to private industry and continued his earlier success, serving in a number of engineering and executive positions. Haupt also wrote several other books, including *Miliary Bridges* (1864*), Tunneling by Machinery* (1876), and the autobiographical *Reminiscences of General Herman Haupt* (1901).

Source: *Dictionary of American Biography.* Vol. 4. Edited by Allen Johnson and Dumas Malone. New York: Charles Scribner's Sons.

Haussmann, Baron George-Eugene (1809-1891)

A French lawyer and administrator, Haussman administered Paris for Napoleon III. He was responsible for managing many aspects of the modern city of Paris, including the sewers, the aqueducts, and street planning. A prime supporter of the Paris Opera House, he made certain that the Opera House's young architect, Charles Garnier, was able to proceed easily and efficiently.

Source: White, Norvel. *The Architecture Book.* New York: Alfred A. Knopf, 1976.

Hennebique, François (1842-1921)

A French structural engineer, Hennebique was impressed with tanks and tubs made of concrete reinforced with wire mesh (devised by Joseph Monier). Determined to apply such a system to construction, Hennebique began producing reinforced floor slabs. In 1892, he patented an entire building system using reinforced structural beams, and his system was demonstrated in the construction of a Paris apartment building. He built several bridges, most notably the Pont de Chatellerault, over the River Vienne in France (1899), a bridge with an extremely low rise-to-span ratio and incredibly thin arches. His work made possible the work of Eugene Freyssinet and other builders in concrete, such as Robert Maillart.

Source: *Encyclopaedia Britannica.* 15th ed. Chicago: Encyclopaedia Britannica, Inc., 1992.

Holland, Clifford Milburn (1883-1924)

Holland, the engineer and builder of the Holland Tunnel in New York, worked his way through Harvard College by doing odd jobs such as teaching evening school, waiting on tables, and reading gas meters. Having passed the New York State civil service examination, he secured an engineering assignment with the Rapid Transit Commission upon graduation in 1906. One of his first assignments was to work with the division constructing the Battery Tunnel between Manhattan and Brooklyn. Holland spent two and a half years doing minor, but important work, such as verifying contract compliance, and in the process acquired a good knowledge of the details of tunnel construction. As Holland moved up in his career, he was charged with the construction of the subway tunnels under the East River in New York City. He was an acknowledged expert in un-

derwater tunneling by the time he left civil service in 1919 to accept the post of chief engineer for the New York State and New Jersey Interstate Bridge and Tunnel Commissions; his main challenge was to design and construct a vehicular tunnel under the Hudson River connecting the two states. There were many new engineering problems involved in designing the proposed tunnel, including how to ventilate the tunnel properly. Holland's eventual design called for pumping massive amounts of fresh air in and out of the tunnel at an extremely rapid rate. The system Holland proposed was so innovative that even some well-intentioned engineers opposed it. However, Holland's reputation and commitment to the system led to the adoption of the plans, with construction to be carried out under Holland's direction. Holland died only two days before the tunnel was "holed through" and it became apparent to all that the project was a success. On 12 November 1924, the tunnel was renamed the Holland Tunnel in honor of its designer and engineer. The success of the tunnel established that long underwater tunnels for vehicular traffic were both possible and practical. Holland probably died of overwork, although the exact cause of his death is uncertain. Before his death, he had gone to Battle Creek, Michigan, to "regain" his health. Despite the stresses of the construction project, Holland was active in professional associations, serving as a member of the board of directors of the American Society of Civil Engineers and as an officer of the Harvard Engineering Society.

Source: *Dictionary of American Biography.* Vol. 5, pt. 1. Edited by Dumas Malone. New York: Charles Scribner's Sons, 1933.

Latrobe, Benjamin Henry (1764-1820)

An architect and engineer, Latrobe designed and built the Philadelphia water supply system, the first in the United States (1799). The system involved a complicated scheme of aqueducts and reservoirs, and served the city well for many years until replaced by a larger system more suitable to Philadelphia's growing population. He also designed the south wing of the White House. Toward the end of his life, Latrobe lost his savings on a scheme to build steamboats for navigation of the Ohio River.

Lilienthal, David E. (1899-1981)

The first director of the Tennessee Valley Authority (TVA), Lilienthal's superb managment skills were exactly right for handling the enormity of the TVA. Lilienthal insisted that each of his managers, when making decisions, consider the project as a whole, rather than simply considering their particular areas of responsibility. Lilienthal wrote about his experiences and the importance of the TVA in his 1944 book *TVA: Democracy on the March.* Among other important projects, Lillienthal was appointed by President Harry S. Truman in 1946 as the first chairman of the Atomic Energy Commission, a position he held until 1950.

Additional Information: Neuse, Steven M. *David E. Lilienthal: The Journey of an American Liberal.* Knoxville: University of Tennessee Press, 1996.

Lin, Maya (1959-)

Creator of the Vietnam Veterans Memorial in Washington, DC, Maya Lin was a senior at Yale University, studying architecture and sculpture, when she designed the monument. She had taken a class in "funerary architecture" and was required, along with her classmates, to submit an entry to the competition. When the choice of her design was announced, several protests were raised, some of which deteriorated to racial attacks (Lin is of Chinese ancestry). H. Ross Perot, who had funded the competition, even flew protesting veterans to Washington to demonstrate against the monument. A group of art critics reviewed the entries and concluded that the original award had been correct. Since the monument was dedicated in 1986, it has come to be recognized as a major artistic achievement, and one that is impressively effective and moving.

Additional Information: *1993 Current Biography Yearbook.* New York: H.W. Wilson, 1993.

Lindenthal, Gustav (1850-1935)

An Austrian-born civil engineer, Lindenthal emigrated to the United States in 1871. Among his early jobs in the U.S., he served as consulting engineer in bridge and railway construction in Pittsburgh, Pennsylvania. In 1902, he was appointed commissioner of bridges in New York City, having already established an international reputation in the engineering community. His most famous work is the Hell Gate Bridge, a steel-arch railroad bridge over the East River in New York City (opened March 1917). He also designed the Queensboro Bridge, a cantilever structure over the East River.

MacAdam, John Loudon (1756-1836)

John MacAdam designed a less-expensive but more serviceable road pavement than that invented earlier by Thomas Telford; MacAdam's roads could carry twice the load that Telford's could. MacAdam was born in Ayr, Scotland, and emigrated at age 14 to New York, where he found financial success working with an uncle in a counting house. He returned to Scotland in 1783 and soon began experimenting with roadway materials. Although well-known and successful as a road engineer, he encountered financial difficulties and was granted £2000 by the British parliament in 1825. His financial difficulties were permanently relieved when he was appointed to the post of surveyor-general of metropolitan roads in 1827. MacAdam's roadways were designed on two principles: the underlying soil beneath a road supports the weight of traffic, and this soil, if kept dry, will carry weight without sinking. McAdam could thus build a road without a heavy foundation, a major improvement over Telford's design. He also concluded that he could cover his roads with rough stone that would be ground tightly by the steel-tired vehicles of his day, helping to make the road watertight. In the twentieth century, the introduction of rubber tires meant that the stone on top of a road would no longer be ground down by traffic, and engineers added a top course of tar to roads built in the MacAdam style, creating the modern roadway material known as tarmac.

Additional Information: Overman, Michael. *Roads, Bridges and Tunnels.* Garden City, NY: Doubleday & Company, Inc. 1968.

Maillart, Robert (1872-1940)

A Swiss engineer, Maillart's startling work in masonry-arch bridge building is significant both for its engineering quality and for its artistic merit. Maillart received a degree in structural engineering in 1894 from the Swiss Federal Institute of Technology. He took several jobs with private engineers, including the innovative Eugene Freyssinet, one of the developers and proponents of the use of reinforced concrete. Maillart's first bridge was the Zuos Bridge in Switzerland, a slender and flat arch that dazzled the engineering community and made it clear that Maillart was a master of his material. He built many other bridges in the Swiss Alps, including the Schwandbach Bridge, an amazingly curved bridge which has been described as ". . . a work of art in modern engineering." Maillart built other structures besides bridges, and temporarily left his native Switzerland to build factories and warehouses in Russia between 1912 and 1919. The Russian Revolution destroyed him financially, but his return to Switzerland revitalized both his career and his finances.

Source: *Encyclopaedia Britannica.* 15th ed. Chicago: Encyclopaedia Britannica, Inc., 1992.

Maybeck, Bernard Ralph (1862-1957)

Designer of the Palace of Fine Arts at the 1915 San Francisco Exposition, Maybeck was born in New York, studied in Paris, and eventually settled in California. Largely unknown to the general public, Maybeck and his work received little attention until, at the age of 89, he received the Gold Medal of the American Institute of Architects. His work was dramatic and matched the California world that he loved, especially its redwood trees. His homes were notable for exposed beams, unpainted finishes, and integration into the surrounding landscape.

Additional Information: http://www.exploratorium.edu/Palace_History/Palace_History.html#Maybeck.

McCullough, Conde Balcom (1887-1946)

The first chief engineer of the Port Authority of New York and New Jersey, McCullough was already well-known when he went to Oregon to work on the Oregon Coast Highway system. His talents as a bridge builder were prodigious and he built a large number of significant bridges, including the Rogue River Bridge, which incorporated the pre-stressing techniques developed by Eugene Freyssinet. The Coos Bay cantilever bridge, built after McCullough's death to complete one of the major links in the Oregon highway system, was renamed the McCullough Memorial Bridge, becoming one of the few engineering structures named in honor of its creator.

Source: Magnusson, Magnus, ed. *Cambridge Biographical Dictionary*. Cambridge, England: Cambridge University Press, 1990.

Moisseiff, Leon S. (1872-1943)

One of the most important engineers of the twentieth century, Moisseiff engineered, designed, or consulted on almost all the major projects of his time, including the Delaware River Bridge, the Golden Gate Bridge, the George Washington Bridge, the Tacoma Narrows Bridge, and the Manhattan Bridge. Through his work on the Manhattan Bridge, Moisseiff successfully established the veracity of the "deflection" theory, which meant that a lighter deck could be used in building bridges. Although other engineers who built failed bridges lost their reputations, the failure of the Tacoma Narrows Bridge did not significantly harm Moisseiff's standing in the profession. The error with the Tacoma Narrows design was almost universally acknowledged to be the result of a lack of general knowledge in the profession rather than a fault of Moisseiff's. Moisseiff died in 1943, three years after the Tacoma Narrows failure.

Additional Information: Petroski, Henry. *Engineers of Dreams*. New York: Alfred A. Knopf, 1995.

Mould, Jacob Wray (1825-1884)

An English-born architect, Mould studied under the renowned architect Owen Jones. He came to New York City in 1853 where he designed several churches and nearby country homes. In the 1860s and 1870s, he worked with Calvert Vaux on several New York City parks and was responsible, along with Vaux, for several well-known masonry-arch bridges, especially in Manhattan's Central Park and Brooklyn's Prospect Park.

Source: *Who Was Who in America, Historical Volume 1607–1896*. rev. ed. Chicago: Marquis Who's Who, 1967, p. 441.

Paine, Thomas (1737-1809)

Philosopher, well-known revolutionary, and author of "Common Sense" and "The Rights of Man," Paine was also a talented designer. He designed a bridge for a Paris competition in 1788; the bridge was based on cast-iron voussoirs that formed an arch bridge with a 20-to-1 span-to-rise ratio. Paine's bridge was redesigned and erected in Sunderland, England, in 1796. The Wear Bridge remained in service until 1858.

Additional Information: Lay, M.G. *Ways of the World*. New Brunswick, NJ: Rutgers University Press, 1992.

Palladio, Andrea (1508-1580)

An Italian renaissance architect, Palladio's influence in using classical forms extended throughout Europe. His "palladian arched truss," a low, nearly-straight curve, was used in early American bridge building. He is considered the founder of modern Italian architecture. His *Four Books on Architecture* (1570) was of significant importance in the two centuries following his death, and was translated into every major European language.

Source: Crystal, David, ed. *The Cambridge Biographical Encyclopedia*. Cambridge: Cambridge University Press, 1994; this encyclopedia is available online at http://www.biography.com.

Palmer, Timothy (1721-1851)

Palmer was the first American to use the arched truss to span long distances with his bridges. Palmer had probably never seen an arched truss, and worked only from drawings, making good use of the Palladian arched truss. Palmer's most important contribution to bridge building was the Permanent Bridge at Philadelphia (1806). The bridge was unusual for the times because it was covered. The bridge lasted far longer than Palmer had expected; before it was destroyed by fire in 1860, it showed no serious deterioration.

Additional Information: Condit, Carl W. *American Building Art: The Nineteenth Century*. New York: Oxford University Press, 1960.

Perronet, Jean Rodolphe (1708-1794)

Known as "The Father of Modern Engineering" and "The Master of the Stone Arch," Perronet was the first builder to realize that, in a multiple-arch bridge, only piers at the ends need to be large and heavy (i.e., horizontal thrusts cancel each other out except at the ends). Thus, Perronet's bridges could be lighter than earlier ones and could allow greater room for waterflow and boats underneath. He also designed a pump to force water out of cofferdams.

Additional Information: Lay, M.G. *Ways of the World*. New Brunswick, NJ: Rutgers University Press, 1992; Steinman, David B. *Famous Bridges of the World*. London: Dover Publications, 1953, 1961.

Pratt, Thomas Willis (1812-1875)

Pratt was an active inventor and civil engineer who invented the "Pratt truss," patented in 1844, and an improved type of steel and wood truss, patented in 1873. Since he was also interested in boats, he invented a new method of hull construction and an improved method of propulsion. Pratt was employed by several railroad lines in his career and built a number of important bridges, the largest over the Merrimac River at Newburyport, Massachusetts.

Source: *Who Was Who in America, Historical Volume 1607-1896.* rev. ed. Chicago: Marquis Who's Who, 1967, p. 494.

John Rennie (1761-1821)

An English engineer, Rennie was the first to consider the importance of protecting underwater pier foundations for bridges. He built bridges over the Thames River in London using a double cofferdam system, with one open box inside another. The space in between the cofferdams was filled with clay, the earth was dug out of the inside box, and pilings were driven in. Timber planking was then placed on top of the piles, and stones for a pier were laid on this platform. Rennie built his first bridge at the age of 24. He also designed many great eighteenth- and early nineteenth-century bridges, including the New London Bridge. He died before the New London Bridge could be constructed, and his work was carried on by his son, Sir John Rennie.

Source: Steinman, David B. *Famous Bridges of the World.* London: Dover Publications, 1953, 1961.

Riquet, Baron Pierre Paul (1604-1680)

An engineer and a prime mover behind the construction of the Canal du Midi in France, Riquet was convinced that a canal could be built that would link the Mediterranean with the Atlantic. After obtaining the support of King Louis XIV, Riquet commenced construction of the important first part of the canal from Toulouse to Sète. This section, which was 260 kilometers (161 miles) in length, required the construction of more than 100 locks, a short tunnel, and a number of aqueducts. Riquet died a few months short of this important segment's completion, but the Canal du Midi, his dream, eventually became a reality.

Source: Magnusson, Magnus, ed. *Cambridge Biographical Dictionary*, Cambridge, England: Cambridge University Press, 1990.

Roebling, John Augustus (1806-1869)
Roebling, Washington (1837 - 1926)
Roebling, Emily Warren (1843–1903)

John is considered the "Father of the Modern Suspension Bridge." He began his career by building aqueducts over rivers in northeastern Pennsylvania; between 1843 and 1850, he experimented with cable structures. His Delaware Aqueduct is the oldest existing cable structure in the United States. In 1851, he constructed the Niagara Bridge, the first suspension bridge in the U.S. designed for railroad traffic. The Brooklyn Bridge was John's greatest accomplishment, but he died two weeks after an onsite accident and before the bridge was completed. Roebling's son, Washington, and Washington's wife, Emily, had been assisting John Roebling, and the two continued the work after John's death. Washington developed a severe case of the bends and was unable to visit the site, but he supervised the project from his Brooklyn apartment, dispatching Emily with instructions several times a day. The acknowledged masterpiece of suspension bridges, the Brooklyn Bridge also became an immediate cultural icon, the object of poetry, song, painting, and photography.

Roberts, Sir Gilbert (1899–1978)

Roberts was the engineer most responsible for the innovations on the suspension bridge over the River Severn in western England. This bridge, which is 972 meters (3,188 feet) long, opened in September 1966. To guard against disaster caused by wind, Roberts used extensive models in a wind tunnel at Great Britain's National Physical Laboratory. He also set two entirely new directions for suspension bridge design in the towers and the road deck. The deck design is especially interesting. Older bridges use a conventional flat deck stiffened longitudinally and laterally by girders, usually trusses. The Severn Bridge is a "torsion box" that is more like a continuous aerofoil. Since hangers are fitted diagonally instead of vertically, the bridge is more stable than older bridges; and since the streamlined torsion box is also cheaper to build, the cost is considerably lower as well.

Source: Overman, Michael. *Roads, Bridges and Tunnels.* Garden City, NY: Doubleday & Company, 1968.

Smeaton, John (1724-1794)

An English engineer, Smeaton was involved in many of the major projects of his day, including bridges, dams, lighthouses, waterwheels, and windmills. He was the first to call himself a "civil" engineer, to distinguish his work from that of the military. His book, *Lives of the Engineers*, went far in establishing the importance of engineering in the consciousness of his society. Although the book has a distinct Victorian approach to biography, in which hard work equals success, he taught his country, and the world, the significance of the engineer or designer behind a project.

Source: Magnusson, Magnus, ed. *Cambridge Biographical Dictionary.* Cambridge, England: Cambridge University Press, 1990.

Steinman, David B. (1887-1960)

An eminent modern bridge builder, Steinman built many great structures, working with most of the other well-known bridge builders of his time, including Othmar Amman, Gustav Lindenthal, and his partner at the end of his career, Holton D. Robinson. At an early point in his career, while a professor of civil engineering in Idaho, Steinman supposedly wrote to Gustav Lindenthal, the acclaimed bridge building master of his period. Steinman, who was probably seeking a new position, told Lindenthal that he wanted to build long-span bridges. Lindenthal, according to the story. wrote back to say that since steel was too expensive, the age of big bridges had ended. Fortunately, Steinman was not discouraged. Among his many notable achievements are the Mackinac Bridge, built in 1957. Two years after writing his first letter, Lindenthal hired Steinman to assist in building two bridges, one of which was the Hell Gate Bridge in New York. In bringing Steinman to New York, Lindenthal put the young engineer in contact with another of his assistants, Othmar Ammann.

Source: Jacobs, David and Neville, Antony E. *Bridges, Canals and Tunnels*. New York: American Heritage Publishing Company, 1968.

Telford, Thomas (1757-1834)

Telford was the British builder of the Menai Straits Bridge in Wales. With a span of 174 meters (571 feet), this structure was the first suspension bridge to extend a longer distance than an arch bridge. The previous span recordholder was a 117-meter (384-foot) timber arch bridge at Wittingen, Germany. The Menai Straits Bridge demonstrated to engineers the great potential of suspension bridges. Telford was also the inventor of a type of road paving (later superseded by the pavement designed by John MacAdam) and the developer of engineering specifications for road grades. His plans for a road from Shrewsbury to Holyhead (opened in 1819) included horizontal and vertical transition curves and carefully thought-out alignments and gradients.

Todt, Fritz (1891-1942)

A German engineer, Todt rose to importance as an inspector of roads. He was the engineer most responsible for the "Reichsautobahnen," or the "Autobahn," as it is now called, the system of German highways that proved that a national system of roads could be successful. In 1937, he engineered the infamous "Siegfried Line." Todt held important positions in the Third Reich. He died in an airplane crash in 1942.

Source: Magnusson, Magnus, ed. *Cambridge Biographical Dictionary*. Cambridge, England: Cambridge University Press, 1990.

Vaux, Calvert (1824-1895)

A noted nineteenth-century landscape architect, Vaux worked with Jacob Wrey Mould in producing some excellent masonry-arch bridges in the parks of New York City, especially in Central Park (Manhattan) and Prospect Park (Brooklyn). He collaborated with the famous Frederick Law Olmsted in the design of those parks. In addition to his work with Olmsted, Vaux was the architect for the first buildings of the New York Museum of Art (now the Metropolitan Museum of Art) and the Museum of Natural History, also in New York. Born in England, Vaux came to the United States at an early age, and was employed as an architect by Jackson Downing of Newburgh, New York. He assisted Downing in several projects, including designing the grounds around the Capitol and the Smithsonian Institution in Washington, DC. In the years before his death, Vaux served as landscape architect for the New York City Department of Public Parks.

Source: *Who Was Who in America, Historical Volume 1607-1896* rev ed. Chicago: Marquis Who's Who, 1967, p. 622.

Weldon, Felix de (1918–)

A famed American sculptor, de Weldon is perhaps best known for his Iwo Jima memorial in Washington, DC. In 1996, de Weldon agreed to provide an AIDS memorial statue for Atlanta's John Howell Park.

Source: McGowan, Cate. "Howell Park to Display Famous Sculptor's 'Hope.'" *Atlanta Magazine* (December 1995), reprinted at http://www.nav.com/atlanta-30306/decfeat/f1.html.

Wernwag, Lewis (1769-1843)

Born in Germany, Wernwag came to the United States in 1786 and worked as an engineer. A major figure in bridge-building history, Wernwag's first bridge was across the Neshaminy Creek in 1810. However, Wernwag's most well-known achievement was the 1812 Colossus Bridge at Philadelphia. This covered bridge, at the time the longest span in the United States at 340 feet (104 meters), was a masterpiece of construction. By the time his career ended, Wernwag had built 29 wagon or railroad bridges in the eastern United States.

Source: *Who Was Who in America, Historical Volume 1607-1896* rev. ed. Chicago: Marquis Who's Who, 1967, p. 643.

Whipple, Squire (1804-1888)

Whipple is known as the "Father of Modern Bridge Building in New York," and a major figure in the history of American bridge construction. Both an engineer and an inventor, Whipple devised a modern truss (the "Whipple truss") that formed the basis of hundreds of bridges in New York and throughout the United States. He used this trapezoidal form of truss in the first long railroad span for the Rensselaer & Saratoga Railroad in 1853. He was responsible for many other bridges and drawbridges throughout New York State. Among his early accomplishments was the invention of a canal lock that could weigh boats as they passed through (1840). Whipple published *A Work on Bridge Building* in 1847, a comprehensive survey of bridge building in the United States, and perhaps an even more important contribution to engineering than the bridges he built. The book was the first to place the problems and solutions of bridge design on a formal, rigorous basis, setting the stage for the development of civil engineering as a professional practice, rather than an activity of craftsmen and artisans.

Source: *Who Was Who in America, Historical Volume 1607-1896* rev. ed. Chicago: Marquis Who's Who, 1967, p. 646.

Wright, Benjamin (1770-1842)

Known mostly as the chief engineer of the Erie Canal, Wright also designed and engineered several other canals, including the Delaware and Hudson Canal and the Chesapeake and Ohio Canal. He began his career as a land surveyor in Oneida and Oswego counties in New York. Long a proponent of building and using canals to transport farm produce, Wright was elected to the New York State Legislature in the late 1790s. When he left the legislature in 1813, he was appointed a county judge in upstate New York. He returned to active engineering in 1816 when he began his work on the Erie Canal.

Source: *Who Was Who in America, Historical Volume 1607-1896* rev. ed. Chicago: Marquis Who's Who, 1967, p. 670.

CHRONOLOGY

c. Twenty-eighth century B.C. Construction of the Sadd el-Kafara (Dam of the Pagans) near what is today Cairo, Egypt. Discovered in 1885, the Sadd el-Kafara is likely the oldest surviving dam in the world.

Twenty-seventh century B.C. The 204-foot-high (62-meter) Step Pyramid of Djoser (Zoser) and its surrounding complex of mortuary buildings were the first major buildings in the world to be constructed entirely of stone.

Twenty-sixth century B.C. Construction of the Great Pyramid of Giza (also known as the Great Pyramid of Khufu or the Great Pyramid of Cheops), which was, until the nineteenth century, the tallest structure in the world, and probably the most massive. Built by the Pharaoh Khufu (or Cheops) as a burial site, the monument took over 20 years to complete and involved the exacting cutting, movement, and placement of more than 2 million blocks of stone.

c. 750 B.C. Marib Dam in Saba (now Yemen) is completed. An irrigation dam, it was built of boulders, without any mortar or other adhesive materials.

c. 600 B.C. The Egyptian pharaoh Necho (d. 593 B.C.) begins a canal between the Nile River and the Red Sea. An early forerunner of the Suez Canal, Necho's waterway was completed by King Darius of Persia sometime between 549 and 486 B.C.

c. 530 B.C. The Samos Water Tunnel on the Greek island of Samos was the first tunnel to bring water into a walled city.

518 B.C. Construction of qanaats begins in Egypt. Qanaats are gently sloping tunnels designed to transport water; more than nine miles of tunnels are completed in the first year of construction. Rediscovered in 1900, the qanaats still carry water.

fifth century B.C. Construction begins on the Grand Canal in China. Almost completed when Marco Polo saw it around 1290, the canal—now 860 miles (1,384 kilometers) long—was and still is used for transportation and shipping.

447-442 B.C. Construction of the Parthenon by the Greek city-state of Athens. An acknowledged masterpiece of Greek architecture, this temple to the goddess Athena was home to a 524-foot-long (160-meter) frieze depicting the Pananthenaic procession held every four years in honor of the goddess. Although parts of the frieze are still in place, large portions were removed to the British Museum in London in 1801 by Lord Elgin.

c. 312 B.C. Construction begins on the Roman Aqua Appia, which eventually will stretch more than 10 miles (16 kilometers). The construction spurs development of support arches for overground aqueducts and other structures.

294-282 B.C. The Colossus of Rhodes, perhaps the best-known statue of ancient times, is constructed at the harbor entrance of the Greek island of Rhodes in the eastern Mediterranean. Constructed to commemorate the resolution of a war, the cost of the 102-foot-high (31-meter) statue was absorbed by selling surplus military equipment left by invaders after peace was made.

c. 246-209 B.C. Emperor Shih Hwang-ti begins erection of the Great Wall of China to protect China from northern invaders. The Wall had various additions made to it throughout the centuries, especially during the period of the

Ming Dynasty (1368-1644). Running for 1,500 miles (2,414 kilometers), the Wall still retains its guard towers and watch stations in some sections. In some areas, the Wall is no more than a dirt mound, but in others is a more serious fortification. In general, the structure was not an effective block to outside invasion.

62 B.C. Lucius Fabricus builds the Pons Fabricus across the Tiber River in Rome; still in service, the Pons Fabricus is the oldest bridge in Rome.

115 A.D. Roman emperor Hadrian begins building a 15-mile (24-kilometer) aqueduct in Athens to bring water to the city. Part of the water supply of modern Athens is still served by the aqueduct.

315 A.D. Emperor Constantine builds the Arch of Constantine in Rome; the structure is now the largest and best-preserved Roman triumphal arch in the world.

c. 1178 Construction of the Pont d'Avignon in France, the longest bridge in medieval Europe. The bridge was supposedly inspired by a vision and a miracle of the shepherd now known as Saint Benezet.

1189 Construction of the Arlesford Dam, England's first recorded dam. The dam and its reservoir still exist, but are no longer in use.

1194-1230 Construction of Chartres Cathedral in France. One of the world's greatest masterpieces of Gothic architecture, the cathedral has a rare artistic unity because of the rapidity with which it was built.

1265-1309 Construction of the Saint Esprit Bridge over the Rhone in southeastern France. For many years, this 2,700-foot-long (823.5-meter) bridge with 26 arches was the longest bridge in the world.

1284 Caernavon Castle in Wales is begun by Master James of St. George. A major achievement in architecture, the castle, actually a complex of buildings, is modeled after the fifth-century walls of the city of Constantinople in an effort to echo Constantinople's connection with the history of Christianity.

c. 1300 Irrigation canals are constructed in the city of Chan Chan, Peru. The canals link the city with water from mountains as far as 40 miles (64 kilometers) away.

1333 First construction of the 657-foot-long (200-meter) Chapel Bridge (Kapellbrucke) across the Reuss River near Lucerne, Switzerland. The Kapellbrucke is the oldest and longest covered bridge in Europe, although it has been rebuilt or refurbished many times.

Fifteenth century The Royal Road, the major road of the Inca Empire, is extended from Chile to Ecuador. Probably the world's longest operating road, the royal road ran from what is now central Chile (Santiago) to what is now northern Ecuador (Quito).

c. 1450 Aqueducts supported by masonry causeways supply water to the Aztec capital of Tenochtitlan, which is built on an island in a lake. Flood protection is supplied by a 9-mile-long (14.5-kilometer) stone-slab dike. All is obliterated by the conquering Spanish in 1521.

1570 Andrea Palladio (see Biographies), an Italian architect, suggests visualizing a rigid, self-supporting system of triangles as a method of bridge construction. The method suggested is what is known today as the truss system for bridge construction. Palladio is credited with first applying the basic concept of transferring both live and dead loads to the abutments or piers via a bridge truss. He built several truss bridges shorter than 100 feet. Early timber truss bridges, based upon the ideas of Palladio and others, were further developed in the eighteenth century by the Grubenmann brothers (see Biographies) of Switzerland.

1611-1613 Construction of the Ponte Alto dam in Italy, the first arch dam in Europc.

1648 The Taj Mahal, one of the most beautiful buildings in the world, is complctcd aftcr 18 years of work. Located outside Agra, India, the Taj Mahal is a monument to the love between Mumtaz Mahal, who died in childbirth in 1629, and Shah Jahan, the Mogul emperor. The building is the finest example of Indian Muslim architecture.

1681 Construction of the Malpas Tunnel in France, the first-ever canal tunnel and the first tunnel excavated using gunpowder.

1692 Completion of the Languedoc Canal in France connects the Bay of Biscay with the Mediterranean, one of the major engineering accomplishments of the period.

1697 Construction of the Frankford Avenue Bridge, the oldest bridge in the United States, and the first stone arch; it still carries U.S. Route 13 over the Pennypack Creek in Philadelphia.

1716 Henri Gautier publishes *Traites des Ponts*, the first treatise devoted entirely to bridge building. A standard reference throughout the eighteenth century, the treatise covered both timber and masonry bridges, as well as foundations, piers, and centering.

1716 France establishes the first national department of transportation, the Corps des Ingenieiurs des Ponts et Chaussees.

1747 Jean-Rodolphe Perronet (see Biographies) is appointed director of the Ecole des Ponts et Chaussees, the oldest academic institution in the world for engineering education.

1758 Opening of the Schaffhausen Bridge in Switzerland. Constructed by Hans Ulrich Grubermann over the Rhine, the bridge had two extremely long trusses for its time. The French army burned the bridge in 1799.

1760 Stephen Riou publishes *Short Principles on the Architecture of Stone Bridges*, the first book on bridge construction written in English.

1761 The first section of James Brindley's (see Biographies) Bridgewater Canal was completed between Worsley and Manchester, England. A tremendous success in transporting coal cheaply, the waterway was the first great accomplishment of England's "canal age"; by 1830, England had 3,639 miles (5,855 kilometers) of canals.

1764-1766 The Grubenmann brothers construct the Limmat River Bridge at Wettingen in Switzerland; the bridge is the first timber span to use a true arch shape.

1779 Opening of Coalbrookdale Bridge across the River Severn in western England; the bridge is the first successful cast-iron arch bridge in the world. Designed by Thomas Farnolls Pritchard and built by Abraham Darby and John Wilkinson to demonstrate the versatility of cast-iron, the success of the 100-foot (30.5-meter) span spurred the use of iron for bridges.

1788 The political philosopher Thomas Paine (author of "Common Sense") (see Biographies) patents an iron bridge design. Paine invented and modeled the first American design of a cast-iron arch bridge with a greater span than ever previously proposed. After being patented in England, the design was reviewed favorably by the French Academy of Sciences, and Paine built and displayed a prototype on Paddington Green in London.

1794 Gaspard Monge establishes the Ecole Polytechnique during the French Revolution. The school set the model for engineering schools and also for the establishment of the United States Military Academy at West Point.

1796 Completion of Sunderland Bridge, the next major bridge after Coalbrookdale. Built by Thomas Wilson across the River Wear at Sunderland, the bridge had an unprecedented arch span of 236 feet (72 meters). For years, it was thought to contain some parts adapted from the bridge Thomas Paine (see Biographies) exhibited in London in 1788.

1800 The 22-mile-long (35-kilometer) Santee Canal, the first canal in the United States, is built in South Carolina and remains in use until 1850.

1801 James Finley (see Biographies) builds the first chain link suspension bridge in the United States over Jacobs Creek near Uniontown, Pennsylvania.

1803 Completion of the 27-mile-long (43-kilometer) Middlesex Canal in Massachusetts establishes canal

transportation as an effective and efficient method for moving goods.

1804 Theodore Burr (see Biographies) completes the Union (Waterford) Bridge across the Hudson; the bridge is the first to employ the Burr truss system (patented 1806), the first bridging system to be widely used in the United States.

1806 The 1,300-foot-long (396.5-meter) Permanent Bridge in Philadelphia may be the first covered bridge in the United States.

1808 Albert Gallatin, secretary of the Treasury, issues his famous "Report on Roads and Canals," setting forth a scheme for developing commercial transportation, especially through the building of canals. The report reinforced the bridge-building trend then sweeping the country.

1811 Thomas Pope publishes his *Treatise on Bridge Architecture*, which is a "comprehensive" review of the world's bridges. He also promoted his "flying pendant lever bridge" with which he proposed to span the Hudson River near New York City. Although his bridge was never built, Pope's suggestion spurred other plans for Hudson bridges.

1812 Lewis Wernwag completes the Colossus Bridge across the Schuylkill at Philadelphia; the 340-foot (104-meter) single-span truss-stiffened arch is the largest timber span in the United States until it burns in 1838.

1816 Wire manufacturers Josiah White and Erskine Hazard complete the world's first wire suspension bridge, a 408-foot (124-meter) temporary footbridge over the Schuylkill in Philadelphia.

1820 Ithiel Town patents the lattice truss (soon known as the Town truss), a bridging system used extensively and successfully in the nineteenth century for aqueducts and highway and railroad structures.

1821 The 408-foot (124-meter) canal tunnel built on the Schuylkill Navigation Canal near Auburn, Pennsylvania is the first tunnel in the United States.

1823 Claude Navier publishes *Rapport et Memoire Sur Les Ponts Suspendus* after two visits to England to study suspension bridges; Navier's work is one of the most influential bridge books of the nineteenth century.

1823 Guilaume Henri Dufour's Saint Antoine Bridge, the world's first permanent suspension bridge, opens to the public in Geneva.

1823 Construction of the Gaunless Viaduct in England, the first iron railroad bridge in the world.

1825 Marc Sequin's double-span Tain-Tournon Bridge, the first bridge ever to carry both vehicles and pedestrian traffic, opens across the Rhone in southeastern France. The bridge's innovative design incorporates several new engineering ideas, including "festoons," multiple cables to carry the deck.

1825 Benjamin Wright (see Biographies) supervises completion of the Erie Canal, running 363 miles (584 kilometers) across New York from Albany to Buffalo. The canal is the longest man-made waterway of its time.

1826 Thomas Telford (see Biographies) completes the 580-foot (177-meter) Menai suspension bridge in Wales, the world's first large-scale suspension bridge. The bridge used flat ends at the eyebar links, with the ends punched with holes to receive iron pins for connections, a design that became standard in England for many years.

1828 Construction of the Danube Canal Bridge in Vienna, Austria; it was the earliest known bridge to use steel instead of iron in the eyebar chains of its suspension system.

1828 Benjamin Wright (see Biographies) connects the Delaware and Hudson Rivers by building the Delaware and Hudson Canal, which runs 106 miles (171 kilometers) from Honesdale, Pennsylvania, to Kingston, New York.

1829 Completion of the Carrollton Viaduct in Maryland, the first railroad bridge in the world, and the first major engineered structure on an American rail line.

1830 Lewis Wernwag (see Biographies) completes the Monocacy River Bridge across the Monocacy River in Maryland; it is the first timber railroad bridge in the United States.

1832 Completion of the Jones Falls Dam, part of the Rideau Canal system and the first arched dam in North America. At the time, it was the highest dam ever built.

1833 Completion of the 900 foot (275-meter) Allegheny Portage Tunnel through a mountain near Johnstown, Pennsylvania; it is the first railroad tunnel in the United States.

1834 The Pennsylvania Canal, a combination of rail lines and canals, runs 400 miles (644 kilometers) to link Philadelphia with Pittsburgh.

1837 While lecturing at West Point, Dennis Hart Mahan writes *Elementary Course of Civil Engineering*, the first modern engineering text.

1839 Opening of Captain Richard Delafield's Dunlap's Creek Bridge, the oldest existing all-metal bridge in the United States, in Brownsville, Pennsylvania.

1840 William Howe patents the Howe Truss design for bridges; the design was used in many railroad bridges because it had simpler framing connections, could be built faster, and allowed for easier field adjustments than other truss configurations.

1840 Earl Trumbull builds the first all-iron truss constructed in the United States to carry a road 70 feet (21 meters) over the Erie Canal at Frankford, New York.

1840 Building of the Cumberland Road, the first major road in the United States constructed with federal funding.

1841 Squire Whipple, (see Biographies) patents the iron bowstring truss. He built a bowstring truss of cast- and wrought-iron over the Erie Canal in Utica, New York. The bridge was the second all-metal truss in the United States, and marked the transition between wooden and iron bridges and the beginning of iron bridge manufacturing.

1842 Charles Ellet, Jr., (see Biographies) completes the Fairmount Suspension Bridge over the Schuylkill in Philadelphia to replace Lewis Wernwag's (see Biographies) Colossus Bridge. The 358-foot (109-meter) Fairmount is the first permanent wire-cable suspension bridge in the United States.

1842 After 18 years of work, the Thames Tunnel in London is completed by using the new and highly significant Brunel Shield invented by Isambard Kingdom Brunel (see Biographies).

1844 Thomas Pratt (see Biographies), an engineer, and his father Caleb Pratt, an architect, patent the Pratt truss, which reverses the position of truss members found in the Howe truss. Used in an all-iron version, the Pratt truss becomes the standard American truss bridge of moderate span for well into the twentieth century.

1845 Opening of Richard Osborne's Manayunk Bridge, the first American all-metal railroad bridge. This Howe truss was built for the Philadelphia and Reading Railroad, and was in service until 1901. One surviving truss of the bridge is exhibited at the Smithsonian Museum of American History.

1845 John Roebling (see Biographies), the future builder of the Brooklyn Bridge, completes his first suspension structure, an aqueduct over the Allegheny River at Pittsburgh to carry both water and the boats of the Pennsylvania Canal.

1846 British iron shipbuilders develop a mechanical riveting process that allows a 12-fold increase in the number of rivets that can be placed per day. The new method is first used extensively on the Britannia Bridge in Wales.

1847 Squire Whipple (see Biographies) publishes *A Work on Bridge Building*, which deals with truss analysis; the book uses parallelograms of forces and force polygons to analyze the sizes of members, and shows methods for determining the strength of trusses needed for both dead loads and moving live loads. Although ignored for many years, the book helped change bridge design from a craft to a profession.

1847   James Milholland builds the first plate-girder bridge, a design that introduces the girder as a practical bridge type for short spans, for the Baltimore & Susquehanna Railroad at Bolton Station, Maryland. The bridge, a 54-foot (16-meter) single-track deck girder, was prefabricated in a shop and then hauled to the site and set in place.

1848   James P. Kirkwood and Julius Adams build the Starrucca Viaduct in Pennsylvania, the first American engineering project to use concrete in an important structural capacity in its foundation.

1848   Development of the Warren Truss, a multiple-system triangular truss with no vertical members in which the diagonals absorb both tensile and compressive stresses.

1848   Charles Ellet (see Biographies) crosses the Niagara Gorge with a 770-foot-long (235-meter) light suspension footbridge.

1848   Construction of the Illinois and Michigan Canal connects La Salle, Illinois, with Chicago, and completes an all-water route from the Gulf of Mexico to Lake Michigan.

1849   John Roebling (see Biographies) completes the Delaware Aqueduct, the largest of four suspension aqueducts built for the Delaware & Hudson Canal at Lackawaxen, Pennsylvania. Now the oldest surviving wire cable suspension structure in the United States.

1849   Charles Ellet (see Biographies) completes the Wheeling Suspension Bridge over the Ohio River at Wheeling, Virginia [now West Virginia]. The bridge becomes the longest span in the world at 1,010 feet (308 meters).

1850   France's Basse-Chaine Bridge collapses when a corroded anchor tears loose during a storm and throws 478 soldiers crossing the bridge into the water, where 226 die. The terrible tragedy prevents the construction of any more suspension bridges in France for 20 years.

1850   Robert Stephenson completes the Britannia Railroad Bridge across the Menai Straits in Wales; the structure becomes the world's longest railroad bridge.

1850-51   Wendell Bollman repairs Lewis Wernwag's (see Biographies for both) bridge at Harpers Ferry, Virginia. The structure was the first all-metal, cast- and wrought-iron Bollman Truss that was capable of supporting more than one ton per linear foot. The strength of the bridge revolutionized railroad bridge construction.

1851   Herman Haupt (see Biographies) publishes *General Theory of Bridge Construction*, which, along with Squire Whipple's earlier *A Work on Bridge Building*, were the two important early books on bridge design.

1852   Wendel Bollman (see Biographies) patents the Bollman Truss. The system uses a composite cast- and wrought-iron Pratt truss, with radiating wrought-iron eyebars that run from the ends of the top chord to each point on the panels from which the deck beams are suspended.

1852   An all-iron railroad bridge crossing the Monongahela River at Fairmont, Virginia is the first major use of Albert Fink's (see Biographies) Fink truss.

1854   Albert Fink (see Biographies) patents the Fink truss, a composite cast- and wrought-iron system distinguished by the absence of a bottom chord. The tension is carried by diagonal braces. It improves upon Wendel Bollman's (see Biographies) truss system, and makes possible spans as long as 300 feet (91.5 meters).

1855   John Roebling's (see Biographies) Niagara Suspension Bridge opens across the Niagara Gorge. The bridge is the first suspension bridge used for railroads; it vindicates Roebling's contention that a railroad could be carried on a suspension bridge.

1855   Opening of the Wabash and Erie Canal, the longest canal in the United States. The 452-mile (727-kilometer) waterway links Toledo, Ohio, and Evansville, Indiana.

1856   Peter Cooper's Trenton Iron Works in Trenton, New Jersey, produces the first I-Beams rolled in the United States. The I-beam revitalized building and railroad construction by making rolled sections stronger, simpler, and cheaper to build.

1859   Albert Fink (see Biographies) builds the longest iron bridge in the United States, a 1,000-foot (305-meter) iron truss bridge across the Green River in Kentucky.

1859   Robert Stephenson's Victoria Bridge over the St. Lawrence becomes, at 5,004 feet (1,526 meters), the longest railroad bridge in the world.

1865   The era of concrete bridge building begins with François Colgnet's construction of a concrete multiple-arch aqueduct to move water from the River Vanne to Paris on a series of reinforced concrete arches.

1866   Opening of Furens Dam in France, the first modern dam in which the engineering took into account the importance of internal stresses.

1867   John and Washington Roebling (see Biographies for both) complete the Cincinnati Suspension Bridge, which, at 1,057 feet (322 meters), becomes the longest bridge in the world. The bridge was a prototype of the Brooklyn Bridge, a later Roebling project.

1867 Opening of Heinrich Gerber's Hassfurt Bridge in Germany, the first modern cantilever bridge.

1869 Completion of the Suez Canal, which was begun by the French engineer Ferdinand de Lesseps (see Biographies). Desired in the area for centuries, the canal crosses northeastern Egypt to link the Mediterranean and Red seas.

1872 Calvert Vaux (see Biographies) completes Cleft Ridge Span, the first concrete arch bridge in the U.S., for Prospect Park in Brooklyn, New York.

1874 Completion of the Eads Bridge across the Mississippi at St. Louis. The 1500-foot (458-meter), triple-arch bridge is the first major use of the cantilever method of construction, the first use of pneumatic caissons in the United States, and the first major use of steel in American bridge construction.

1876 With the loss of 76 lives, the collapse of the Ashtabula Bridge is the worst American bridge disaster of the nineteenth century. An investigation by the American Society of Civil Engineers blamed the use of composite cast- and wrought-iron members, and led to the use of all wrought-iron trusses with riveted connections.

1876 After more than 20 years of construction, the Hoosac Tunnel is finally completed in Massachusetts; the longest American railroad tunnel of its time, the Hoosac provided invaluable lessons in tunnel engineering to later builders.

1879 The success of the Eads Bridge at St. Louis led William Sooy Smith to built the first all-steel truss in the U.S. for the Chicago & Alton Railroad across the Missouri River at Glasgow, Missouri. The five river-spans of the Glasgow Bridge were each 300-foot-long (91.5 meter), pre-connected Whipple trusses.

1879 In December, Scotland's Tay Bridge collapses carrying 75 people in a mail train to their deaths. When the bridge opened in 1877, it was considered one of the seven wonders of the modern world. The board of inquiry speculated that the failure was the result of ignorance of metallurgy, uneven manufacturing standards resulting in defective castings, and lack of aerostatic stability. The board concluded that cast-iron bridges should not be built in the United Kingdom.

1883 Washington Roebling completes the project begun by his father John (see Biographies for both), the Brooklyn Bridge, a 1,505-foot (459-meter) span over the East River connecting Manhattan with Brooklyn. The bridge becomes on completion the longest suspension bridge in the world and is considered by many to be the preeminent engineering accomplishment of the nineteenth century.

1883 Completion of the Niagara Cantilever Bridge, the first important American cantilever; designed to carry heavy locomotives, the 495-foot (151-meter) span set the precedent for later long-span cantilevers over the Firth of Forth (Firth of Forth Bridge), the St. Lawrence River (Quebec Bridge), and the Hudson River (Poughkeepsie Railroad Bridge).

1884 Consulting engineer J.A.L. Waddell publishes *The Designing of Ordinary Highway Bridges*; Waddell advocates an unusual idea for the time—engineers, rather than bridge fabricating companies, should design bridges.

1884 California's Bear Valley Dam becomes the first arch dam in the United States.

1887-1889 Construction of the Eiffel Tower to mark the opening of the 1889 International Exhibition of Paris. The tallest building in the world until 1930, the 300-meter (984-foot) tower has become one of the world's most recognized symbols of France. Nearly torn down as outdated in 1909, the tower's destruction would now be unthinkable.

1889 Completion of the San Mateo Dam in California, the first dam to be built entirely of concrete.

1890 Scotland's Firth of Forth Bridge opens. Designed by John Fowler and Benjamin Baker, the span of 1,710 feet (522 meters) was a world record for 27 years. The world's first all-steel long-span bridge, the Firth of Forth ushered in the age of great cantilevers and accelerated the demise of wrought-iron bridges.

1892 Construction of the Vyrnwy Dam in Wales, the first large masonry dam in Great Britain, and the first use of a dam in Wales to provide water to a city in England (Liverpool).

1893 Completion of Austin Dam in Texas, the first dam specifically built to produce hydroelectric power.

1894 The world's first vertical-lift bridge, the South Halsted Street Bridge, is constructed by J.A.L. Waddell in Chicago, Illinois.

1897 The Topeka Bridge, a 125-foot (381-meter) span over the Kansas River, is the first major concrete bridge in the United States; it was built by Edwin Thatcher, one of the first engineers to develop completely scientific design theories for reinforced concrete.

1902 The 6,400-foot-long (1,952-meter) Aswan Dam across the Nile River in Egypt becomes the longest dam in the world.

1902 Congress passes the Reclamation Act to encourage and promote the irrigation of the arid and semi-arid lands of the American West. The U.S. Bureau of Reclamation, which is established under the act, becomes the controlling authority for hydroelectric projects that produce electricity to be sold to cities. One of the Bureau's first projects, Theodore Roosevelt Dam on the Salt River east of Phoenix, Arizona is the largest masonry dam to date. Roosevelt Dam began selling electricity to Phoenix in 1911.

1903 Completion of New York City's Williamsburg Bridge, the first large suspension bridge to be built with steel towers, rather than towers made of iron, concrete, or stone.

1905 Completion of the 295-foot (90-meter) stone arch bridge at Plauen, Germany, the largest bridge ever built in stone masonry.

1906 After 33 years of delays and engineering difficulties, the Hudson Tubes connecting New York and New Jersey finally open; the tunnels are still in use as the PATH (Port Authority of New York and New Jersey) system linking the two states.

1907 The collapse of the uncompleted Quebec Bridge, which costs the lives of 80 workers, spurs investigation of engineering techniques and a re-evaluation of the requirements of the bridge-building process.

1907 Completion of New York's Bronx River Parkway, the first modern parkway in the United States.

1909 Gustav Lindenthal's (see Biographies) 1,182-foot-long ((361-meter) Queensboro Bridge, which crosses the East River to connect Manhattan with Queens, becomes the longest cantilever in the United States.

1910 Completion of the Buffalo Bill Dam (formerly Shoshone Dam) in Wyoming, the first concrete dam to be built higher than it is wide.

1912 Eugene Freyssinet (see Biographies) perfects a prestressing method for concrete.

1914 After 33 years of sporadic construction, the Panama Canal, engineered by George Washington Goethals (see Biographies), opens, connecting the Atlantic and Pacific Oceans across the Isthmus of Panama; the largest engineering project completed by any government to date, the canal cost the U.S. over $336 million.

1915 The Delaware, Lackawanna, & Western Railroad completes A. Burton Cohen's Tunkhannock Viaduct, a structure made of permanent, low-maintenance concrete near Nicholson in northeastern Pennsylvania.

1915 Gustav Lindenthal (see Biographies) completes the Hell Gate Bridge, a magnificent steel arch bridge across the East River between Manhattan and Long Island.

1917 Gustav Lindenthal, David B. Steinman, and Othmar H. Ammann (see Biographies for all three) complete the longest continuous truss in the world, the 1,550-foot (473-meter) Sciotoville Bridge over the Ohio River in southern Ohio.

1917 The 1,800-foot (549-meter) (second) Quebec Bridge across the St. Lawrence opens to become the longest, single nonsuspension span in the world.

1923 A Pan American Highway System to link all the capitals of all the American nations is proposed. Today the system is almost complete.

1927 The world's largest canal tunnel, the 4.5-mile-long (7.2-kilometer) Rove Tunnel opens in southern France.

1927 The first major automobile tunnel in the world, John Holland's Holland Tunnel, runs under the Hudson River to connect New York and New Jersey.

1927 The Moffat Tunnel near Denver, Colorado, becomes the first tunnel to cross under the Continental Divide.

1929 The 8-mile-long (13-kilometer) New Cascade Tunnel through the Cascade Mountains in Washington becomes the longest tunnel in the United States.

1929 Opening of David B. Steinman's (see Biographies) Grand-Mère Bridge in Quebec; the Grand-Mère is the first bridge to use twisted-strand cables rather than parallel wire cables.

1930-1931 Construction and opening of the Empire State Building in New York City. Representing the essence of the skyscraper, this building was intentionally planned in the 1920s to be the tallest building in the world. By 1931, the exuberance of the previous decade had diminished greatly as the Great Depression spread throughout the United States and the world. The Empire State Building has since become one of the most illustrious addresses in the world, and one of the country's most desirable locations.

1931 Othmar H. Ammann's (see Biographies) 3,500-foot (1,067.5-meter) George Washington Bridge across the Hudson between New York City and New Jersey becomes the longest suspension bridge in the world by almost doubling the span length of the previous recordholder.

1931 The 1,652-foot (504-meter) Bayonne Bridge across the Kill van Kull between Staten Island, New York, and New Jersey becomes the longest single-span steel-arch in the world.

1932 Opening of Virginia's Mount Vernon Memorial Highway, the first parkway to be built and maintained by the United States government.

1933 Congress creates the Tennessee Valley Authority (TVA) to stimulate and integrate development in the Tennessee River Valley; the TVA will construct many dams and hydroelectric power plants throughout the southeastern United States, bringing jobs, power, and industry to the region.

1933   Construction begins on the Autobahns, Germany's high-speed road system. Carefully constructed and engineered to the most modern specifications, the Autobahns set the standard for modern road construction. Following World War II, other roads were modeled on the principles of the Autobahn, including the United States Interstate Highway System.

1934   Opening of one of the largest underwater road tunnels in the world, the 2-mile-long (3.2-kilometer) Mersey Tunnel at Liverpool, England.

1936   Opening of the Hoover Dam, a 1,244-foot-long (379-meter) dam on the Colorado River between Nevada and Arizona.

1936   Opening of the 8-mile (13-kilometer) San Francisco-Oakland bridge-tunnel complex.

1937   Opening of the Golden Gate Bridge in San Francisco, California. Engineered by Joseph B. Strauss, with assistance from Charles Ellis, Clifford Paine, and Leon Moisseiff (see Biographies for all three), the 4,200-foot (1,281-meter) Golden Gate was, at completion, the longest suspension span in the world.

1937   The Lincoln Tunnel opens, connecting New York and New Jersey. The later addition of two tubes made the Lincoln the only three-tube vehicular tunnel in the world.

1940   The Tacoma Narrows Bridge opens across Puget Sound in Washington in July. Known almost immediately as Galloping Gertie because of its undulations in wind, the 2,800-foot (854-meter) suspension bridge collapses four months after its opening. Its replacement opens in 1950. The collapse spurred research into the effect of aerodynamic forces on suspended decks.

1940   The Pennsylvania Turnpike, running between Harrisburg and Pittsburgh, becomes the first superhighway built in the United States.

1942   The Grand Coulee Dam, a massive hydroelectric project near Spokane, Washington, becomes the world's largest concrete structure.

1942   Sando Bridge over the Angerman River in Sweden becomes the world's longest concrete arch at 866 feet (264 meters).

1944   Opening of the 105-mile (169-kilometer) Delaware Aqueduct, the world's longest tunnel, which delivers water to New York City.

1949   Opening of the Walnut Lane Bridge in Philadelphia, the first major prestressed and precast concrete girder bridge.

1951   Completion of the 3.5-mile (5.6-kilometer) Delaware Memorial Bridge connecting New Jersey and Delaware; a twin to this bridge is opened alongside it in 1968.

1957   David B. Steinman (see Biographies) completes the Mackinac Bridge, connecting the two peninsulas of Michigan; the bridge, one of the longest clear-span suspension bridges in the world, is, at 8,614 feet (2,627 meters) the longest span from anchorage to anchorage.

1964   Completion of the 23-mile (37-kilometer) Chesapeake Bay Bridge-Tunnel, which passes both above and below the bay to connect Norfolk, Virginia, with Virginia's Eastern Shore.

1964   Opening of the Verrazano Narrows Bridge connecting Brooklyn and Staten Island; at the time, it had the longest clear span of any type of bridge at 4,260 feet (130 meters).

1970   Completion of the Aswan High Dam on the Nile in Egypt; although politically and socially controversial, the project is a major engineering feat of the twentieth century.

1984   Dedication of the Vietnam Veterans Memorial Wall, which was designed by Maya Ying Lin, then a student at Yale University. An incredibly moving and stark memorial to the war dead of Vietnam, the Wall is one of the most visited and most beloved landmarks in Washington D.C., and in the nation.

1985   The 33.5-mile-long (54-kilometer) Honshu-Hokkaido Tunnel, which links two of Japan's main islands, becomes the world's longest railroad tunnel.

1988   At nearly 8 miles (13 kilometers) long, Great Seto Bridge, connecting the Japanese islands of Honshu and Shikoku, becomes the world's longest double-decker bridge; it is capable of carrying both railroads and automobiles.

1989-1995   Construction and opening of the Denver International Airport in Colorado. The Denver Airport was the largest public works project in the world after the Channel Tunnel between Great Britain and France.

1990s   Construction begins on the Boston Central Artery/Tunnel, also known as "The Big Dig." Half tunnel and half roadway, this structure is the largest highway project ever undertaken in a major United States city.

1991   Itaipú Dam in Brazil is completed. It is the largest dam in South America and one of the largest hydroelectric complexes in the world, generating up to 12,600 megawatts of power and supplying about one-fourth of Brazil's power requirements, and almost three-quarters of Paraguay's.

1995   Completion of the Channel Tunnel or "Chunnel" between Great Britain and France. Proposed many times since the beginning of the nineteenth century, the tunnel was the largest public works project in the world. Two high-speed rail lines make the 31-mile (50 kilometer) trip several times each day, making interaction between Great Britain and the rest of Europe as easy as a daily commute to work.

1996 The Petronas Towers, the tallest building in the world, opens in Kuala Lumpur, Malaysia. A last-minute decision to add 20 feet (6.1 meters) of spires to the towers made the building a record breaker.

1999 Expected completion date of the Chongqing Tower in Chongqing, China. This tower will be the tallest building in the world at a height of 460 meters (1,509 feet), but will be equaled by the Millennium Tower in Tokyo, Japan, which is due to open in the first decade of the twenty-first century.

2001 Expected completion date of the Shanghai World Financial Center in Shanghai, China. Planned to be the tallest building in the world, the Center is intended to demonstrate Shanghai's prosperity and to symbolize the "hopes and lofty ideals" of the Chinese peoples. The building will equal the height of the Chongqing Tower to be completed in Chongqing, China, in 1999, but will be surpassed by the Millennium Tower scheduled to open in Tokyo in the first decade of the twenty-first century.

Before 2010 The Millennium Tower in Tokyo, Japan, will be completed before the end of the first decade of the twenty-first century. The building will be an amazing 800 meters (2,624 feet) high, nearly twice as high as any other building ever constructed.

GLOSSARY

Abutments The supports for a bridge at either end.

Acre-foot A measurement of volume, usually used in discussing the capacity of dams and reservoirs. One acre-foot is the volume that can be held by a theoretical container one foot deep by one acre wide.

Aqueduct A bridge designed to carry water between two points. Some aqueducts might also have a pedestrian roadway in them. *See also* Bridge.

Arch A typically curved structural member spanning an opening and serving as a support for a wall or other weight above the opening. Arch shapes are used most magnificently in bridges, aqueducts, and dams. The force of the arch is toward its ends. When concrete arches are constructed, the arch is built in two sections, and the center or "keystone" piece of material is inserted between the two sections, forcing the arch to push toward its ends. For an excellent brief summary of the history of arch theory, *See* Lay, M.G. *Ways of the World*. New Brunswick, NJ: Rutgers University Press, 1992, pp. 274-275. *See also* Spandrel; Voussoir.

Arch Bridge A bridge built in the shape of an arch, usually no longer than about 1,700 feet (518 meters). The roadway of an arch bridge may lie on top of the arch, or underneath it. When the roadway is on top, the weight of the roadway is carried by vertical supports underneath called spandrel columns. If the roadway is beneath the arch, it is hung from girders or beams that are also connected to the abutments of the arch, preventing the arch from separating. *See also* Bridge.

Art Deco A design style introduced in the 1920s and popular again in the 1960s. Art Deco is characterized by geometric patterns, curved forms, and strong outlines. The name is shortened from the style's formal presentation at the 1925 Paris Exposition Internationale des Arts Decoratif et Industrials Modernes.

Baroque A style in both art and architecture that is characterized by excessive use of ornamentation. Complex and fascinating at its best, the baroque style can, at its worst, descend to the grotesque.

Bascule Bridge A "lift bridge" in which a weight at one end of a span or section can be lowered to raise the other end of the section. *See also* Bridge; Drawbridge.

An example of a single bascule bridge.

Beam Bridge Also known as a "girder bridge." The simplest type of bridge, akin to an arrangement where a beam or length of material is supported at the ends, as in a log or piece of timber laid across a stream or on top of supporting blocks. Beam bridges can extend as far as 1,000 feet (300 meters). *See also* Bridge; Truss Bridge.

Beaux Arts   An architectural style associated originally with the Ecole des Beaux Arts in Paris, France. Popular in France at the end of the nineteenth century, and in the United States and elsewhere at the beginning of the twentieth century, the style makes broad use of earlier French architectural styles, resulting in complicated effects that are frequently imposing and, sometimes, cumbersome. In a Beaux Arts style, symmetrical designs and patterns are common.

Bends   A condition which, until properly understood at the beginning of the twentieth century, affected tunnelers and bridge workers excavating foundations. A too-rapid ascent from a deep location under pressure causes nitrogen bubbles to form in the blood; the resulting condition is painful and can cause lasting and permanent effects or even death. Workers coming to the surface are now required to spend time in a "decompression chamber" where pressure is adjusted slowly and the nitrogen gases have an opportunity to dissipate harmlessly. Limitations are also placed on the amount of time a worker may spend under pressure. The "bends" is also a concern for scuba divers, underwater salvagers, and others working in highly pressurized environments.

Bollman Truss   Developed by Wendell Bollman for the Baltimore and Ohio Railroad, this truss uses cast-iron compression members and wrought-iron tension members. Each panel point is supported by two diagonal tension members connected to the top of both end posts so that each floor beam of the truss is suspended from the cast-iron end posts at each abutment. In such a system, the failure of a diagonal tension member will cause the collapse of only a single floor beam and not bring down the entire structure. For additional information, *See* Jackson, Donald C. *Great American Bridges and Dams*. Washington, DC Preservation Press, 1988, pp. 121-22. *See also* Howe Truss; Lenticular Truss; Pony Truss; Pratt Truss; Through-Truss; Town Truss; Truss; Truss Bridge.

Box Girder   A type of beam bridge in which the girders are fashioned into a rectangular, box shape. *See also* Beam Bridge.

Bridge   A structure carrying a roadway or pathway over a depression or obstacle. *See also* the various types of bridges, e.g., arch, beam, cable-stayed, cantilever, suspension.

Cable-Stayed Bridge   This newest type of large bridge has cables connected directly to the deck of the bridge from the bridge's towers. The stresses on the cables are greater than the stresses on the cables of suspension bridges, and only recently has technology developed the appropriate materials for such cables. *See also* Bridge; various types of bridges.

Caisson/Pneumatic Caisson   An underwater structure designed to allow workers and materials to remain dry while construction is proceeding. Depending on the type of work being done (e.g., underwater excavation, concrete casting), caissons can be large and technically sophisticated. A pneumatic caisson is kept dry inside by using a pneumatic pump to remove water.

Caisson Disease   *See* Bends.

Cantilever Bridge   This type of bridge has a support arrangement in which a horizontal member is supported at the outside end by an anchor arm, and is then free toward the center end. In cantilever bridges, the roadway is supported at either end, and the central section is connected but not directly supported. Most cantilever bridges will have two anchor arms, one on each side, and a center span. As of 1997, the 1800-foot (549-meter) Quebec Bridge is the longest cantilever bridge in the world. *See also* Bridge; various types of bridges.

Cast Iron   An iron formed by casting, a process in which the iron is melted and poured into a mold while still in a liquid state. Cast iron is strong, but also brittle and easily fractured. *See also* Wrought Iron.

Chicago-type Bridge   A style of bridge used throughout the United States, and used almost exclusively in Chicago since the construction of the Cortland Street Bridge in 1902. This type of bridge rotates around its axle (trunnion) while the arms of the bridge lift upward about 76 degrees beyond horizontal. The lifting mechanism involves the use of both powerful motors and massive counterweights. For an extensive description of the history of the Chicago-type bridge, *See* Schodek, Daniel L. *Landmarks in American Civil Engineering*. Cambridge, MA MIT Press, 1987. *See also* Bridge; various types of bridges.

Cofferdam   A cofferdam is a box-shaped base for a bridge abutment or foundation. A cofferdam, sometimes with a cutting edge on the bottom, is forced downward until it rests on a solid footing, and the foundations of the structure above are built upon it. *See also* Dam; various types of dams.

Compression   The pushing force on a support or member. Forces in bridges and other structures are analyzed in terms of compressive forces and tension. *See also* Tension.

Concrete Arch Dam   Concrete arch dams are frequently built in steep canyons; the walls of the canyon take up some of the pressure of the dammed-up water. Because of the strength of the arch, the dam wall is usually relatively thin compared to other constructions. *See also* Arch; Dam; various types of dams.

Concrete Buttress Dam   Usually low, concrete buttress dams are supported on the air face side by a series of buttresses (or supports). *See also* Dam; various types of dams.

Concrete Gravity Dam   In a gravity dam, water is held back by the dam wall itself. The weight of the dam, held down by gravity, is sufficient to withstand the pressure of water. Unlike rockfill and earthfill dams, a gravity dam's water-facing wall is perpendicular to the water (straight up and down). Gravity dams require a solid base of rock. Older gravity dams, some built as late as the turn of the last century, were made of rock instead of concrete. *See also* Dam; various types of dams.

Continuous Bridge   Besides the supports at its ends, a continuous bridge is supported at one or more additional points. A continuous structure can be longer than a structure supported only at its two end points. A continuous bridge does have some disadvantages, including the problem of keeping all supports at the appropriate level. *See also* Bridge; various types of bridges.

Corduroy   A type of road construction in which closely placed timbers are laid across a cleared area or path. In an "enhanced corduroy" road, beams are laid lengthwise on top of the timbers on each side of the road. For more information, *See* Lay, M.G. *Ways of the World*. New Brunswick, NJ Rutgers University Press, 1992.

Course (of masonry)   A line of bricks or other masonry within a structure. In a brick wall, for example, a single horizontal line of bricks is a "course."

Cutwater   A device at the base of a bridge that diverts water away from the bridge support, or minimizes the flow of water against the support.

Dam   A structure designed to hold back or control moving water. Types of dams are determined by the function they serve, or by their basic material and structure. Functional descriptions include storage dam, diversion dam, and detention dam. Descriptions based on material include earthfill dams, concrete gravity dams, concrete arch dams, and concrete buttress dams. *See also* various types of dams.

Damper   A safety mechanism on a bridge to absorb some of the energy of vibration and sway.

Dead Load   The constant weight that the supports of a structure can bear. In a bridge, for example, the weight of the deck, towers, cables (and sometimes tollbooths) are dead loads. *See also* Live Load.

Dead Weight   The weight of a structure itself, rather than the weight of what the structure carries, such as vehicular traffic or a railroad. *See also* Live Weight.

Detention Dam   A dam built to hold water back from its natural course in special circumstances. A detention dam may be built in a storm area, for example, to lessen the effects of flooding from rain. Less permanent detention dams might be built to allow bridge supports to be installed in a location usually underwater. *See also* Dam; various types of dams.

Diversion Dam   A dam built to move water out of its usual path, such as into an irrigation canal. *See also* Dam; various types of dams.

Drawbridge   A bridge in which all or part of the span can be lifted to allow traffic to pass below. Bascules are considered drawbridges, but drawbridges are of other types as well, including the vertical lift bridge, in which the movable span is lifted upwards from both ends, keeping the span in a horizontal plane. *See also* Bascule Bridge; Bridge; various types of bridges.

Earthfill Dam   Modern earthfill dams tend to be massive; the special technology involved in building these dams centers around the ability to excavate, move, and rearrange large amounts of rock and earth. An earthfill dam usually has a clay core and an impervious membrane installed on the water face, as well as a water cut-off wall extending down from the base of the dam to the underlying bedrock. To prevent against erosion from water flowing over its top, an earthfill dam is usually connected to a spillway, which provides a relief system. *See also* Dam; Rockfill Dam; various types of dams.

Eyebar   A structural member of a truss in the form of a metal bar with an enlargement at either end to create an "eye."

Falsework   Support material, usually scaffolding, used in the construction of a bridge or a building. In some cases, such as bridges over deep waters, ways have to be found to construct the bridge without falsework.

Finial   An ornamental feature at the top of a structure or portion of a structure. The round ball sometimes found on top of a flagpole is a finial.

Fresnel Lens   Invented by French physicist Augustin Fresnel in 1822, the Fresnel Lens is used in lighthouses to amplify the light that is beamed out. Although a normal light from an open flame loses 97 percent of its light, the Fresnel lens is so efficient that up to 83 percent of the light produced by a lamp is radiated outward. The lenses were manufactured in gradients relating to their efficiency. The strongest light came from a First Order Fresnel lens. For additional information, *See* http://zuma.lib.utk.edu/lights/fresnel.html.

Girder   A horizontal support beam in a structure, such as a bridge or building.

Girder Bridge   *See* Beam Bridge.

Gothic Architecture   A style of construction that originated in twelfth-century France, exemplified by such cathedrals as Notre Dame de Paris and Chartres. The name was originally a scornful one, first applied during the Renaissance and meant to call up the image of the barbarian

Notre Dame Cathedral in Paris, a famous example of Gothic architecture. Photo courtesy of C. and B. Antonelli.

Goths who had overrun Rome in the fifth century. The Gothic style reflects a command of engineering concepts, especially of pointed supporting arches. Gothic constructions are taller than had been possible before and have fewer interior supporting walls and columns. Consequently, interior space is greater and better lighted because large outer walls contain impressive expanses of stained glass windows. In addition to the stained glass windows, Gothic exteriors are massive presences and are highly ornamented with statues, gargoyles, and other decorations.

Gravity Dam   *See* Concrete Gravity Dam.

Heading   A working entrance to a tunnel under construction. Frequently, tunnels have at least two headings, one at each end. Additional headings are sometimes created for a tunnel by digging a vertical shaft down to tunnel depth, and then digging horizontally toward the headings at the end.

Hectare   A metric measurement of area, approximately equal to 2.47 acres.

Helical   A spiral- or helix-shaped pattern.

Howe Truss   A well-known method of supporting a bridge, named after William Howe who patented the design in 1840. The Howe truss was attractive to railroad bridge builders because it could be constructed easily and could be adjusted easily to the terrain requirements of the project. The original design used wooden diagonal compression members and threaded vertical wrought-iron rods to deal with compression. *See also* Bollman Truss; Lenticular Truss; Pony Truss; Pratt Truss; Through-Truss; Town Truss; Truss; Truss Bridge.

Keystone   *See* Arch.

Lenticular Truss   An iron truss distinguished by polygonal top and bottom trusses, based on patents secured in 1878 and 1885 by William O. Douglas. The lenticular truss used less iron than its popular counterpart, the Pratt truss, and was the primary bridge type used by the highly successful Berlin Iron Bridge Company. The lenticular truss is also called the Pauli truss. *See also* Bollman Truss; Howe Truss; Pony Truss; Pratt Truss; Through-Truss; Town Truss; Truss; Truss Bridge.

Lift Bridge   *See* Bascule Bridge.

Live Load   The variable weight that a bridge or other structure can bear, most commonly the weight of the traffic (trains, cars, people) moving across the bridge. *See also* Dead Load.

Live Weight   The variable weight that a bridge or other structure must carry, e.g., vehicular or railroad traffic. *See also* Dead Weight.

Meter   A unit of length. One meter is approximately equal to 3.3 feet.

Mortising   A woodworking method for making "joints" or connections in which the pieces of wood or timber to be connected are cut so as to make them meet in a way that provides more strength than simply butting them together. Mortise joints come in several varieties. One of the impressive aspects of Lewis Wernwag's Colossus Bridge was that the connections between timbers were not mortised.

Muleback Bridge   A type of arch bridge in which the roadway follows the curve of the upper portion of the arch. Examples include the bridge that once stood in Mostar, Bosnia, and the Kintai Bridge in Japan. *See also* Bridge; various types of bridges.

Neoclassical   A style of architecture popular in the late eighteenth and early nineteenth centuries. The style made use of Greek architectural models to create typically geometric structures that appeared simple and presented an appearance of strength. Low reliefs in facades and other decorations were common.

Nonoverflow Dam   Designed to hold a certain amount of water, a nonoverflow dam would be endangered by water flowing over its top. Earthfill dams usually cannot tolerate long periods of water flowing over their tops. *See also* Dam; various types of dams.

Oakum   Loose fiber used for caulking seams. The fiber is usually treated with some material to make it water resistant. Originally, oakum was made by picking apart old weathered ropes.

Open Crib Dam   A dam arrangement intended less for storing water than for diverting water to a channel. The water level is raised because the water is forced to flow within a limited area. *See also* Dam; various types of dams.

Overflow Dam   A dam that is specifically designed to allow water to flow over its top; because of the likelihood of erosion, these dams usually are made of manufactured materials. *See also* Dam; various types of dams.

Penstock   An enclosed area for holding water before it flows into a hydroelectric facility. Using a penstock allows the operators of a facility to control water flow through the generators, rather than having the water flow depend solely on what flows over or around a dam.

Pontoon   A floating structure, usually used as a temporary support for a bridge or part of a bridge's construction, although some bridges have been built by making use of permanent pontoons.

Pony Truss   A truss bridge in which the deck is on top of the trusswork, as opposed to a "through-truss" in which the deck goes through the trusswork. *See also* Bollman Truss; Howe Truss; Lenticular Truss; Pratt Truss; Through-Truss; Town Truss; Truss; Truss Bridge.

Pound Locks   Canal locks with gates on either side. With both gates closed, a vessel can be raised or lowered by adding water to or removing water from the closed lock.

Pozzolino   A type of cement used by Roman builders, pozzolino was strong enough to hold under the strains of arches and other structures. Pozzolino was also useful in underwater situations since it set hard and strong in water.

Pratt Truss   A type of truss or bridge support invented by Thomas Willis Pratt and patented in 1844. *See also* Bollman Truss; Howe Truss; Lenticular Truss; Pony Truss; Through-Truss; Town Truss; Truss; Truss Bridge.

Prestressed Concrete Developed to a level of commercial use by the French engineer Eugene Freyssinet, and further developed by Robert Maillart and Francois Hennebique, prestressed concrete has steel wires, bars, or plates inside which are under permanent stress, providing an extremely strong beam that can bear much more tension than unstressed steel alone. Two methods are used in making prestressed concrete. In one method, the steel is placed in forms and stressed before the concrete is poured. A second method (post-tensioning) involves passing the steel through cured concrete, stressing the steel at that point, and then filling up spaces with concrete. Post-tensioning is done directly at a construction site. *See also* Reinforced Concrete.

Railroad Bridge A bridge designed to carry railroad train loads, rather than vehicular traffic. Because the live weight of these bridges was usually higher than vehicular bridges, developments in railroad bridges became the basis for developments in vehicular bridges as the number of automobiles on the road increased. An explosion in the weight of railroad cars, especially locomotives, meant that many nineteenth-century railroad bridges had to be replaced, despite their status as superb engineering achievements. *See also* Bridge; various types of bridges.

Reinforced Concrete Concrete that is strengthened by the laying of steel rods inside its core. The steel is laid in the concrete form before the concrete is poured. *See also* Prestressed Concrete.

Rockfill Dam Similar to an earthfill dam, but uses rock instead of earth. As in the earthfill dam, erosion from the force of water is a danger. The rockfill dam might have concrete, steel, or water-resistant earth on its upside face. *See also* Dam; Earthfill Dam; and various types of dams.

Sandhog A tunnel worker, specifically one who is involved in the day-to-day operations of digging and blasting. For more information, *See* Paul E. Delaney. *Sandhogs: A History of the Tunnel Workers of New York.* Longfield Press, 1983.

Scour The destructive effect of flowing water on a dam or bridge piers. Moving water, if strong enough, literally scours or wipes away the material in its path. Many foundation failures and related bridge collapses have been linked to the effect of scour, especially in locations where the course of a river has changed after a bridge has been constructed. To protect against scour, protective buffers are sometimes constructed around the outside of a pier, or various methods of diverting water flow are used.

Shield A machine used in building tunnels that is designed to protect the tunnel workers while they are working; the first shield was the Brunel Shield used to build a tunnel under the Thames River in London.

Shop Work Work done off-site or work on a portion of a project that is performed away from the site.

Sluice A structure with a gate made for holding back water. Usually an adjunct to a dam, the gates of a sluice can be opened to lower water pressure against a dam.

Span A length of a bridge between supports, or the distance between two supports of a bridge, although occasionally used to describe an entire bridge's length. A bridge can have a main span, side spans, or multiple spans.

Spandrel The sometimes decorated space between the right and left exterior curve of an arch and an enclosing right angle. *See also* Arch; Voussoir.

Stilling Basin An area into which water for a dam or aqueduct is channeled so that the flow of water at the structure is at an angle to the original flow of the water. The basin protects the structure from the destructive force of the water in its original channel.

Storage Dam A dam designed to hold water for later use, such as water supply, hydroelectric power, or irrigation. *See also* Dam; various types of dams.

Stress Limit The amount of force of one type or another that a structure or material is able to withstand without damage. The French Ban Dam, for example, had a higher "compressive stress limit" than did the Furens Dam; The Ban Dam was able to withstand a higher amount of force from water than was the Furens.

Suspension Bridge A bridge with a deck supported by vertical cables descending from a catenary cable suspended from towers. This bridge type is ideal for long lengths; when adjusted properly, the towers are stressed in only a vertical

This "spinning wheel" was used during cable-spinning operations on the Delaware Memorial Bridge. The wheel held four wires, and raced across the bridge at 710 feet per minute slowing only at the ends to allow workers to grab the wires and place them in the proper position. Photo courtesy of the Bethlehem Steel Corporation.

direction. Horizontal stresses are carried to the outer supports of the bridge. *See also* Bridge; various types of bridges.

Swing Bridge A bridge in which the central span pivots around a central point in a horizontal plane to allow ships or other traffic to pass beneath it. *See also* Bridge; various types of bridges.

Tension   The pulling force on a support or member. In a suspension bridge, the force of the weight of the deck acting upon the cables is one of tension. *See also* Compression.

Through-Truss   A truss bridge in which the deck goes through the trusswork, as opposed to a pony truss in which the deck is on top of the trusswork. *See also* Bollman Truss; Howe Truss; Lenticular Truss; Pony Truss; Pratt Truss; Town Truss; Truss; Truss Bridge.

Tied Arches   An arrangement of at least two arches supporting a bridge where the arches are connected at the base. *See also* Arch.

Town Truss   A truss bridge type designed and patented by Ithiel Town in 1820. A Town truss used wooden planks crossed in a diamond pattern and fastened with wooden pins. The design was popular because it was simple, made of readily available materials, and within the capacity of a good carpenter. The Town truss was popular in the early nineteenth century for railroad and highway bridges. *See also* Bollman Truss; Howe Truss; Lenticular Truss; Pony Truss; Pratt Truss; Through-Truss; Truss; Truss Bridge.

Trial-load Analysis   A complicated mathematical method for determining the abilities of a proposed arch dam to resist various stresses and forces. Used extensively in the United States in the first half of the twentieth century, the method successfully aided the design of such dams as the Buffalo Bill Dam and the Hoover Dam. In contrast, many dams outside the United States were designed using a modeling method, in which actual models of the proposed structure were made and mathematical adjustments for differences in materials were applied to the results of tests made on the models. For additional information, *see* Smith, Norman. *A History of Dams*. Secaucus, NJ The Citadel Press, 1972.

Truss   A support structure, in various types of arrangements, for a bridge or other type of overpass. For a variety of technical reasons, a beam bridge that needs to support more than occasional light usage will require support to be supplied by various combinations of I and T sections arranged into "trusses." The development of various truss arrangements, especially in the nineteenth and early twentieth centuries, led to the large number of different trusses used in different situations (e.g., a railroad bridge needs to bear more weight than a footbridge). Trusses are frequently named after the people who invented them (e.g., Bollman, Pratt, Whipple).

See also Bollman Truss; Howe Truss; Lenticular Truss; Pony Truss; Pratt Truss; Through-Truss; Town Truss; Truss Bridge.

Truss Bridge   A bridge supported by trusses. Numerous types of truss systems have been designed for increased support and strength, but all trusses depend upon the inherent strength and stability of triangular components. Truss bridges are usually no longer than 1,000 feet (300 meters). *See also* Bollman Truss; Howe Truss; Lenticular Truss; Pony Truss; Pratt Truss; Through-Truss; Town Truss; Truss.

Two-Hinged Arch   A system for metal arch construction developed in the 1870s and 1880s. The bottom ends of the arch are pinned into their abutments, preventing the arch from rotating relative to the abutment; this allows the arch itself to be made of lighter members. A two-hinged arch sometimes refers to a system in which the ends of the arch rest on large wheels called "rolling pins." The rolling pins provide flexibility for times when the arch is compressed under a heavy load or when the stress is relieved. A "three-hinged" arch can be constructed in which a third pin is placed at the top of the arch where its two halves meet; a single-hinge arch can be built in which only one pin is used at the top of the arch. For additional information, *See* Silverberg, Robert. *Bridges*. Philadelphia: Macrae Smith Company, 1966. *See also* Abutments; Arch.

Vertical-lift Bridge   A movable bridge in which the movable portion is lifted up and down between towers. *See also* Bascule Bridge; Drawbridge.

Viaduct   Usually a bridge designed for pedestrian traffic. Because it supports less live weight, a viaduct usually can be more attractively designed than bridges designed for car or railroad traffic. *See also* Bridge; Live Weight.

Voussoir   Any of the pieces of an arch, cut and designed to resemble a wedge with the point cut off. *See also* Arch; Spandrel.

Warren Truss   A highly popular and successful truss patented in 1848. If constructed properly, it is possible for the builder to determine if the various parts of the truss are in tension or compression, and thus if the truss is sufficiently strong.

Weir   A small diversion dam, usually used as part of a larger dam project. An example of its use includes the slight reorienting of a stream or creek. *See also* Diversion Dam.

Wrought Iron   A form of iron with little or no carbon in it; wrought iron is more durable and less brittle than cast iron. *See also* Cast Iron.

BIBLIOGRAPHY

Books

Academic American Encyclopedia, electronic version. Danbury, CT: Grolier Electronic Publishing, 1991.

Allen, Richard Sanders. *Covered Bridges of the Middle Atlantic States*. Brattleboro, VT: Stephen Greene Press, 1959.

Bartoli, Georges. *Moscow and Leningrad Observed*. Translated by Amanda and Edward Thomson. New York: Oxford University Press, 1975.

Beaver, Patrick. *A History of Tunnels*. Secaucus, NJ: Citadel Press, 1973.

Bishop, Gordon. *Gems of New Jersey*. Englewood Cliffs, NJ: Prentice-Hall, 1985.

Branner, Robert. *Chartres Cathedral*. New York: Norton, 1969.

Brown, David J. *Bridges*. New York: Macmillan, 1993.

Bruce, Moffat. *Forty Feet Below: The Story of Chicago's Freight Tunnels*. Interurban Press, 1982.

Chao, Lorraine S., ed. *Let's Go: The Budget Guide to Italy*, 1993 edition. New York: St. Martin's Press, 1993.

Chippindale, Christopher. *Stonehenge Complete*. London: Thames and Hudson, 1983.

Clucas, Philip. *England's Churches*. New York: Crescent Books, 1984.

Condit, Carl W. *American Building Art: The Nineteenth Century*. Vol. 1. New York: Oxford University Press, 1960.

——. *American Building Art: The Twentieth Century*. Vol. 2. New York: Oxford University Press, 1961.

Crystal, David, ed. *The Cambridge Biographical Encyclopedia*. Cambridge: Cambridge University Press, 1994; this encyclopedia is available online at http://www.biography.com.

Cudahy, Brian J. *Rails Under the Mighty Hudson: The Story of the Hudson Tubes, the Pennsy Tunnels and Manhattan Transfer*. Brattleboro, VT: Stephen Greene Press, 1975.

Cupper, Dan. *The Pennsylvania Turnpike*. Lebanon, PA: Applied Arts, 1990.

DeLony, Eric. *Landmark American Bridges*. Boston: Bullfinch Press, 1993.

Demby, William. *The Catacombs*. Boston: Northeastern University Press, 1991.

Diamonstein, Barbaralee. *The Landmarks of New York*. New York: Harry N. Abrams, Inc., 1993.

Dictionary of American Biography. New York: Charles Schribner's Sons.

Ditzel, Paul C. *How They Build Our National Monuments*. New York: Bobbs-Merrill, 1976.

D'Onofrio, Cesare. *Castel S. Angelo: Images and History*. English Edition. Rome: Roma Società Editrice, 1984.

Duffield, Judy; Kramer, William; and Shephard, Cynthia. *Washington D.C.: The Complete Guide*. New York: Vintage Books, 1988.

Dunlop, Fiona. *Fodor's Exploring Paris*. 3rd ed. New York: Fodor's Travel Publications, 1993.

Dunn, Andrew. *Tunnels*. New York: Thompson Learning, 1972.

Eicher, David. *Civil War Battlefields: A Touring Guide*. Dallas: Taylor Publishing Company, 1995.

Ellis, Hamilton. *The Pictorial Encyclopedia of Railways*. Prague, Czechoslovakia: Paul Hamlyn, 1968.

Encyclopaedia Britannica. 15th ed. Chicago: Encyclopaedia Britannica, Inc., 1992.

Bibliography

Epstein, Sam and Epstein, Beryl. *Tunnels*. Boston: Little Brown & Company, 1985.

Farb, Peter. *The Story of Dams*. Irvington-on-Hudson, NY: Harvey House, 1961.

Franzoi, Umberto. *Canal Grande*. English Edition. New York: Vendome Press, 1993.

Freedgood, Seymour. *The Gateway States*. New York: Time, 1967, 1970.

Gies, Joseph. *Adventure Underground: The Story of the World's Great Tunnels*. Garden City, NY: Doubleday & Company, 1962.

Gillespie, Angus K. and Rockland, Michael Aaron. *Looking for America on the New Jersey Turnpike*. New Brunswick, NJ: Rugers University Press, 1989.

Goldsmith, Edward and Hildyard, Nicholas. *The Social and Environmental Effects of Large Dams*. San Francisco: Sierra Club Books, 1984.

Hadfield, Charles. *World Canals*. Newton Abbot, England: David & Charles, 1986.

Hamilton, George Heard. *The Art and Architecture of Russia*. 3rd ed. Pelican History of Art Series. New York: Penguin Books, 1983.

Harrington, Lynn. *The Grand Canal of China*. Chicago: Rand McNally & Company, 1967.

Haskins, Frederic J. *The Panama Canal*. Garden City, NY: Doubleday, Page & Company, 1913.

Hawkes, Nigel. *Structures: The Way Things Are Built*. New York: Macmillan, 1990.

Hayward, R.A. *Cleopatra's Needles*. Buxton, England: Moorland Publishing Company, 1978.

Holler, Anne. *Florencewalks*. rev. ed. New York: Henry Holt and Company, 1993.

Hopkins, H.J. *A Span of Bridges: An Illustrated History*. New York: Praeger, 1970.

Jackson, Donald C. *Great American Bridges and Dams*. Washington, DC: The Preservation Press, 1988.

Jacobs, David and Neville, Antony E. *Bridges, Canals and Tunnels*. New York: American Heritage Publishing, 1968.

Karp, Walter. *The Center: A History and Guide to Rockefeller Center*. New York: American Heritage Publishing Company, 1982.

Keates, Jonathan and Hornak, Angelo. *Canterbury Cathedral*. London: Scala/Philip Wilson, 1980.

Kinross, Lord John Patrick Douglas Balfour. *Between Two Seas: The Creation of the Suez Canal*. New York: William Morrow and Company, 1969.

Kirby, Richard Shelton et al. *Engineering in History*. New York: Dover Publications, 1990.

Knowles, Rebecca, ed. *Let's Go: The Budget Guide to France,* 1993 edition. New York: St. Martin's Press, 1993.

Kost, Mary Lu. *Milepost I-80: San Francisco to New York*. Milepost Publications, 1993.

Kubler, George. *Building the Escorial*. Princeton, NJ: Princeton University Press, 1981.

Lay, M.G. *Ways of the World*. New Brunswick, NJ: Rutgers University Press, 1992.

Louis, Victor and Louis, Jennifer. *Complete Guide to the Soviet Union*. New York: St. Martin's Press, 1991.

MacIvor, Iain. *Edinburgh Castle*. London: B.T. Batsford/Historic Scotland, 1993.

Magnusson, Magnus, ed. *Cambridge Biographical Dictionary*. Cambridge, England: Cambridge University Press, 1990.

Matthys, Levy and Salvadori, Mario. *Why Buildings Fall Down*. New York: W.W. Norton & Company, 1992.

McCullough, David. *The Great Bridge*. New York: Simon and Schuster, 1972.

McNeil, Ian, ed. *An Encyclopedia of the History of Technology*. London: Routledge, 1990.

Michelin Paris. Paris: Michelin et Cie, 1992. [Contains a detailed description of the Paris sewer system.]

Montgomery-Massingberd, Hugh. *Blenheim Revisited: The Spencer-Churchills and Their Palace*. London: Bodley Head, 1985.

Morris, Richard B., and Irwin, Graham W., eds. *Harper Encyclopedia of the Modern World*. New York: Harper & Row, 1970.

National Geographic Society. *Exploring America's Scenic Highways*. Prepared by the Special Publications Division. Washington, DC: National Geographic Society, 1985.

O'Connor, Colin. *Roman Bridges*. Cambridge, England: Cambridge University Press, 1993.

Oggins, Robin S. *Castles and Fortresses*. New York: Michael Friedman Publishing Group, 1995.

Outerbridge, Davis. *Bridges*. Photographs by Graeme Outerbridge. New York: Harry N. Abrams, 1989.

Overman, Michael. *Roads, Bridges and Tunnels: Modern Approaches to Road Engineering*. Garden City, NY: Doubleday & Company, 1968.

Patton, Phil. *Open Road: A Celebration of the American Highway*. New York: Simon and Schuster, 1986.

Pearson, John. *Arena: The Story of the Colosseum*. New York: McGraw-Hill, 1973.

Petroski, Henry. *Engineers of Dreams*. New York: Alfred A. Knopf, 1995.

Pietrangeli, Carlo et al. *The Sistine Chapel: A Glorious Restoration*. New York: H.N. Abrams, 1994.

Platt, Colin. *The Castle in Medieval England and Wales*. New York: Barnes & Noble, 1996.

Railway Directory & Year Book from web page.

Reisner, Marc. *Cadillac Desert: The American West and Its Disappearing Water*. New York: Penguin Books, 1987.

Robins, Anthony. *The World Trade Center*. Englewood, FL: Pineapple Press, 1987.

Rose, Albert C. *Historic American Roads: From Frontier Trails to Super Highways*. New York: Crown Publishers, 1976.

Russell, Solveig Paulson. *The Big Ditch Waterways*. New York: Parents' Magazine Press, 1977.

Saalman, Howard. *Haussmann: Paris Transformed*. New York: G. Braziller, 1971.

Salway, Peter. *Roman Britain*. New York: Oxford University Press, 1981.

Sandak, Cass R. *Bridges*. New York: Franklin Watts, 1983.

——. *Canals*. New York: Franklin Watts, 1983.

——. *Dams*. New York: Franklin Watts, 1983.

——. *Tunnels*. New York: Franklin Watts, 1984.

Sandstrom, Gosta E. *Man the Builder*. New York: McGraw-Hill, 1970.

——. *Tunnels*. New York: Holt, Rinehart and Winston, 1963.

Schodek, Daniel L. *Landmarks in American Civil Engineering*. Cambridge, MA: MIT Press, 1987.

Schreiber, Herman. *The History of Roads: From Amber Route to Motorway*. Translated by Stewart Thomson. London: Barrie and Rockliff, 1961.

Shaw, Ronald E. *Canals for a Nation: The Canal Era in the United States, 1790-1860*. Lexington: University Press of Kentucky, 1990.

Silverberg, Robert. *Bridges*. Philadelphia: Macrae Smith Company, 1966.

Smith, Norman. *A History of Dams*. Secaucus, NJ: Citadel Press, 1972.

Spangenburg, Ray and Moser, Diane K. *The Story of America's Canals*. New York: Facts on File, 1992.

Steinman, David B. *Famous Bridges of the World*. London: Dover Publications, 1953, 1961.

Steinman, David B. and Watson, Sara Ruth. *Bridges*. New York: Dover Publications, 1941, 1957.

Trachtenberg, Marvin. *Statue of Liberty*. New York: Viking Press, 1976.

Wacher, John. *Roman Britain*. London: J.M. and Sons, 1978.

Waldron, Arthur. *The Great Wall of China: From History to Myth*. Cambridge: Cambridge University Press, 1990.

Walker, Charles. *Wonders of the Ancient World*. New York: Crescent Books.

Wallis, Michael. *Route 66: The Mother Road*. New York: St. Martin's Press, 1990.

Walton, Guy. *Louis XIV's Versailles*. Chicago: University of Chicago Press, 1986.

Wetterau, Bruce. *The New York Public Library Book of Chronologies*. New York: Stonesong Press, 1990.

White, Norvel. *The Architecture Book*. New York: Alfred A. Knopf, 1976.

Who Was Who in America, Historical Volume 1607-1896. rev. ed. Chicago: Marquis Who's Who, 1967.

World Almanac and Book of Facts 1993. New York: Pharos Books, 1992; this almanac can be accessed at http://booksrv2.raleigh.ibm.com:80/ cgibin/bookmgr/bookmgr.cmd/BOOKS/ BUILDNGS/CCONTENTS.

Young, Edward M. *The Great Bridge: The Verrazano Narrows Bridge*. New York: Ariel/Farrar, Strauss and Giroux, 1965.

Journal, Magazine, and Newspaper Articles (including electronic versions)

Austin, Teresa. "U.S. Toll Project Update." *Civil Engineering* 64:2 (February 1994), p. 58.

Brooke, James. "Tackling a Crucial Highway to Keep Skiers Happy." *New York Times* (23 December 1995), P:7.

Brown, Patricia Leigh. "Driving, Buying, Reading and Remembering Route 66." *New York Times* (24 August 1995), A:1.

Cagley, James R. "Restoring Freedom at the Capitol Dome." *Civil Engineering* 64:6 (June 1994), pp. 57-59.

Caile, Bill. "Mechanical Systems Match Complexity of Denver Airport." *Engineered Systems* 12 (1 August 1995), p. 74; see also C.A. Merriman in http://infodenver.denver.co.us/-aviation/ factrvia.html.

Chiles, James R. "'Remember, Jimmy, Stay Away from the Bottom of the Shaft!' (Water Tunnel Number Three, New York, New York)." *Smithsonian* 25 (July 1994), p. 60.

Colt, George Howe. "Raising Alexandria." *Life* (April 1996), p. 70.

"Croat Offers to Rebuild the Bridge at Mostar." *New York Times* (13 March 1994), I:1.

Darnton, John. "Ferris Wheel Will Look Down on Big Ben." *New York Times* (16 April 1996), I:7.

Davis, Tony. "Managing to Keep Rivers Wild." *Technology Review* (May/June 1986).

"Edifice Complex" [Nina Tower and other buildings in Asia]. *Asia, Inc. Online* (August 1995), see http://www.asia-inc.com/archive/ 0895edifice.html.

Faber, Harold. "Hardenbergh Journal: One of New York's Covered Bridges Gets a Reprieve." *New York Times* (4 May 1992), B:5.

——. "Sunday Outing: On a Covered Bridge, Reverence Is Required (Horse and Buggy Optional)." *New York Times* (22 March 1992), I, pt. 2, p. 40.

Failla, Kathleen Saluk. "A Decade After Disaster State Rebuilds at Rapid Pace." *New York Times* (29 June 1993), 13CN:1.

Bibliography

Fairweather, Virginia. "The Channel Tunnel: Larger Than Life, and Late." *Civil Engineering* 64:5 (May 1994), pp. 42-46.

"Flood Is Called Right Tonic for Grand Canyon," *New York Times*, I:32.

Foderaro, Lisa W. "Strolling in a Bucolic New Jersey Corner." *New York Times* (24 November 1995), C1:1.

Fowler, John P., II, and Thompson, Kurt R. "The Toll Road That Wouldn't Die." *Civil Engineering* 65:4 (April 1995), pp. 48-51.

Gessner, George and Selvaratnam, Selva. "Bridge Over the River Karnali." *Civil Engineering* 66:4 (April 1996), pp. 48-51.

Goldberg, Carey. "Golden Gate Bridge to Institute Suicide Patrols." *New York Times* (25 February 1996), A:32.

Goldstein, Harry. "Triumphant Arches." *Civil Engineering* 65:7 (July 1995), pp. 48-49.

Gordon, Michael R. "Despite Cold War's End, Russia Keeps Building a Secret Complex." *New York Times* (16 April 1996), A:1.

Greenberger, Robert S. "U.S. Opposes Aid for Firms to Help Build China Dam." *Wall Street Journal* (13 October 1995), A.

Gray, Christopher. "Streetscapes: The University Heights Bridge; a Polite Swing to Renovation for a Landmark Span." *New York Times* (25 February 1990), X:9.

Harriman, Stephen. "A Day Trip to Bridge on River Kwai." *The Virginian-Pilot* (12 November 1995), see at http://scholar3.lib.vt.edu/VA-news/VA-Pilot/issues/1995/vp951112/9511100073.html.

Heidengren, Charles R. "Japan Builds 21st Century Monuments." *Civil Engineering* 64:8 (August 1994), pp. 57-59.

Janofsky, Michael. "New Toll Road Offers Glimpse at Future." *New York Times* (29 September 1995), A:16.

Keizo, Shimizu. "Sky-High Ambition." *Asia, Inc. Online* (August 1995), see http://www.asia-inc.com/archive/0895sky.html.

Kinzer, Stephen. "Leipzig Plans to Restore Once-Regal Train Station." *New York Times* (7 April 1996), I:10.

Lindgren, Hugo. "Visible City: Sears Tower." *Metropolis* (November 1994), reprinted at http://www.enews.com/magazines/metropolis/archive/1994/110194.1.html.

Lindgren, Kristina. "Denver Airport: A Beauty with a Few Flaws." *Los Angeles Times* (10 March 1995), Business:7.

Maitland, Leslie. "Moses, 90, Nostalgic About Whitestone Bridge, 40." *New York Times* (30 April 1979), B:5.

Mazie, David. "The Glory That Was Greece; Restorers Battle Time and Elements to Save Ancient Acropolis." *Los Angeles Times* (7 February 1993), A:10.

McGowan, Cate. "Howell Park to Display Famous Sculptor's 'Hope'." *Atlanta Magazine* (December 1995), reprinted at http://www.nav.com/atlanta-30306/decfeat/f1.htm.

Mellor, William. "Mirage or Vision? Asia's Suez Canal." *Asia, Inc.* (December 1995), reprinted at http:\\198.111.253.144/archive/1295thaisuez.html.

"Mexico, U.S. and Canada Linked by 4.6 Mile Road Completing I-5." *New York Times* (13 October 1979).

"News: Nordic Neighbors to Sink Link." *Civil Engineering* 65:11 (November 1995), pp. 20-22.

"News: Record-Setting French Bridge Set to Open." *Civil Engineering* 65:2 (February 1995), p. 18.

"News: Roosevelt Dam Reaches a New Height of Safety." *Civil Engineering* 66:5 (May 1996), p. 10; see also http://donews.do.usbr.gov/dams/TheodoreRoosevelt.html.

"News: Study Shows Aluminum Deck Resists Corrosion." *Civil Engineering* 65:9 (September 1995), p. 10.

"News: Unconventional Bascule Bridge Crosses Superfund Site." *Civil Engineering* 65:12 (December 1995), pp. 12-14.

O'Kon, James A. "Bridge to the Past." *Civil Engineering* 65:1 (January 1995), pp. 62-65.

Pacelle, Mitchell. "U.S. Architects in Asia: Only Way to Go Is Up." *Wall Street Journal* (21 March 1996), B:1.

Perry, George. "The New Opera House" from *The Complete Phantom of the Opera*. Wordsworth Books, excerpted at http://phantom/skywalk.com/operahouse/opera-house.html.

Petzold, Ernst H. "Cables Over the Mississippi." *Civil Engineering* 64:2 (February 1994), pp. 62-65.

Pollak, Axel J. and Lalas, V. Peter. "Boston's Third Harbor Tunnel." *Civil Engineering* 66:3 (March 1996), pp. 3A-6A.

Robert Reinhold. "Opening New Freeway, Los Angeles Ends Era." *New York Times* (14 October 1993), A:16.

Robison, Rita. "Boston's Home Run." *Civil Engineering* 66:7 (July 1996), pp. 36-39.

——. "Malaysia's Twins: High Rise, High Strength." *Civil Engineering* 64:7 (July 1994), pp. 63-65.

Searles, Denis. "Debate Flies Over Denver's Big, New Airport. . . . ." *Los Angeles Times* (19 July 1992), A:2.

Sierpina, Diane Sentementes. "Stamford; A (Sort of Purple) Iron Bridge Gathers Champions." *New York Times* (4 June 1995), XIIICN:2.

Smith, Malcolm G. "The Multi-million Dollar Bridge." *Coastal Beacon* (online version) (14 December 1995), see http://www.mbeacon.com/archive/121495dir/b.html.

South Dakota Bureau of Tourism. "Mount Rushmore: The Four Most Famous Guys in Rock." See online at http://www.state.sd.us/tourism/rushmore/rushmore.html.

Svitil, Kathy A. "The Coming Himalayan Catastrophe." *Discover* 16:7 (July 1995), pp. 80-84.

Terry, Don. "Chicago's Well-Kept Secret: Tunnels." *New York Times* (15 April 1992), D:27.

"Travel Advisory: Lucerne's Rebuilt Chapel Bridge Open Again." *New York Times* (29 May 1994), V:3.

"Travel Advisory: Normandy Bridge Cuts Travel Time to North." *New York Times* (26 March 1995), V:1.

Williams, Daniel. "Art Sleuths Find 'Mona Lisa's' Landscape." *Rockland Journal-News* (15 December 1995), A:17.

Internet Websites

Although every effort has been made to include current WWW sites in this listing, some addresses may have changed or some sites may have been replaced. If you cannot find a site listed here, you may need to undertake an Internet search to find the new site.

Akashi-Kaikyo Bridge: http://www.kobe-u.ac.jp/hyogo/akashi.html

Aki Tunnel: http://www.rtri.or.jp/japanrail/JapanRail_E.html

Alaska Highway: http://alaskan.com/bells/alaska_highway.html

Alexander Column: http://www.spb.su/fresh/sights/palacesq.html; http://www.spb.su/lifestyl/136/where.html

Antioch Bridge: http://www.dot.ca.gov/dist4/calbridgs.htm#ab

Arc de Triomphe: http://www.focusmm.com.au/~focus/fr_re_02.html; http://www.paris.org/Monuments/Arc.html

Arch of Constantine: http://harpy.uccs.edu/roman/html/archconslides.html

Arlington National Cemetery: http://www.dgsys.com/-mwardell/cem.html

Arts and Industries Building: http://www.si.edu/organiza/museums/artsind/homepage/artsind.html

Ashland Bridge: http://n1-44-225.macip.drexel.edu/top/bridge/CBAshl.HTML; http://william-king.www.drexel.edu/top/bridge/CBAshl/html

Ataturk Dam: http://gurukul.ucc.american.edu/ted/ataturk.htm; American University's "Trade and Environment Database" at http://www.ids.ac.uk/e/did/data/d0s1/e02135.html

Baiyoke Sky Hotel: http://www.asia-inc.com/archive/0895edifice.html

Bank of China Tower: http://bcwww.cityu.edu.hk/b-s/boc.html

Barrackville Covered Bridge: http://www.wvonline.com/post/Barrackville.htm

Bartlett Dam: http://donews.do.usbr.gov/Denver/tsc/Concdams/Buttress/Bartlett.h tml

Bath-Bath #28 Bridge: http://vintagedb.com/guides/covered3.html

Battle Monument, West Point Military Academy: http://tamos.gmu.edu/-marcus/battle.html

Bay Area Rapid Transit (BART): http://www.transitinfo.org/BART/

Big Walker Mountain Tunnel: http://199.183.146.20/gorp/activity/byway/va_bigwa.htm

Boston Central Artery/Tunnel: http://www.tiac.net/users/kat/CAT.html

Bridge Over the River Kwai: http://scholar3.lib.vt.edu/VA-Pilot/issues/1995/vp951112/9511100073.html

Bridgeport Covered Bridge: http://www.websterweb.com/bridgeport.html

Buckingham Palace: http://www.londonmail.co.uk/palace.html; http://www.buckinghamgate.com/events/past_features/palace/palace.html; http://cafeonternet.CD.uk/bucks.htm

California State Water Project: http://tiger-2.water.ca.gov/dir-state_water_project/State_Water_Project.html

Campton-Blair Bridge #41: http://vintagedb.com/guides/covered3.html

Canterbury Cathedral: http://www.lonelyplanet.com.au/dest/eur/eng.htm

Cape Cod Canal: http://www.virtualcapecod.com/towntext/bourne.html

Cape Hatteras Lighthouse: http://zuma.lib.utk.edu/lights/banks1.html

Capilano Suspension Bridge: http://www.bendtech.com/attractions/capilano/index.html; http://worldtel.com/vancouver/susbridg.html; http://capbridge.com/fastfax.htm; http://www.capbridge.com/history.htm

Centennial Bridge: http:william-king.www.drexel.edu/top/bridge/CBCent.html

Central Plaza Building: http://bcwww.cityu.edu.hk/b-s/cenplaz.html

Century Freeway: http://www.tmn.com/iop/introduction.html; http://www.tmn.com/iop/decree.html

Channel Tunnel: http://www.doc.gov/JTP.Mosaic.Materials/JTP/JTLB/JTP.JTLB.JTLB23/4.Chunnel.html; http://www.starnetinc.com/eurorail/eurostar.Chunnel.html

Chartres Cathedral: http://member.aol.com/detechmendy/capital.html; http://www.designbase.com/was/038.htm

Bibliography

Cheesman Dam: http://www.water.denver.co.gov/cheesman.htm

Chehalis River Riverside Bridge: http://wsdot.wa.gov/eesc/environmental/Bridge-WA-111.htm

Chesapeake and Ohio Canal: http://www.fred.net/kathy/canal.html

Chongqing Office Tower: http://www.asia-inc.com/archive/0895edifice.html

Church of the Bleeding Saviour: http://www.spb.su/fresh/sights/bleeding.html

Cimitiére du Pére Lachaise: http://www.io.org/cemetery/Lachaise/lachaise.intro.html; Harley, Adam. "Cimitière du Père Lachaise" at http://www.users.globalnet.co.uk/~lilth/tcimit.htm

Clifton Suspension Bridge: http://pyt.avonibp.co.uk/Gallery/Clifton/Bridge.html

CN Tower: http://ppc.westview.nybe.north-york.on.ca/Ernest/CNTower.html; http://www.banfdn.com/cntower.htm

College Park Airport: http://www.wp.com/avianet/cgs.html

Colossus of Rhodes: http://pharos.bu.edu/Egypt/Wonders/colossus.html

Confederation Bridge: http://www.peinet.pe.ca/SCI/br_d.html

Cornish-Dingleton Hill Bridge #22: http://vintagedb.com/guides/covered2.html

Dale Dyke Dam: http://pine.shu.ac.uk/~engrg/victorian/DDDam/Flood1.htm

Delaware and Hudson Canal: http://www.mhroc.org/kingston/kgndah.html

Denver International Airport: http://infodenver.denver.co.us/-aviation/factrvia.html

Dismal Swamp Canal: http://www.albemarle-nc.com/camden/history/canal.htm

Drift Creek Bridge: http://william-king.www.drexel.edu/top/bridge/CBDrift.html

Dulles Greenway: http://www.his.com/~cwealth/greenway/facts.html

Dwight D. Eisenhower Interstate and Defense Highway System: http://www.tfhrc.gov/pubrds/summer96/p96su28.htm

Edinburgh Castle: http://www.ibmpcug.co.uk

Eiffel Tower: http://www.paris.org/Monuments/Eiffel/html

Elephant Butte Dam: http://www.usbr.gov/power/elephant.html

Empire State Building: http://theinsider.com/topthing/2EMPIRE.htm; http://www.hotwired.com/rough/usa/mid.atlantic/ny/nyc/city/midtown.html; http://www.stern.nyu/~hlee10/empstat.htm

Engineer's Castle: http://www.spb.su/fresh/sights/engineer.html

European Castles: http://www.nectec.or.th/rec-travel/europe/castles; http://www.cde.com/~alpsboy/alpsboy.htm

Extraterrestrial Highway: http://www.csicop.org/si/9609/highway.html; http://www.pcap.com/extrater.htm

First Owens River—Los Angles Aqueduct: http://www.ladwp.com/aboutwp/facts/supply/supply.htm

Flatiron Building: http://www.hotwired.com/rough/usa/mid.atlantic/ny/nyc/city/midtown.html

Forbidden City: http://www.ihep.ac.cn/tour/bj.html

Fredericksburg and Spotsylvania National Military Park: http://woodstock.mro.nps.gov/frsp/natcem.htm

Friant Dam: http://www.usbr.gov/cdams/friant.html; http://nps.gov/jeff/arch-home/default.htm

Gateway Arch: http://www.st-louis.mo.us/st-louis/arch/structur.html; http://nps.gov/jeff/arch-home/default.htm

Glen Canyon Dam: http://www.pagehost.com/lakepowell/gcdam.htm

Grand Central Station: http://www.hotwired.com/rough/usa/mid.atlantic/ny/nyc/city/midtown.html

Grand Palace: http://www.asiatour.com/thailand/e-03bang/et-ban70.htm; http://www.asiatour.com/thailand/e-03bang/et-ban71.htm

Great Pyramid of Giza: http://pharos.bu.edu/Egypt/Wonders/pyramid.html

Great Sphinx: http://sunship.com/egypt/giza/photor1.html; http://www.idsc.gov.eg/cgi-win/pser3.exe/10

Guavio Dam: http://www.eia.doe.gov/emeu/cabs/colombia.html

Hampton Court Palace: http://www.buckinghamgate.com

Hanging Gardens of Babylon: http://pharos.bu.edu/Egypt/Wonder.Home.html

Harlech Castle: http://www.wp.com/castlewales/harlech.html

Harpole Bridge: http://wsdot.wa.gov/eesc/environmental/Bridge-WA-133.htm

Hartland Covered Bridge: http://www.mi.net/tge.html

Haverhill-Bath Bridge #27: http://vintagedb.com/guides/covered3.html

Heceta Head Light: http://zuma.lib.utk.edu/lights/heceta.html

Henniker-New England College Bridge #63: http://vintagedb.com/guides/covered4.html

High Coast Bridge: http://swe.connection.se/hoga-kusten/uk/hk.html

Highland Light: http://zuma.lib.utk.edu/lights/cod8.html

"Hope" Statue: http://www.nav.com/atlanta-30306/decfeat/fl.htm

Horse Mesa Dam: http://www.srp.gov/aboutsrp/water/srplakes.html

Horseshoe Dam: http://www.srp.gov/aboutsrp/water/srplakes.html#hs_dam

Inistearaght Lighthouse: http://zuma.lib.uk.edu/lights/eagle/eagle10-30.html

Itaipú Dam: http://pharos.bu.edu/Egypt/Wonders/ Modern/itaipn.html

Jami Masjid: http://www.lonelyplanet.com.au/dest/ ind/cal.htm

Kra Canal: http://198.111.253.144/archive/ 1295thaisuez.html

Kinzua Viaduct: Phillips, D. Harvey. "The Kinzua Viaduct and the Man Who Built It" at http:// www.bradford-online.com/blsvduct.html

Kremlin: http://www.russia.net/country/moscow/ sights/kremlin.html

Kurushima Oohashi Bridge: http:// www.webcity.co.jp/mi/tour/hashiE.html

Lachine Canal: http://www.cam.org/~fishon1/ oldmtl.html

Lake Powell: http://www.pagehost.com/lakepowell/ man-made.htm

LaSalle Street Tunnel: http://cpl.lib.uic.edu/ 004chicago/timeline/tunneltrffc.html

Lighthouse at Alexandria: http://www.newton. cam.ac.uk/egypt/news.html

Lincoln Highway: Lin, James. *Lincoln Highway History* at http://www.ugcs.caltech.edu/~jlin/ lincoln/history/part1.html

Lion's Gate Bridge: http://www.b-t.com/liongate.htm; http://www.bctour.com/bctour/lowmain/stypk.htm

Louvre Museum: http://www.focusmm.com.au/ ~focus/fr_re_02.html; http://sunsite.unc.edu/wm/ paris/hist/louvre.html

Marble Palace: http://www.spb.su/fresh/museums/ marble.html

Marib Dam: http://www.ceng.metre.edu.tr/~e78199/ Walid/Marib.html

Mausoleum of Halicarnassus: http://pharos.bu.edu/ Egypt/Wonders/mausoleum.html

Maybeck, Bernard Ralph: http:// www.exploratorium.edu/Palace_History/ Palace_History.html#Maybeck

Mica Dam: http://www.intorchg.ube.ca/ behydro...ronment/recreation/site=mica1.html

Millennium Tower: http://www.asia inc.com/ archive/0895sky.html

Moffat Tunnel: Smith, Christopher. "You Can Buy Historic Rail Tunnel—Hole Sale" in online edition of *Salt Lake Tribune*, 1 January 1997 at http://ftp.sltrib.com/97/jan/01/tci/00341717.htm

Mount Rushmore National Memorial: http:// www.state.sd.us/tourism/rushmore/ rushmore.html

Mount Vernon Memorial Highway: http:// www.cr.nps.gov/phad/ncrobib.html

National Statuary Hall: http://acs5.bu.edu:8001/ ~pviles/Hall2.html; http://xroads.virginia.edu/ ~CAP/CAP_home.html

Neuschwanstein Castle: http://www.nectec.or.th/rec-travel/europe/castles; http://www.cde.com/ ~alpsboy/alpsboy.htm

Nevsky Prospect: http://www.spb.su/fresh/sights/ nevsky.html

New York State Barge Canal: http:// www.history.rochester.edu:80/canal/index.htm

Nina Tower: http://www.asia-inc.com/archive/ 0895edifice.html

North Woodstock Bridge-Clark's Trading Post #64: http://vintagedb.com/guides/covered3.html

Notre Dame de Paris: http://www.focusmm.com.au/ ~focus/fr_re_02.html

Oland Island Bridge: http://www.destination-oland.se/engelska/index-e.htm

Oldfield Covered Bridge: http://www.mi.net/users/ alstondj/bridges.html

Olympian Zeus: http://pharos.by.edu/Egypt/Wonders/ zeus.html

Olympic Tower: http://www.centrepoint.com.au/ tower/index.html

Oresund Link: http://www.oresund.com/bron/ broinfo.htm

Ostankino Moscow Tower: http:// www.centrepoint.com.au/tower/index.html

Owyhee Dam: http://www.usbr.gov/cdams/dams/ owyhee.html.

Palace of Fine Arts-Exploratorium: http:// www.exploratorium.edu/Palace_History/ Palace_History.html#Maybeck

Panama Canal: http://iaehv.iaehv.ni/users/grimaldo/ canal.html

Paris Opera House: http://phantom.skywalk.com/ operahouse/opera_house.html

Parthenon: http://www.elibrary.com/cgi-bin/hhweb/ hhfetch?32411045X0Y670:QOO2:D013

Peace Bridge: http://www.grasmick.com/ peacepoe.htm

Peak Forest Canal: http://www.blacksheep-org/canals

Pentagon: http://www.dgsys.com/-mwardel/pent.html

Petronas Towers: http://www.jaring/my/petronas/ compro/twintwrs/twintwrs.html; http:// www.jaring/my/petronas/latest/press2.html

Pons Fabricus: http://www.iol.it/tci sdp/html/ eo706.htm

Portland Bridge(s): http://www.mbeacon.com/archive/ 121495dir/b.html; http://www.portland.com/ bridge/012497.htm

Poughkeepsie Railroad Bridge: http:// www.academic.marist.edu

Quebec Bridge: http://www.civeng.carleton.ca/ECL/ files/ecl.270txt

Queen Emma Pontoon Bridge: http:// www.interknowledge.com/curacao/ancpn01.htm

Qutab Minar: http://www.lonelyplanet.com.au/dest/ ind/cal.htm; http://sun10.v52.bsme.hu/ ~s7809sak/qutab.html

Radio City Music Hall: http://www.hotwired.com/ rough/usa/mid.atlantic/ny/nyc/city/midtown.html

Red Fort: http://www.lonelyplanet.com.au/dest/ind/ del.htm

Bibliography

Royal Gorge Bridge: http://bechtel.colorado.edu/ Graduate_Programs/Sesm/Courses/cven3525/ royal/royal.html

Saint Basil's Cathedral: http://www.adventure.com/ library/encyclopedia/ka/rfibasil.html (Knowledge Adventure Encyclopedia)

Saint Paul's Cathedral: http:// www.lonelyplanet.com.au/dest/eur/lon.htm

Saint Peter's Church: http://www.iol.it/tci.sdp/html/ eo706.htm

Salisbury Cathedral: http://www.eluk.co.uk/spireweb/ cinfo.htm

Saint Peter's Square: http://www.iol.it/tci.sdp/html/ eo706.htm

Salt River Project: http://www.srp.gov:80/aboutsrp/ water/srplakes.html

Sears Tower: http://bechtel.colorado.edu/ Gradua..Cven3525/Sears/Overview/facts .html; http://cpl.lib.uic.edu/004chicago/timeline/ searstower.html; http://www.enews.com/ magazines/metropolis/archive/1994/ 110184.l.html

Selkirk Lighthouse: http://www.maine.com/lights

Seoul Tower: http://www.centrepoint.com.au/tower/ index.html

Seven Wonders of the Ancient World: http:// pharos.bu.edu/Egypt/Wonders/list.html

Shasta Dam: http://donews.do.usbr.gov/dams/ Shasta.html

Shubenacadie Canal: http://ttg.sba.dal.ca/nstour/ halifax/points/downdart.html

Skarnsunder Bridge: http://www.mti.unit.no/~eriks/ pros...e/skarnsundet/skarnsundbridge.html

South Fork Newaukum River Bridge: http:// wsdot.wa.gov/eesc/environmental/Bridge-WA- 112.htm

Space Needle: http://spaceneedle.com/whatis.html; http://spaceneedle.com/histfaq.html

Step Pyramid of Djoser (Zoser): http:// ccat.sas.upenn.edu/arth/zoser.html

Sydney Opera House: http://www.anzac.com/aust/ nsw/soh.html

Sydney Tower: http://www.centrepoint.com.au/tower/ index.html

Taj Mahal: http://www.bucknell.edu/departments/ library/multimedia/new/taj_mahal.html

Temple of Artemis (Diana): http://pharos.bu.edu/ Egypt/Wonfer.Home.html

Theodore Roosevelt Dam: http://donews.do.usbr.gov/ dams/TheodoreRoosevelt.html

Thomas Mill Bridge: http://william- king.www.drexel.edu/top/bridge/CBTmill.html

Tien An Men Square: http://www.ihep.ac.cn/tour/ bj.html

Tokyo Tower: http://www.centrepoint.com.au/tower/ index.html

Tower of London: http://www.millersv.edu/~english/ homepage/duncan/medfern/tower.html

Transco Tower: http://www.centrepoint.com.au/ tower/index.html

United States Capitol: http://www.acc.gov/history/ cap_hist.html

Vietnam Veterans Memorial: http://www.nps.gov/ nps/ncro/nacc/vvm.html

Wabasha Street Bridge River Crossing: http:// bridges.sppaul.gov/WabINFO.html

Waco Suspension Bridge: http://ohare.easy.com/waco/ bridge.html

Walnut Street Bridge: http://bertha.chattanooga.net/ chamber/walnut.html

Washington Monument: http://sc94.ameslb.gov/ TOUR/washmon.html

Washington Square: http://www.libertynet.org/iha/ _tomb.html

Washington Street Tunnel: http://cpl.lib.uic.edu/ 004chicago/timeline/tunneltrffc.html

Wertz's Bridge: http://william-king.www.drexel.edu/ top/bridge/CBWertz.html

West Gate Bridge: http://www.civeng.carleton.ca/ ECL/reports

Westminster Palace: http://www.parliament.the- stationery-office.co.uk

White House: http://www.whitehouse.gov

Wiman Mek Palace: http://www.asiatour.com/ thailand/e-03bang/et-ban72.htm

Winchester Mystery House: http://www.netview.com/ svg/tourist/winchester/story.html

Winchester-Coombs #2 Covered Bridge: http:// vintagedb.com/guides/covered1.html

Windsor Castle: http://www.londonmail.co.uk/ windsor/default.html

Woolworth Building: http://www.hotwired.com/ rough/usa/mid.atlantic/ny/nyc/city/midtown.html

Yerba Buena Tunnel: http://www.dot.ca.gov/dist4/ calbridgs.htm#ab

Zook's Mill Bridge: http://william- king.www.drexel.edu/top/bridge/CBZook.html

GEOGRAPHICAL INDEX

Geographical Index

World Landmarks

Geographical Index

SUBJECT INDEX

by Michelle B. Graye

Numbers in **boldface** refer to a full entry. Illustrations are indicated by (il)